Advanced Array Systems, Applications and RF Technologies

Signal Processing and its Applications

Books in the series
P. M. Clarkson and H. Stark, *Signal Processing Methods for Audio, Images and Telecommunications* (1995)

R. J. Clarke, *Digital Compression of Still Images and Video* (1995)

S-K. Chang and E. Jungert, *Symbolic Projection for Image Information Retrieval and Spatial Reasoning* (1996)

V. Cantoni, S. Levialdi and V. Roberto (eds.), *Artificial Vision* (1997)

R. de Mori, *Spoken Dialogue with Computers* (1998)

D. Bull, N. Canagarajah and A. Nix (eds.), *Insights into Mobile Multimedia Communications* (1999)

D. S. G. Pollock *A Handbook of Time-Series Analysis, Signal Processing and Dynamics* (1999)

Advanced Array Systems, Applications and RF Technologies

NICHOLAS FOURIKIS

Phased Array Systems,
Ascot Park, Australia

ACADEMIC PRESS

A Harcourt Science and Technology Company

San Diego • San Francisco • New York • Boston
London • Sydney • Tokyo

Academic Press
A Harcourt Science and Technology Company
32 Jamestown Road, London NW1 7BY, UK
http://www.academicpress.com

Academic Press
A Harcourt Science and Technology Company
525 B Street, Suite 1900, San Diego, California 92101-4495, USA
http://www.academicpress.com

ISBN 0-12-262942-6

A catalogue for this book is available from the British Library

Transferred to digital printing 2005
Printed and bound by Antony Rowe Ltd, Eastbourne

Contents

Series Preface

Signal processing applications are now widespread. Relatively cheap consumer products through to the more expensive military and industrial systems extensively exploit this technology. This spread was initiated in the 1960s by the introduction of cheap digital technology to implement signal processing algorithms in real-time for some applications. Since that time semiconductor technology has developed rapidly to support the spread. In parallel, an ever increasing body of mathematical theory is being used to develop signal processing algorithms. The basic mathematical foundations, however, have been known and well understood for some time.

Signal Processing and its Applications addresses the entire breadth and depth of the subject with texts that cover the theory, technology and applications of signal processing in its widest sense. This is reflected in the composition of the Editorial Board, who have interests in:

(i)   Theory – The physics of the application and the mathematics to model the system;
(ii) Implementation – VLSI/ASIC design, computer architecture, numerical methods, systems design methodology, and CAE;
(iii) Applications – Speech, sonar, radar, seismic, medical, communications (both audio and video), guidance, navigation, remote sensing, imaging, survey, archiving, non-destructive and non-intrusive testing, and personal entertainment.

Signal Processing and its Applications will typically be of most interest to postgraduate students, academics, and practising engineers who work in the field and develop signal processing applications. Some texts may also be of interest to final year undergraduates.

Richard C. Green
The Engineering Practice,
Farnborough, UK

To my parents

Preface

This book is dedicated to radar engineers and scientists with a pronounced interest in phased array systems. More explicitly, we consider current systems and their precursors as well as systems that are in the proposal/planning/developmental stages. While the first two sets of systems help the reader reconstruct the vector of evolution, the consideration of future systems provides the reader with a philosophy of minimum surprises. Future systems are based on proposals that promote the innovation process and prototypes bridge the gap between proposals and production models.

The focus of the book, however, is in the processing of the signals derived from active array systems at their radio/intermediate frequency, RF/IF. While Chapters 1 and 2 cover systems and applications, Chapters 3, 4, and 5 are dedicated to the array antenna elements, transmit/receive (T/R) modules, and beamformers, respectively. The book therefore complements other books that are devoted to the processing of the signals derived from phased arrays at a baseband. Active phased arrays offer the array designer new and exciting opportunities for signal processing at the RF/IF, or baseband. Polarization processing and multibeaming are but two areas of great promise.

The book has the following defining characteristics:

(i)   It accepts a holistic view of array systems performing the radar/electronic warfare (EW) or communications functions as well as systems dedicated to the derivation of radiometric images of celestial sources or of the Earth's surface.
(ii)   The gap is bridged between the idealized phased arrays, covered in several textbooks, and working phased arrays.
(iii)   Conventional active phased arrays, offering several degrees of freedom to the designer, are now affordable. Furthermore, 'intelligent' or self-focusing/self-cohering arrays hold the promise for the realization of even more affordable or cheaper arrays.

In the book we demonstrate that all phased arrays share the same theory and that there is a continuum between radar arrays having short integration times and radiometric/radioastronomy arrays having short or long integration times. While the instantaneous beam derived from a densely populated radar array can have low sidelobes and the corresponding beam of a sparsely populated radioastronomy array is useless because it has high sidelobes, the latter sidelobes can be lowered to an acceptable level. The linkages between arrays performing a variety of functions not only enrich and deepen one's understanding of array systems but also allow a designer imbued in one discipline to utilize the techniques and approaches developed by another discipline. Lastly, the defense of high-value platforms or the efficient operation of airports depends on array systems that derive interdependent and interrelated data. At a technological level it is impossible to design robust radars without knowledge of the capabilities of electronic warfare (EW) systems designed to disable it.

It is hard to imagine another discipline, apart from radar, that has continued to offer so much to an ever-increasing number of communities since its genesis in the 1930s. Ingenious scientists and engineers orchestrated and staged its many metamorphoses, and with every metamorphosis another community harnessed and enjoyed a new set of benefits.

Radar array systems similar to PAVE PAWS/COBRA DANE, airborne early warning (AEW) operating at microwaves, and over the horizon radars (OTHRs) operating in the high-frequency band contribute to the defense of many countries. Recently the latter radars have been used in several applied science applications such as the monitoring of wind patterns over vast stretches of the ocean. These early military radar systems were so highly valued that a veil of secrecy covered their critical parameters; in addition, the cost of realizing these systems was treated as an independent variable. By contrast, array cost is not treated as an independent variable throughout this book, a reflection of the post-cold-war environment.

The designers of the flying phased array radar, better known as synthetic aperture radar (SAR) took the next evolutionary step. An airplane equipped with a SAR system yields all-weather, high spatial resolution maps of the Earth's surface. Typically very large aerial maps having half a meter spatial resolution or better on the Earth's surface are used for topography, resource monitoring/management, pollution/waste threat control, the mitigation of natural hazards, and surveillance.

The originators of interferometric SAR took the final step in the evolutionary ladder. Deformations on the Earth's surface amounting to a few millimeters can now be measured and monitored over several hundreds of square kilometers from orbiting satellites instantaneously and on a routine basis by interferometric SARs. Some of the branches of science that benefit from interferometric SAR maps include topography, hydrology, ecology, tectonics/erosion/earthquake studies and global volcano monitoring.

While rudimentary bistatic radar systems have been used in the past, it is only recently that bistatic radars are reaching their full potential. The transmitters of FM/broadcast stations and those on board the US/Russian global positioning systems (GPSs) have been used as illuminators of opportunity in bistatic radar systems that detect several diverse targets. The importance of these experiments cannot be over-emphasized because the resulting systems have low probability of interception (LPI) capabilities.

Proposals for pseudo-satellites powered by ground-based array systems constitute just one of the many applications of wireless power transmission (WPT) and can be used to provide communications and/or radar surveillance over an area of 1000 km in diameter on the Earth's surface.

Phased arrays also contributed significantly to the development of: (i) communication systems utilizing satellites on the geostationary orbit as well as constellations of satellites that ushered in the wireless revolution; and (ii) radioastronomy/radiometric systems.

We can state without reservation that radioastronomy became a respectable branch of science only after it adopted phased array-based radiotelescopes. Radioastronomy unraveled a radio universe that turned up to be as interesting as the universe explored by optical astronomers since antiquity. The complementary knowledge derived from observations taken at different wavelengths helped us assemble the wondrous mosaic of the cosmos we live in.

Modern radiometric systems that meet diverse sets of requirements essentially evolved from radioastronomy systems. More explicitly, meteorologists, oceanographers, environmentalists, climatologists and the military require measurements of rain rates, soil moisture content, land/sea surface temperatures and atmospheric water vapor content under all weather conditions over the entire Earth at sufficient spatial resolution to understand how the weather engine works and to predict future conditions.

At the end of Chapter 1 we conclude that the information obtained from modern array systems is invaluable not only for our survival on planet Earth, the defense of high value assets, and the maintenance of peace, but also for the understanding of the cosmos and our position in it.

In the early sections of Chapter 2, we erect the theoretical infrastructure necessary to understand array systems. Naturally, array sidelobe and grating lobe control are assigned considerable importance. While the well-known conventional amplitude/space tapers are briefly delineated, genetic algorithms and other modern approaches used to meet various array requirements are treated in some detail. More importantly, systems having low array sidelobes and grating lobes are considered when real-world considerations are taken into account.

The later sections of the same chapter are dedicated to the characterization and realization aspects of working phased arrays. Heat management, for instance, and the power supplies used in conjunction with arrays—topics often not treated in phased arrays books—are assigned the importance they deserve in this chapter. Chapter 2 ends with an extended discussion on the realizability aspects of

- wideband arrays with particular emphasis on broadband, multifunction shared aperture arrays;
- second-generation radioastronomy arrays;
- pseudo-satellites; and
- LPI radars utilizing illuminators of opportunity.

The remaining three chapters are devoted to the RF technologies of phased arrays comprising the array antenna elements, T/R modules and beamformers, respectively. These are the subsystems that constitute the sensors of the systems we have considered. The new thrusts for sensor developments come from the users who want to know more and to know it sooner.

After continuous research and development spanning over several decades, high-quality, low-cost, microstrip conformal antenna elements can meet the diverse requirements of current and future arrays. More explicitly, microstrip narrowband, wideband and dual-/multiple-band antenna elements are now available to the array designer. At the other end of the requirement spectrum, we consider offset antennas that offer the designer multiple staring beams over several frequency bands. These antennas are eminently suitable as antenna elements for future wide field of view radioastronomy arrays or the radio equivalents of the Schmidt optical telescopes.

While the cost of the array antenna elements and beamformers is relatively low, the T/R modules are expensive subsystems. Here again continuous R&D thrusts over the last two decades have resulted in affordable T/R modules. Additionally, the cost of monolithic microwave integrated circuit (MMIC)-based T/R modules is bound to decrease further as the associated manufacturing processes are better understood with the passage of time. We are now witnessing the widespread usage of phased arrays in many applications because affordable MMIC-based T/R modules that exhibit an unprecedented uniformity of the critical module parameters are now available. In a parallel research thrust, we have witnessed a renaissance of interest in vacuum tube technology culminating in the realization of multiple-beam tubes, gyro-klystrons, electrostatic amplifiers, and the microwave power modules (MPMs) that combine the attractions of solid-state devices and vacuum tubes. In the not too distant future, vacuum microelectronic devices (VMDs) might offer the array designer alternative transmitters.

The beamformer of a typical passive array can be bulky and its insertion loss affects the overall array noise temperature. By contrast, the beamformer of an active array is transparent to the system and volumetrically attractive. These two beamformer characteristics lend the designer several degrees of freedom, which we have explored in Chapter 5. Although some of the beamformers we have considered are well established, most of the beamformers explored are at different stages of development. The crossbar beamformers, for example, considered in this chapter have unprecedented versatility, are volumetrically attractive, dissipate low powers, and are capable of deriving several simultaneous and independent agile beams. Megalithic MMIC-based realizations of these beamformers have already been reported.

While more R&D is required to attain the full potential of the beamformers we have considered, there is little doubt that some promising candidates are emerging. What is more significant is that R&D in the area of beamforming has lately increased significantly, mainly owing to the challenging requirements that emerged from recent systems and the application of MMIC/digital/photonic technologies to beamformers suitable for active phased arrays utilizing MMIC-based T/R modules. Finally, a variety of low-cost, 'intelligent' arrays with unique capabilities are considered in the last sections of Chapter 5.

It is a pleasure to acknowledge the help that the following colleagues/friends extended to me: Nicholas Shuley, Harry Green, Chris Hemmi, Tom Brukiewa, Holger Meinel, Robert Westphal, Darla Wagner, Vladimir Vanke, Stig Rehnmark, Masato Ishiguro, Staffan Jonsson, Rene Grognard, Marvin Cohen, James C. Wiltse, Dianne Davis, Bob Alper, and Eli Brookner. Special thanks are due to Robyn Fourikis and to Josephine Smallacombe. I commissioned Constantine Tsitsinaris to draw the original drawings of the book and Maureen Prichard to produce the basic book cover illustration, which was further processed by Academic Press. The illustration is based on interferograms of the Earth's surface in the vicinity of fault lines and were kindly supplied by JPL/NASA.

Nicholas Fourikis
Adelaide, South Australia,
September 1999

Abbreviations

AAR	active array radar
A/D	analog-to-digital
ADC	analog-to-digital coverter
AESA	active electronically scanned array
AEW	airborne early warning
AGF	array geometric factor
AI	artificial intelligence
AM	amplitude modulation
AMRAAM	advanced medium-range air-to-air missile
AMTI	airborne moving target indication
AOA	angle of arrival
AQ	amplitude quantization
ARM	anti-radiation missile
ASA	annular synthesis antenna
ASAI	ASA type I
ASAII	ASA type II
ASAA	advanced shared aperture array
ASAP	Advanced Shared Aperture Program
ASCM	antiship cruise missile
ASDE	airport surface detection equipment
ASIC	application-specific integrated circuit
ASR	airport surveillance radar
ATC	air traffic control
AWACS	airborne warning and control system
BER	bit error rate
BFN	beamforming network
BJT	bipolar junction transistor
BPF	bandpass filter
BRG	Bragg reflection grating
BWO	backward-wave oscillator
C^3I	command, control, communications intelligence
CAD	computer-aided design
CAM	computer-aided manufacturing
CAT	computer-aided testing
CBIR	crossbeam interferometer

CCTWT   coupled-cavity TWT
CDM   circular disk monopole
CDS   cyclic difference set
CFA   cross-field amplifier
CFT   cross-field tube
CIF   clutter-improvement-factor
CIM   computer-integrated manufacturing
C/N   carrier-to-noise ratio
COBE   cosmic background explorer
COMINT   communications intelligence
COPS   complex operations per second
COTS   commercial off-the-shelf
CW   continuous wave
CWESA   cyclotron wave electrostatic amplifier

DAR   distributed array radar
DBF   digital beamformer
DEW   directed-energy weapon
DF   direction finding
DME   distance-measuring equipment
DMPM   decoy microwave power module
DOA   direction of arrival
DR   dynamic range
DRO   dielectric resonator oscillator
DSB   double sideband
DSN   Deep Space Network
DSP   digital signal processing

EA-FEM   electrostatic-accelerator free-electron maser
ECCM   electronic counter-countermeasures
ECM   electronic countermeasures
EDFA   erbium-doped fiber amplifier
EFIE   electric field integral equation
EIKA   extended-interaction klystron amplifier
EIKO   extended-interaction klystron oscillator
EIRP   effective isotropic radiated power
ELINT   electronic intelligence
EM   electromagnetic
EMC   electromagnetically coupled
EMI   electromagnetic interference
EMM   electromagnetic missile
EMP   electromagnetic pulse
EOM   electrooptic modulator
EPC   electronic power conditioner
ERP   effective radiated power
ESA   electrostatic amplifier
ESM   electronic support measures

ESR	equivalent series resistance
ESTAR	electronically scanned thinned array radiometer
ETHER	Energy Transmission toward High altitude long endurance airship ExpeRiment
EW	electronic warfare
E–W	East–West

FEA	field emitter array
FET	field-effect transistor
FGP	fiber grating prism
FH	frequency hopping
FMA	ferrite microstrip antenna
FMCW	frequency modulated continuous wave
FMICW	frequency modulated interrupted continuous wave
FOV	field of view
FPA	final power amplifier
FSS	frequency-selective surface

GA	genetic algorithm
GBR	ground-based radar
GEO	geostationary Earth orbit
GLONASS	GLObal'naya NAvigatsionnaya Sputnikovaya Sistema
GMTI	ground moving-target indication
g.o.	geometric optics
GPS	global positioning system

HALROP	high-altitude long-range observing platform
HBT	heterojunction bipolar transistor
HDMP	high-density microwave packaging
HEMT	high-electron-mobility transistor
HGA	high-gain antenna
HPA	high-power amplifier
HPBW	half-power beamwidth
HPF	high-pass filter
HRR	high range resolution
HTS	high-temperature superconductor
HWSIC	hybrid wafer-scale integrated circuit

IC	integrated circuit
ICBM	intercontinental ballistic missile
IF	intermediate frequency
IFB	instantaneous fractional bandwidth
IFF	identification-friend-or-foe
iMEMS	integrated MEMS
IMPATT	impact avalanche transit time
INSAR	interferometric SAR
IRD	image rejection downconverter

ISAR	inverse synthetic aperture radar
ISL	intersatellite link
JEM	jet engine modulation
JORN	Jindalee Operational Radar Network
JPL	Jet Propulsion Laboratory
J/S	jammer-to-signal ratio
KGD	known good die
LAN	local area network
LBT	linear-beam tube
LCC	life-cycle cost
LEO	low Earth orbit
LHC	left-handed circular
l/rhs	left-/right-hand side
LNA	low-noise amplifier
LO	low observable or local oscillator
LOS	line of sight
LPI	low probability of interception
LPIR	LPI radars
LSA	large self-cohering array
LTA	lighter-than-air
MAFET	microwave and analog front-end technology
MAG	maximum available gain
MAM	multibeam array model
MBA	multibeam array
MBK	multiple-beam klystron
MCA	multichip assembly
MCM	multichip module
MEMS	microelecromechanical systems
MEO	medium Earth orbit
MESAR	multi-function electronically scanned adaptive radar
MESFET	metal–semiconductor field effect transistor
MHDI	microwave high-density interconnect
MIMIC	millimeter and microwave integrated circuit (program)
ML	main lobe
MLS	microwave landing system
MMA	millimeter array
MMACE	Microwave & Millimeter-wave Advanced Computational Environment
MMIC	monolithic microwave integrated circuit
MMPM	millimeter microwave power module
MMT	Multiple Mirror Telescope
M/NRA	minimum-/null-redundancy array
MODFET	modulation-doped FET
MOSAR	Maquette Orientee pour un Systeme d'Analyse de Resonances

MOSFET	metal oxide semiconductor field-effect transistor
MPM	microwave power module
MRA	minimum-redundancy array
MSAG	multifunction self-aligned gate
MSP	microwave signal processing
MSSL	mean-square sidelobe level
MTBCF	mean time between catastrophic failure
MTBF	mean time before/between failure
MTI	moving-target indicator
NF	Noise figure
NMA	Nobeyama Millimeter Array
NNEMP	non-nuclear electromagnetic pulse
NOAA	National Oceanic and Atmospheric Administration
NPD	noise power density
NRA	null-redundancy array
NRAO	National Radio Astronomy Observatory
NRE	nonrecurring engineering
NRL	National Research Laboratory
NTT	new technology telescope
OBK	one-beam klystron
OBP	onboard processing
OMT	orthomode transducer
OTHR	over-the-horizon radar
OVLBI	orbiting VLBI
PAE	power-added efficiency
PAR	precision approach radar
PBA	perfect binary array
PCA	printed-circuit antenna
PCB	printed-circuit board
PD	pulsed Doppler
PHEMT	pseudomorphic HEMT
PIN	positive-intrinsic-negative
PM	pseudomorphic
POC	proof of concept
POI	probability of interception
PRF	pulse repetition frequency
PSM	polarization scattering matrix
RADANT	RADome ANTenna
RAM	radar absorbing material
RAMP	radar modernization program
RBF	radial basis function
RCS	radar cross section
Rectenna	RECeive anTENNA

RF	radiofrequency
RFI	radiofrequency interference
RHC	right-handed circular
RIN	relative-intensity noise
ROTHR	relocatable over the horizon radar
RSLL	relative sidelobe level
RWR	radar warning receiver

SA	surveillance array
SAA	shared aperture array
SAG	self-aligned gate
SAR	synthetic aperture radar
SAW	surface acoustic wave
S/C	signal-to-scatter (ratio)
SDH	selectively doped heterostructure
SETI	search for extraterrestrial intelligence
SFDR	spurious-free dynamic range
SHARP	stationary high-altitude relay platform
SIR	spaceborne imaging radar
SKA	square kilometer array; also large-N SKA where N is equal to 1000 to 2000
SL	sidelobe level
SLIC	system-level integrated circuit
SMM	spiral-mode microstrip
SMPS	switched-mode power supply
SNR	signal-to-noise ratio
SOIC	small outline integrated circuit
SS	solid-state
SSB	single sideband
SSD	solid-state device
SSPA	solid-state power amplifier
Starllite	surveillance targeting and reconnaissance satellite

TDMA	time domain multiple access
TDRSS	Tracking and Data Relay Satellite System
TDU	time delay unit
TDWR	terminal Doppler weather radar
TEG	two-dimensional electron gas
TEM	transverse electromagnetic mode
TOA	time of arrival
TOI	third-order intercept
T/R	transmit/receive
TSA	tapered slotline antenna
TWS	track while scan
TWT	traveling-wave tube
TWTA	TWT amplifier

UAV	unmanned aerial vehicle
UHF	ultrahigh frequency
URR	ultrareliable radar
UWB	ultra-wide bandwidth
VCO	voltage-controlled oscillator
VLA	Very Large Array
VLBA	very long-baseline array
VLBI	very long-baseline interferometer
VM	vector modulator
VMD	vacuum microelectronic device
VPB	vacuum power booster
VSWR	voltage standing-wave ratio
WDM	wavelength division multiplexing
WPT	wireless power transmission
YBCO	$YBa_2Cu_3O_7$
YIG	yttrium–iron garnet

1

Systems and Applications

Boy you guys in sales are all the same. You remind me of the farmer in 1850. If you asked him what he wanted, he would say he wanted a horse that was half as big and ate half as many oats and was twice as strong. And there would be no discussion of a tractor.

D. T. Kearns and D. A. Nadler, *Prophets in the Dark*

In this book we have adopted a holistic view of array systems that perform the radar, electronic warfare (EW), radiometric, and communications functions. As all array systems share the same theory, this fundamental premise seems eminently reasonable. At a real-world level, it is well known that robust radar systems cannot be realized without knowledge of EW systems that can disable them. Additionally, the survival of a modern high-value platform, such as a ship or aircraft, depends on its radar, EW, and communication systems. Similarly, the efficient management of a busy airport depends on radar and communication systems that monitor meteorological phenomena and manage the traffic on its tarmac and within its airspace. Lastly, a systems designer familiar with these diverse systems can respond to the challenges imposed by the recent trend toward the integration of systems that derive interelated and interdependent information.

While there are many similarities between array systems used by different communities of researchers, the differences between them are also considered and emphasized. It is hoped that our approach enriches and deepens the reader's understanding of array systems and allows a designer imbued in one discipline to utilize the techniques and approaches developed by another discipline.

The communications community pioneered the use of phased arrays to improve the communication links between the United States and the United Kingdom for the first time in the late 1930s [1]. Next the radar and radioastronomy communities adopted array systems and we can now state without any reservations that:

- Arrays have redefined radar functions [2].
- Radioastronomy became a respectable branch of science only after it adopted array-based radiotelescopes.
- Phased arrays continue to provide the impetus for new and significant developments in other sciences.

In our coverage of systems we include:

- Defense radars such as the PAVE PAWS/COBRA DANE, over-the-horizon radars (OTHRs) and airborne early warning (AEW) systems.

- Affordable target identification systems.
- Flying phased arrays, better known as synthetic aperture radars (SARs) and inter-ferometric SARs.
- Bistatic/multistatic radars and low probability of intercept (LPI) radars.
- EW and polarimetric systems.
- First-generation radioastronomy array systems, radiometric systems, and proposals for second-generation radioastronomy arrays.
- Proposals for distributed array radars (DARs), and pseudo-satellites that can provide communication/surveillance functions over an area of 1000 km in diameter on the Earth's surface and are powered by ground-based array systems.
- Proposals for wideband, truly multifunction arrays that can perform: (i) the radar/radar-related and other functions essential for the survival of high-value plat-forms; or (ii) the functions essential for the efficient management of busy airports.

Some of the branches of science that benefit from the products of the above systems include: topography, hydrology, ecology, tectonics/erosion, meteorology, oceanography climatology and radioastronomy/planetary radar astronomy. Additionally, the products of some of the systems considered contribute to environmental monitoring, the man-agement of Earth resources, earthquake studies, and global volcano monitoring.

Can we delineate a typical or notional phased array? What are the defining character-istics of arrays? What makes phased arrays so versatile? What are the radar/EW, communication functions and radioastronomy aims? For a book dedicated to phased arrays, the answers to these questions constitute an appropriate starting point.

Throughout this chapter, the drivers for future phased array systems are clearly identified and a strong case is made for the evolution of novel systems that meet the challenges of our era.

1.1. PHASED ARRAYS AND WHAT THEY OFFER

The collecting area of a conventional filled aperture is a continuous surface, usually a paraboloid, whereas the collecting area of a typical array is made up of several antenna elements that take many forms. If we assume a receive mode of operation and divide the surface of the paraboloid into many contiguous segments, the powers received by all the segments are added constructively at the focus of the paraboloid. Similarly, the powers received by the many array antenna elements are added constructively at a convenient summing point. A corresponding arrangement holds if we assume a transmit mode of operation, and the spatial resolution attained for both systems is governed by diffrac-tion.

While the antenna segments of a filled reflector are contiguous, the antenna elements of the array are separated by a distance that can range from a few millimeters to thou-sands of kilometers. All antenna elements, however, are 'electrically' connected and the received signals are suitably phased before they are combined coherently.

Despite the similarities between the two types of apertures, there are fundamental differences between them. Let us assume that the gain, G, of the filled aperture of diameter d is taken as a reference. The gain of a densely populated array of diameter d

is G_1 and is approximately equal to G. The gain, G_2, of a sparsely populated array of diameter d, however, is reduced in proportion to its collecting area so that $G_2 << G$ and equation (2.78) in section 2.5.5.2, defines the gain of thinned apertures.

The diameter of radar array is a few tens of meters and is either densely populated or lightly thinned. By contrast, a typical radioastronomy array has a diameter d_{max} that can be comparable to the Earth's diameter. The same array is highly thinned and its gain, compared to that of an aperture having a diameter comparable to the Earth's diameter, is many orders of magnitude lower. It is a simple trade-off between realizable radar array systems having maximum gain and moderate spatial resolutions and radioastronomy systems having very fine spatial resolutions and low gains—see sections 1.5.3 and 1.7.8, Themes 2 and 3.

From optics we know that when a single slit is illuminated by a coherent source, the resulting diffraction pattern on a screen has a main lobe and sidelobes. If, for instance, the illumination function is uniform across the slit, the resulting pattern is a $\text{sinc}^2 x$ function, where x is the linear distance along the screen. We also know that the illumination function and the diffraction pattern constitute a Fourier transform pair. Similarly we recall from optics that if a linear diffraction grating is illuminated by a coherent source, the resulting diffraction pattern consists of several grating lobes repeated along the screen.

A filled aperture and an array yield far-field patterns that can be likened to the diffraction patterns corresponding to a single slit and a diffraction grating, respectively. Arrays, therefore, can have grating lobes and sidelobes, while the filled apertures have sidelobes only. As sidelobes and grating lobes introduce target ambiguities in the spatial domain, approaches to minimize the sidelobe level and the elimination of grating lobes of arrays assume considerable importance.

The last difference between these two types of apertures is subtle. Radar systems utilizing filled apertures or densely populated arrays can yield beams having low sidelobes; these apertures therefore perform the many radar functions efficiently. The highly thinned radioastronomy arrays on the other hand can have beams with unacceptably high sidelobe/grating lobe levels and the conventional approach to lowering these levels is to integrate the received signals emanating from weak celestial sources over long periods of time—see section 1.7.4. A different approach, however, is used when the imaging of a strong variable source like the active Sun is required. This approach calls for the lowering of the resulting sidelobes in quasi-real time before the imaging of the active Sun is undertaken—see section 2.4.2.

From the foregoing considerations it is clear that, despite the differences we have considered between arrays, there is a continuum between radar array systems having short integration times and radiometric/radioastronomy arrays having short or long integration times.

1.1.1. Typical Phased Arrays

An active radar array consists of m_1 antenna elements that take the form of conventional dipoles, slotted waveguides, horns, or microstrip antennas tightly spaced and each antenna element is connected to a transceiver. If the number of transceivers is n_1, $n_1 = m_1$ for an active array; by contrast, $m_1 > n_1$ in a passive array. A circular/elliptical aperture populated with antenna elements or transceivers yields a pencil or fan beam, respectively;

the latter beam is wide in the direction of the shortest dimension of the ellipse and narrow in the direction of its longer dimension.

A typical or notional radar phased array operating at a wavelength λ has an aperture of $50\lambda \times 50\lambda$ and yields a pencil beam of 1° [3]. If costs are not constrained, the aperture is densely populated by 10 000 antenna elements, the maximum number of elements spaced $\lambda/2$ apart. This interelement spacing insures that the array grating lobes are eliminated—see section 2.2.1.1. If costs are constrained, the aperture can be populated by a smaller number of antenna elements commensurate with the available funds. Let us assume that the number of antenna elements of the typical radar phased array is between 5000 and 10 000 with the understanding that some arrays will be larger or smaller.

In another realization, the array antenna elements take the form of conventional reflectors that are either located on terra firma or are spaceborne. These are the classical radiotelescope realizations where a few tens of parabolic reflectors are scattered over very large areas. A stable and sensitive receiver selected from a suite of receivers usually follows each antenna and this arrangement enables the radiotelescope to receive radiation from celestial bodies of interest at the frequency band usually defined by the receiver.

Compact radioastronomy arrays (or radiotelescopes) extend over some tens of kilometers and all antenna elements are connected by low-loss transmission lines such as coaxial cables, oversized waveguides, or optical fibers. The spatial resolution attained by these arrays is of the order of one arc second, comparable to the spatial resolution attained by Earth-bound optical telescopes.

The spatial resolution of compact arrays can be increased if the antenna elements are further separated, in which case the connection between antennas is not implemented by low-loss transmission lines. The antennas of these arrays can be dispersed in one or more continents, forming a very long-baseline interferometer, or VLBI network. The signals received by each antenna of the VLBI network are recorded together with exceptionally stable time signals derived from an atomic clock or hydrogen maser. After the observations are completed, the signals recorded by the network elements are coherently processed in one central processor.

If all antennas used are Earth-bound, the spatial resolution attained is of the order one milli-arc second; when satellite-borne antennas are used in conjunction with a VLBI network of antennas, the attained resolution is even finer—see section 1.3.3.

1.1.2. Multiple Beams/Nulls

Solar concentrators operate in a 'light bucket' mode because the phase of the contributions received by individual collectors is not preserved; by contrast, the phases of radiation received by the many antenna elements of a phased array are preserved. In what follows we shall assume that the prerequisite conditions derived in section 2.2.1.2 are satisfied so that we are justified in considering the phases between the wavefront of the incoming radiation and the antennas instead of the corresponding time delays.

In Figure 1.1, we illustrate an active linear phased array, operating in the receive mode and its cross-bar beamformer. The received signals are first amplified in low-noise amplifiers (LNAs) before entering the beamformer, which consists of several programmable vector modulators connected between the antenna lines and the summing ports where the derived beams are formed. Depending on the computer commands X and

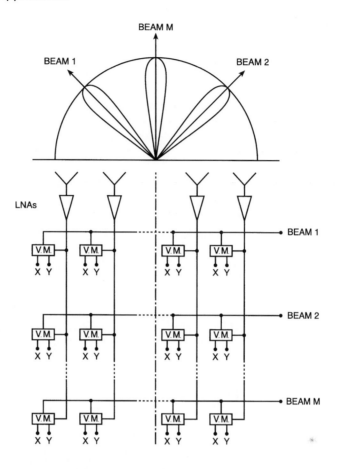

Figure 1.1 The antenna elements of a linear phased array operating in a receive mode followed by a cross-bar beamformer. The resulting beams, shown at the top of the figure, are available at the summing ports shown on the right-hand side of the figure. The vector modulators, VM, shown are computer controlled by commands applied at ports X and Y.

Y, each vector modulator can insert any phase angle between any antenna and any summing port. The synthesized beam M shown on the top of the figure, for instance, is formed when all the phases of the vector modulators (VMs) connecting the antennas and the summing port where the beam M is formed, are equal. Similarly, beams 1 and 2 are formed at 45° on either side of the antenna boresight axis when the VMs associated with these beams are suitably programmed to provide the appropriate phases; for a given array, the calculations of the required phases are straightforward because the interelement spacing and wavelength of operation are known.

As the programmable VMs are computer controlled, each beam can be independently scanned and some beams may be stationary or staring. The array can therefore 'see' a volume of space that is defined by the radiation patterns of its antenna elements and the constraints outlined in section 1.1.5.

The question that naturally arises is: How many independent antenna beams can be generated from an N-element planar array? Theoretically, the maximum number of

independent antenna beams realizable is related to the number of non-redundant baselines between the antenna elements [4,5]. For some planar arrays the number of
independent beams can be as high as $N^2/2$ or even N^2 (when N is large). In the absence
of any knowledge of the spacings between antenna elements, we shall conservatively
assume that an N-element planar array can yield an equal number of independent
beams. For the typical array we have defined, the number of independent beams ranges
from 5000 to 10 000, and is considerable.

Each of the derived beams is often referred to as 'inertialess', to discriminate it from
the beam, derived from a conventional paraboloid, that can be scanned mechanically.
While the beams derived from both apertures can be scanned, the paraboloid has a non-
zero inertia.

Unlike the ideal beams shown in Figure 1.1, the synthesized beams have sidelobes
the magnitude of which is dependent on many factors such as the array illumination. We
have already noted that phased arrays, like diffraction gratings, have grating lobes that
have the same shape as the array main beam and are located at regular intervals of angular distance from the main beam—see Figure 2.1b. The positional ambiguity of targets
is reduced if (a) the array sidelobes are minimized and (b) either the grating lobes are
eliminated or their amplitudes are minimized.

We have already considered an active array operating in the receive mode. Every
antenna element is followed by a LNA, which in turn defines the array noise figure—
see section 2.5.3. A final power amplifier (FPA) is connected to the antenna of an
active array operating in the transmit mode, so that losses associated with transmission
lines or waveguides connecting the antenna and its transmitter are minimized. The
same array exhibits the following attractive characteristics:

- The powers generated by individual modules at the array face are low.
- The power density on a target is high and meets the system requirements.

So far it has been convenient to assume that VMs are simply phase-shifters; in reality VMs can provide quantized phase and amplitude settings between antenna elements
and beams under computer control. If all VMs are set appropriately, one can generate
nulls toward M directions. Similarly, one can have M_1 beams toward M_1 targets and M_2
nulls toward M_2 sources of interference, or jammers, so that spatial filtering is attained
with ease; here we have assumed that $M_1 + M_2 = M$, the total number of beam/nulls generated by the array.

In the radar context, the flexibility offered by phased arrays to generate scanning or
staring beams over a wide surveillance volume, coupled with the provision of spatial filtering toward jammers in an ever-changing, dynamic environment, provides invaluable
array characteristics.

1.1.3. Graceful Degradation of Performance

The conventional passive radar array utilizes one high-power transmitter, which is usually a tube amplifier or oscillator, to feed its many array elements. If the tube or 'bottle'
fails, the whole system is down, a hard-failure condition. In an era marked by demands
for continuous 24-hour, 365-day operation of systems, these failures are not tolerated.
Some systems utilize redundant transmitters but their bulk restricts their domain of

applicability to fixed installations. Other more attractive realizations of passive arrays are described in section 2.5.1.

Active phased array systems, on the other hand, utilize many hundreds or even thousands of solid-state transceivers or transmit/receive (T/R) modules, and this approach has the following attractive characteristics:

- The failure of a small number of modules causes an insignificant degradation in the system's performance: a fail-soft condition. The degradation of performance is mainly assessed by the increase in the sidelobe level of the array, while the decrease of the array gain is slight.
- For conventional arrays, self-calibration and self-healing procedures described in section 4.2.2 ensure that the array operates efficiently at all times—see also Theme 16 in section 1.6.15. For unconventional variants of arrays, delineated in section 1.8.1, a 'hot maintenance' capability can meet the current requirements for continuous system availability—see section 4.2.1.

Operationally these characteristics of radar phased arrays are very attractive.

1.1.4. Radiation Patterns on Demand

We have already seen that an array can yield high-resolution beams or nulls toward the directions of targets or jammers, respectively. These are but a couple of the many radiation patterns an array can generate.

A target is often tracked by a radar array with the aid of a sum and two difference beams. The sum beam has a maximum toward the direction of the target, while the difference beams have minima or nulls toward the target. One difference beam has two lobes on either side of the target in elevation, while the lobes of the other difference beam are along the azimuth direction; mathematically, a difference beam has a radiation pattern resembling the Rayleigh (squared) function.

So long as the target remains along the boresight of the radar, no error signals are generated by the difference beams; if the target slips from the radar's boresight axis, the power of the sum beam will decrease slightly but the power received by the difference beam is increased and is in turn used to generate an error signal that guides the sum/difference beams toward the target again. Receive phased arrays can easily generate sum and difference beams and the four difference beams are clustered around the required position. Satellite communication systems use sum and difference beams also. The term 'monopulse tracking' of targets is used to indicate that the target's position is derived from one radar pulse with the aid of sum and difference beams—see also section 2.2.3.3.

While one requires high-resolution beams to locate and track targets, airborne radars having low-resolution beams are required for terrain following/avoidance maneuvers; these maneuvers are often used by an aircraft to penetrate a surveillance territory. In section 1.6.8 we will examine in some detail the reasons why conventional radars are not effective in tracking targets that fly close to the ground or sea. Phased arrays can yield lower-resolution beams by introducing the appropriate phase gradient across the array; the comparison is made here with respect to the beam defined by diffraction. More generally, the same array can easily yield beams of the required shape by the introduction of appropriate phasing and implementation of simple signal processing operations.

1.1.5. The Unattractive Characteristics

Our account of the defining characteristics of phased arrays would be incomplete if we emphasized the suite of significant attractive characteristics and omitted any mention of their two unattractive characteristics. From another point of view, an early statement of these two characteristics will set the scene for the approaches taken to offset and/or minimize their influence on array systems.

We have already mentioned that phased arrays have grating lobes that are generally undesirable. For a receive array, the presence of grating lobes results in positional ambiguities of targets. For a transmit array, power is wasted toward many directions. In the radar context, the conventional approach to eliminating grating lobes is to set the distance, d, between antenna elements to half the wavelength of operation—see section 2.2.1.1. For thinned arrays, the same spacing is randomized to eliminate grating lobes— see section 2.2.2.

The other unattractive characteristic of a phased array, when compared to a conventional aperture, is that the shape of the resulting beam varies as a function of the direction in which the beam is pointed—see section 2.2.1.2. As the array area, projected toward the direction in which its beam is pointed varies, the shape of its beam also varies—see section 2.3. By contrast, the half-power beam width (HPBW) of a conventional aperture, mounted on a fully steerable structure, is at least theoretically invariant as the direction of the antenna boresight axis varies.

To overcome this unattractive characteristic of phased arrays, one or more phased arrays are used to match the required azimuth/elevation coverage of the array. The maximum azimuth coverage of a phased array is usually ±60°, so a hemispherical coverage requirement is met by three arrays if no overlap in coverage is required or by four arrays if some overlap is required.

As can been seen, the above unattractive characteristics have a tolerable cost penalty only when compared to the suite of their attractive characteristics that we have already outlined. We shall consider other array attractors resulting from flexibility, which in turn affords the designer several degrees of freedom as we develop the appropriate frameworks.

If phased arrays can offer so much to the system designer, why is it that phased arrays are not widespread? The answer is complex, but we shall try to identify the essential isues here.

Conventional radars utilizing a filled aperture performed the functions required at the time. By contrast, the first phased array radars were realized for national defense and costs were allowed to be independent variables. With the passage of time, the requirements increased and the challenges multiplied to such an extent that conventional radars could no longer be considered for future systems. On the other hand, the costs of phased array systems (i) are better understood, and (ii) have decreased significantly through the evolution of affordable mass-produced monolithic microwave integrated circuits (MMICs) that are used for the T/R modules, the most costly building blocks of phased arrays.

As cost is no longer considered as an independent variable for many systems, we shall re-visit this topic at every opportunity; the cost of MMIC-based T/R modules is considered in detail in Chapter 4.

Although the communications community first used phased arrays, the impetus for phased array developments came mainly from the radar and radioastronomy communities; it is therefore appropriate to consider the radar and radioastronomy fundamentals before we explore the related applications.

1.2. RADAR FUNDAMENTALS AND APPLICATIONS

A monostatic radar transmits and receives electromagnetic (EM) radiation either through a common antenna or through adjacent antennas. The EM radiation illuminates targets or objects of interest and the scattered radiation is further processed by an elaborate receiving system that can discern targets from non-targets. Objects of interest can be clouds, aircraft, swarms of birds or insects, cars, or ships. We shall refer to objects of interest as targets thereafter. The distinguishing characteristic of an active sensor, like radar, is its ability to measure the range, R, of a target, i.e. the radar-to-target distance. The comparison is made here with respect to passive sensors.

The radar cross section (RCS) of a target is defined as the ratio of the power of radio waves that a target scatters back in the direction of the radar to the power density of the radar's transmitted waves at the radar's range. The aim of armaments manufacturers is therefore to reduce the RCS of ships, aircraft, and armed vehicles, while the aim of radar manufacturers is to realize systems with increased sensitivity. The target's RCS is usually measured in m^2—see section 1.6.2.1.

A bistatic radar uses antennas for transmission and reception at sufficiently different locations that the angles or ranges to the target are significantly different. Essential transmitter information, however, such as location, pointing data, and waveforms, is required at all receivers—see section 1.6.6.

Let us consider a monostatic pulsed radar that consists of an aperture of diameter D, and that operates at a frequency f or wavelength λ. Its pulse period is T, the pulse duration is τ, pulse repetition frequency PRF = $1/T$ and bandwidth is B. The resulting system beamwidth, $\theta = b\lambda/D$, where b is a beam broadening factor—see section 2.2.1.2.

Let us assume that Δt ($\Delta t < T$) is the time interval between the transmit and receive pulses. If the target is moving, the resulting Doppler frequency shift is Δf and the radar–target range rate is R'. The radar system, aided by the following equations, can measure the target's range, R, and range rate R', with range and azimuth resolutions ΔR and $\Delta\alpha$, respectively.

$$R = \frac{\Delta t}{2} c \tag{1.1a}$$

$$R' = \frac{\Delta f}{2} \lambda \tag{1.1b}$$

$$\Delta R = \frac{c\tau}{2} = \frac{c}{2B} \tag{1.1c}$$

$$\Delta\alpha = \theta R \tag{1.1d}$$

where c is the velocity of light. In equation (1.1c) we have used the relation $B = 1/\tau$, on the assumption that the receiver is well matched to deliberate modulations of the signal often used for range resolution purposes, e.g. pulse compression, within τ. Pulses lasting 1, 0.1, and 0.01 μs, for instance, yield range resolutions of 150, 15, and 1.5 m and the corresponding radar bandwidths are 1, 10, and 100 MHz respectively.

For a moving target, its Doppler shift is used not only to determine the target's range rate but also as a discriminant against clutter or returns resulting from stationary masses of land or sea. In more general terms, clutter constitutes any unwanted returns that usually mask the target returns. Comprehensive studies of the effects of clutter on radars constitute a significant body of knowledge that allows the user to discriminate targets from clutter.

1.2.1. Main Radar Functions

Conventional radars are used to detect targets through clouds, mist, smoke, rain, and haze; after detection, the targets are tracked. Pro tem the main radar functions are:

- To survey a certain volume of space frequently; typically the surveillance function is performed over a period, T_2, that varies between 2 and 10 s and the surveillance coverage is usually hemispherical. Airport radars that consist of a reflector rotating in azimuth are typical examples of conventional radars.
- To detect and track as many targets as possible; if the radar is used to defend a high-value asset, the task involves the detection of targets as soon as possible so that defensive actions can be undertaken. Similarly, early detection of targets within an airport's airspace facilitates the efficient management of operations.

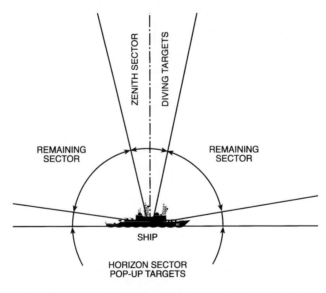

Figure 1.2 The surveillance volume of a ship's radar can be divided into the zenith and horizon sections where diving targets and sea-skimming missiles are expected. Other targets are expected in the remaining surveillance sector.

- To determine as many target parameters as possible. The determination of range, bearing and range rates for every target constitutes a minimal requirement.

We shall augment the above list of radar functions as we develop the required framework.

In the military context, it is often convenient to divide the hemispherical surveillance volume of a radar into three sectors as illustrated in Figure 1.2. A high-value platform such as a ship needs protection from diving targets in the zenith sector and from sea-skimming missiles that pop-up from the horizon. Other targets and seekers occupy the remainder of the surveillance volume. While the zenith sector is of marginal importance to airport radars, the same volume of space is of crucial importance to a ship's survival. As we shall see in section 1.6.8, sea-skimming missiles pose another significant threat to ships because conventional radars are unable to detect them.

The radar range for pop-up targets extends to the ship's horizon; the range for diving targets can be medium, while the range for targets occupying the remaining surveillance volume is long.

In Table 1.1 we list the time spent by a four-face phased array operating in the 4–8 GHz band [6] to perform the listed radar functions as a percentage of T_2. The radar under consideration operated in a peace-time environment, where the few targets detected were tracked. Naturally the proportion of time spent performing the different radar functions is scenario-dependent.

Table 1.1 Time spent as percentage of T_2 for the different radar functions [6]

Radar function	Time spent as percentage of T_2
Surveillance	
Long range	50
Horizon	26
Medium	13
Short	3
Tracking	7
False alarm	1

In a dense multitarget environment, a radar has to detect and track up to one thousand targets at a time. With the passage of time, and as the density of air travel increases, the number of targets at any time is bound to increase.

In Table 1.2 we list the commonly used designations of frequency bands. The 4–8 GHz band, for instance, is designated as C-band. In this book we shall mainly use the radar band designations. In peace time, spectrum management regulatory bodies usually assign the frequency bands available for different regions for different applications.

Following the Gulf incident when an Iranian Airbus was mistakenly identified as an F-14 fighter and was downed [7], radar users are no longer satisfied with seeing blips on radar screens; they would like to recognize friendly from unfriendly targets. The systems used for the identification of targets range from elaborate and expensive imaging systems to affordable target identification systems, considered in sections 1.2.2.4 and 1.6.11, respectively. The identification function is therefore another important radar function and is usually performed by a dedicated radar.

Table 1.2 Designation of frequency bands

Frequency	Wavelength λ	NATO	Frequency band	Radar band designation
250 MHz	1.2 m	A	MF HF VHF 230	
500 MHz	60 cm	B	UHF	P
750 MHz	40 cm	C		
1000 MHz	30 cm			
2 GHz	15 cm	D		L
3 GHz	10 cm	E	3 GHz	S
4 GHz	7.5 cm	F		
5 GHz	6 cm	G		
6 GHz	5 cm			C
7 GHz	4.3 cm	H		
8 GHz	3.75 cm		UHF 8	
9 GHz	3.33 cm	I		X
10 GHz	3 cm		12	K_u
		J	18	
20 GHz	1.5 cm		30 GHz 27	K
		K		
40 GHz	0.75 cm		42 $\frac{40}{}$	K_a
		L	54 $\frac{46}{}$	Q
60 GHz	0.50 cm		92 $\frac{62}{}$	V
		M		
100 GHz	0.3 cm		96	W
			137	
			143	D
300 GHz	1 mm		300 GHz	

(Frequency band column middle-to-lower region labelled vertically: Millimeter waves EHF)

The above functions are required during benign conditions and in the presence of rain, clutter, chaff, and intentional and unintentional jammers. While intentional jammers need no elaboration, unintentional jammers comprise transmissions from broadcasting stations, other radars, or communication systems.

An EW system typically detects radar/communications systems by their transmissions, measures their pertinent characteristics (e.g. frequency and bearing), and then emits high jamming powers toward them. If the radar/communication systems have no in-built defenses against jammers, the systems will be disabled.

We shall consider EW systems in some detail in section 1.6.13 because (i) it is impossible to design radar systems able to operate efficiently in realistic environments without a knowledge of the capabilities of EW systems; and (ii) it is impractical to adapt and/or modify radars designed to operate in benign environments so that they cope with the challenges imposed by realistic environments.

1.2.2. Applications

It is no secret that the early radars were used to defend several countries at different epochs and that the same practice continues to this day; high-value assets such as aircraft and ships also use radars for their defense. Here we shall consider three archetypical radars of that era, but many more types of radars still in operation are described in reference [8].

If all the elements of a phased array are connected in phase, the resulting beam is along the boresight axis of the aperture. When this phase array is gimbaled in two dimensions, the resulting beam can be steered. Most US fighter planes such as the F-18s utilize arrays of this kind mounted on the nose of the aircraft; the radar performs a multitude of functions to ensure the survival of the fighter. The generic characteristics of these radars are as follows [9]:

- Antenna size is 750 mm $\times$ 500 mm.
- Its MTBF (mean time between failures) of 150–200 hours is low owing to the mechanical movement of the aperture.
- The weight is approximately 140 kg.
- Although the inertia of the system is low compared to that of a paraboloid, it still represents a fundamental limitation when the radar has to track multiple high-velocity targets and/or when the radar has to perform several functions.
- The acquisition cost is around $1.8M (with limited spares), while the lifetime cost is typically 2–3 times the acquisition cost over a 15–20-year period [10].

The nose radar of the B-1B consists of an X-band steerable phased array used for navigation, penetration, weapons delivery (see section 1.2.2.1), and air-refueling tasks [11]. This radar does not share the limitations of the mechanically steerable radars.

The S-band AN/APY-1 radar is part of the E-3A AWACS (airborne warning and control system) that consists of a 707 civilian aircraft on top of which the antenna system, enclosed in a rotodome, is mounted. The oblate ellipsoidally shaped radome rotates with the antenna in azimuth and the antenna is electronically scanned in elevation. Electronically steerable arrays in one dimension are relatively easier to realize and less expensive, compared to arrays that are electronically steerable in two dimensions. The E-3A is designed for airborne warning and control of airborne assets in conflict; more specifically, the radar is able to track low-flying aircraft that cannot be tracked by ground-based radars.

The radar array has the following characteristics [9]:

- The antenna consists of a planar waveguide slot array of 4000 slot radiators and its size is 7.5 m $\times$ 1.5 m.
- The acquisition cost of an E-3A aircraft system is about $268M, 80% of which is attributed to the radar system.
- Its sidelobes are ultra-low at the −40 to −50 dB level in azimuth.

The last archetype radar array we shall consider is the PAVE PAWS phased array system that consists of two phased arrays each having a 120° coverage so that the total surveillance volume of the system is 240° in azimuth [12]. The system is designed to provide early warning of attacks by submarine-launched ballistic missiles and to aid in tracking satellites at the UHF band.

Figure 1.3 The antenna elements of the PAVE PAWS radar system. (Courtesy Dr. Eli Brookner.)

Each face has a diameter of 31.09 m and 1792 antenna elements followed by solid-state T/R modules each of which can generate 400 W. As the total number of antenna element positions is 5354, the array is not fully populated.

Assuming that a target has a radar cross section of 10 m^2, the radar range is about 5556 km; PAVE PAWS systems have been built at four locations in the United States. The array elements, shown in Figure 1.3 are bent dipoles and the array can select its polarization.

With the passage of time, radars have been used for many applications of significant military/applied science value. Here we shall outline some representative radar applications to illustrate the range of possibilities; important as these applications might be, the promise of some as yet unrealized proposals is of considerable import.

1.2.2.1. Fire control/Mid-course Guidance

A radar is often required to guide a missile launched to intercept a target [13]. In one realization a radar tracks the designated target and a beam-riding missile, launched to intercept the target, follows or 'rides' the radar's beam. Eventually the radar's beam guides the missile to its target. In another realization the radar illuminates the target and the missile is guided to its target by the reflections off the target.

Some missiles have their own radar that guides them to their targets; others are guided to their target by detecting the radiations emanated by the targets; these emanations are often jamming powers.

1.2.2.2. Airport Traffic Control

A suite of radars is used in civilian or military airports and airbases for air traffic control (ATC) [14] to manage traffic flow, mobility, and safety. While the airport surveillance radar (ASR) performs the surveillance and tracking of airplanes in the airport's airspace, the airport surface detection equipment (ASDE) radar monitors the movement of equipment around the airport area. The precision approach radar (PAR) guides aircraft during final approach, and the terminal Doppler weather radar (TDWR) monitors meteorological phenomena and provides automatic detection of microbursts and low-level wind shear [15]. Wind shear induced by microbursts can result in a sudden, significant change in an aircraft's airspeed that has disastrous effects during the take-off or landing stages.

Bistatic radars can measure the full vector windfield throughout the complete surveillance volume [16]. Vector windfields are useful to many professional groups including meteorologists, weather forecasters, aviation safety groups, users of weather maps, and academics.

1.2.2.3. Navigation and Space Flight

In the civilian sector, radars provide aircraft with coarse navigation capability; measurement of heights above ground (altimeters); positional information through distance-measuring equipment (DME) systems; and search and rescue capability.

In the military sector, radars are used for blind low-altitude flight to avoid radar interceptions—see section 1.6.8; early warning and surveillance—see section 1.6.9.2; radar map-matching; and beacons used for rendezvous with aerial tankers.

The availability of the global positioning systems (GPSs) (NAVSTAR and GLONASS, GLObal'naya NAvigatsionnaya Sputnikovaya Sistema) has greatly simplified the navigation tasks of all mobile platforms.

While conventional radars can offer coarse spatial resolution that is useful for navigation, there is a large set of applications where high spatial resolution is a requirement. We have already included in our list of radar functions the requirement for high-resolution images of targets so that friendly airplanes can be recognized and distinguished from others. Synthetic aperture radars (SARs) and inverse SARs (ISARS) can provide the required spatial resolution.

As the scope of space missions is broadened, radars have been used to perform a number of critical tasks including landing aids; rendezvous; and mapping/sounding. The wider role that radar plays in planetary explorations is outlined in section 1.4.

1.2.2.4. High-resolution systems: SARs and ISARs

Airborne radars having fine spatial resolutions are required for many military and civilian applications such as mapping of large areas of land that are normally under cloud cover; surveillance and reconnaissance of war zones or borders; and monitoring of the sea-state of oceans.

To increase the radar's range resolution one has to increase the radar's bandwidth; similarly, azimuth resolution can be increased by an increase of the aperture's size; or by a decrease of the wavelength of operation. However, there are constraints related to

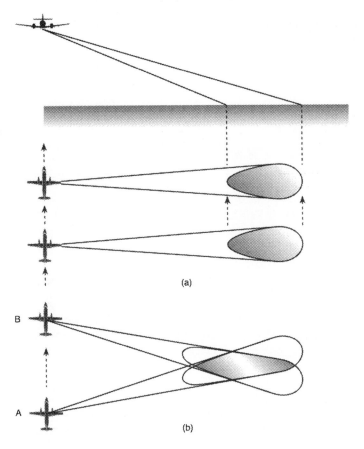

Figure 1.4 Airborne SAR systems: (a) conventional SAR mode; (b) spot-SAR mode.

the size of the airborne aperture and the system's frequency of operation. The latter constraint is imposed by the available power from current transmitters and the atmospheric attenuation at millimeter wavelengths. Furthermore, inspection of equation (1.1d), implies that, as the radar range increases, the attained azimuth resolution deteriorates. Ideally we require that the spatial resolutions attained be range-independent.

We have already seen that an aperture can be synthesized by a phased array that has antenna elements populating an area commensurate with the required spatial resolution. Similarly, an aperture can be synthesized by moving an airborne antenna pointed toward a direction perpendicular to the line of flight, as illustrated in Figure 1.4a; it is here assumed that the aircraft flies along a straight line at constant speed and altitude. As can be seen in the figure, the radar antenna on board the aircraft 'looks' sideways, and the illuminated ground is shown as a shaded area.

As the aircraft occupies different positions, the airborne antenna occupies the positions that the antenna elements of a phased array would occupy, and thus a phased array is simulated. This arrangement, however, cannot be used to attain infinitesimally fine resolutions because the real antenna beam ceases to illuminate the same area on the ground as the airborne platform follows its line of flight.

If ℓ is the length of the antenna, the minimum azimuthal resolution distance $d_{\text{azim}}|_{\text{min}}$ is given by the equation

$$d_{\text{azim}}\Big|_{\text{min}} = \frac{\ell}{2}$$

Thus the shorter the airborne antenna is, the wider is its beamwidth, the longer the antenna illuminates the same area on the ground, and the smaller is $d_{\text{azim}}|_{\text{min}}$. The other factor that affects the dimensions of the antenna is the requirement to attain an acceptable system signal-to-noise ratio, SNR. As the antenna size decreases, the transmitted effective power and received power are decreased and the system SNR is diminished. For an antenna 3 m in length (a typical dimension for airborne SARs), $d_{\text{azim}}|_{\text{min}}$ is 1.5 m. As can be seen, $d_{\text{azim}}|_{\text{min}}$ is independent of the frequency of operation and the radar range.

In our considerations we have assumed that:

(i)   the SAR coherently processes the returns from the area it illuminates;
(ii) the aircraft's motion is compensated; and
(iii) the SAR is focused, so that geometric distortions are taken into account [13].

Synthetic arrays like phased arrays have sidelobes and grating lobes; the minimization of the sidelobes of a synthetic array is achieved by suitably weighting the contributions of its elements, while its grating lobes are eliminated by the suitable selection of the radar's pulse repetition frequency (PRF). More explicitly, the PRF selected to minimize the grating lobe when the aircraft travels with a velocity V_R, is given by

$$\text{PRF} = \frac{V_R}{\delta} \qquad \text{where} \qquad \delta = n\, \frac{\lambda}{2 \sin \theta_N}\,, \qquad n = 1$$

and θ_N is the null-to-null azimuth beamwidth of the real antenna [13].

In Figure 1.4b the operation of SAR operating in the spotlight mode is shown; the SAR's antenna beam is directed to illuminate the same area on the ground as the aircraft moves from A to B; with this arrangement the attained d_{azim}, is independent of ℓ. This being the case, a longer antenna can be used to increase the radar's SNR.

The SAR frequency is selected to meet diverse requirements; L-band operation is used if some penetration into the ground or foliage is required and for studies related to aspects of biomass and soil moisture. X-band operation is used for studies related to short vegetation, crops, leaves, and twigs. Other frequency bands are used to map different vegetations, bush fires, or oil slicks. For remote sensing of Earth resources and crops, the aircraft usually carries a suite of SARs, each having a dual polarization capability and operating at a different frequency band. The NASA/JPL airborne SAR system (popularly known as AIRSAR) is a typical system that operates at the P-, L- and X-bands with a full polarimetric capability [17]—see section 1.6.12 and the comments relating to Table 1.3.

Reference [18] is a comprehensive summary of the current status and future directions for spaceborne SARs. Applied science areas that have benefited from SAR imaging include ecology, hydrology, marine science, studies of ice sheets and glaciers, solid earth sciences, and topography.

Table 1.3 The essential characteristics of a representative sample of spaceborne imaging radar (SIR) SARs [18,19]

Satellite names	Frequency (GHz)	Year	Polarization[a]	Data	SAR technology	Antenna beam
SEASAT	1	1978	HH	Analog	Central Tx/Rx	Fixed
SIR-A	1	1981	HH	Analog	Central Tx/Rx	Fixed
SIR-B	1	1984	HH	Analog	Central Tx/Rx	Mechanical beam steering
SIR-C	1 and 5	1994	HH, HV, VH, VV	Digital	Distributed T/R modules	Electronic beam steering
SIR-C/X		1994	HH, HV, VH, VV	Digital	Distributed T/R modules	Electronic beam steering
L- and C-bands	1 and 5					
X-band	10		VV		Central Tx/Rx	Mechanical beam steering

[a] H, horizontal; V, vertical.

Table 1.3 lists the essential characteristics of spaceborne imaging radars (SIR) SARs. In the fourth column the system's polarimetric capabilities are listed and the notation used needs some elaboration. A system that transmits and receives horizontal (H) polarization, is assigned the designation HH; similarly, a system that transmits horizontal polarization and receives vertical (V) polarization is given the designation HV. In section 1.6.12, dedicated to polarimetric radars, we shall deduce that a system attains full polarimetric capabilities when it can be designated as HH, HV, VH and VV. The trend to systems having full polarization capabilities at many frequency bands is evident in the table and digital technology is increasingly adopted.

Future generations of SAR systems require dual-polarized antennas with steerable beams, a requirement that can be met by steerable, dual-polarized phased arrays [20]. The availability of affordable monolithic microwave integrated circuits (MMICs) is crucial to the realization of these systems [21]. To limit the level of azimuth and range ambiguities in the SAR signal, the physical size of the antenna used cannot be smaller than certain prescribed limits. A typical size for a spaceborne L-band SAR is 10×2 m [18]. In reference [22] the potential for SAR to be used in a number of operational areas is outlined. Representatives of realistic opportunities foreseen for the operational exploitation of SAR [22] include mapping and charting; resource monitoring and management; pollution and waste threats; mitigation of natural hazards; ocean and ice; and law enforcement and surveillance. A proposal has been made for a spaceborne radar capable of performing airborne moving target indication (AMTI), ground moving target indication (GMTI), and SAR functions [23]. The authors of this proposal claim that one multibeam cellular radar can perform these functions optimally.

While the operation of a SAR depends on the motion of the airborne radar with respect to the areas of interest, the operation of the inverse SAR, or ISAR, depends on the motion of the target with respect to the radar; the other important difference between the two systems is related to the frequency of operation. The SAR's frequency of operation is fixed, while that of the ISAR's varies continually within certain bounds.

An ISAR provides an inexpensive means of identifying a selected target. In one realization the ISAR has a single aperture and an instantaneous bandwidth, Δf, the center frequency of which, f_c, changes continually. The image is formed by utilizing stored information taken at different frequencies f_c. While Δf is defined by $\Delta f = c/2\rho$, the total bandwidth B is limited by the instrumentation available. In this relationship, c is the velocity of light and ρ is the range depth of the target's ensemble of scattering elements.

We have used the terms range and azimuth resolution for radars; in more general terms the same resolutions are often referred to as slant range and cross range resolutions. The slant and cross range resolutions, Δr_s and Δr_c, attainable by an ISAR system are given by [24]

$$\Delta r_s = \frac{c}{2B} \quad \text{and} \quad \Delta r_c = \frac{\lambda}{2\omega T_1}$$

where ω and T_1 are the target's rotation rate and integration time, respectively. As can be seen, the slant and cross range resolutions decrease when the total system bandwidth is increased and when the wavelength of operation decreases, respectively. Operation at millimeter wavelengths, therefore, holds the promise of improved slant and cross resolutions because of the inherent increased absolute bandwidth potential of these systems.

The limiting value of T_1 cannot exceed the coherence time of the target and typical values between 2 and 20 s have been reported [25]; thus the cross range resolution cannot be increased indefinitely. In references [25] and [24] typical applications of ISARs are reported.

1.2.2.5. SAR Interferometry

SARs usually record the amplitude and phase of each pixel and SAR interferometry is a technique that involves the interferometric phase comparison of successive SAR images of the Earth's surface; this mode of operation is better known as repeat-pass mode.

SAR interferometry allows the detection of subtle changes in the Earth's surface over a period of days to years with a scale, accuracy, and reliability that are unprecedented. More specifically, the coverage is global and movements of the order of a few millimeters are measured under all weather conditions. These measurements are possible provided the following conditions apply [26]:

- The topography of the region is accurately known.
- The position and orientation of the antennas during each image acquisition is known.
- Changes within a pixel between passes are not larger than one wavelength of operation.
- Within a pixel, the position of radar scatterers has not changed to any significant degree but the ensemble of scatterers has moved up, down, or sideways in some correlated fashion.

Let us consider the basic equations on which the SAR interferometry methodology is based. With reference to Figure 1.5, let us assume that two antennas A_1 and A_2 are pointed to a point P on the ground and that one antenna transmits and receives while the other receives only. ϱ and $\delta\varrho$ are the ranges of point P from the antennas A_1 and A_2 and the vector B is the baseline between the two antennas. The ordinate of point P, P_y is given by

$$P_y = h - \varrho \cos \theta \tag{1.2}$$

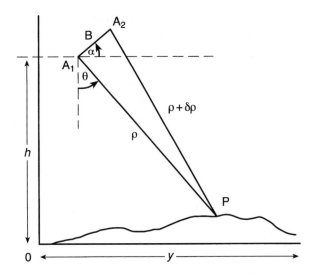

Figure 1.5 The basic imaging geometry of a SAR interferometer. A_1 and A_2 represent two antennas viewing the same point on Earth or a single antenna viewing the same point on two separate passes.

The relation

$$(\varrho + \delta\varrho)^2 = \varrho^2 + B^2 + 2\varrho B \sin(\alpha - \theta)$$

can be deduced by applying the law of cosines to the triangle A_1A_2P. The angle ϕ, the measured phase difference between the two antennas, is given by

$$\phi = \frac{\lambda}{2\pi}\,\delta\varrho$$

Additional techniques such as 'phase unwrapping' [27] are required to solve for the integer number of wavelengths and obtain absolute range. Lastly, an accurate estimate of vector $\boldsymbol{B}$ can be attained with the aid of the differential GPS system. The approximation $(\delta\varrho)^2 \to 0$ is made and for spaceborne geometries the term $(B^2/2\rho)$ can be ignored. From these basic equations, ground images are derived for interferometric SARs operating in many modes including the repeat-pass mode [26].

The limits and potential of radar interferometry are outlined in [28] and the airborne/spaceborne topographic applications are described in references [29] and [30], respectively.

Figure 1.6a illustrates the observed coseismic interferogram for the Landers earthquake (24 April–7 August 1992) and Figure 1.6b is a synthetic interferogram for the same earthquake [26,31]. The images extend over an area of 90 km × 110 km and one cycle of shading represents 28 mm of change in range; a comparison of the two images indicates that there is good agreement between the calculated (synthetic) and observed interferograms. As can be seen, SAR interferometry records reveal minute surface changes over a phenomenally wide area.

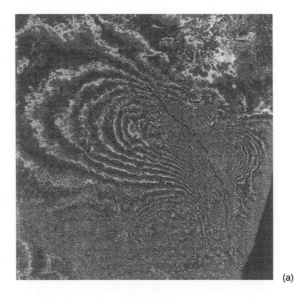

(a)

(b)

Figure 1.6 Interferograms of the Landers earthquake, 24 April–7 August 1992. (Courtesy JPL/NASA.) (a) The observed coseismic interferogram taken by using the interferometric SAR technique. (b) The calculated synthetic interferogram. The images extend over an area of 90 km by 110 km and one cycle of shading represents 28 mm of change in range [26]. The correlation between the calculated and observed interferograms is good.

The scientific applications of SAR interferometry are limited only by the imagination of the users but the following list demonstrates the exceptionally large range of applications [26]: earthquake studies; global volcano monitoring; hydrology, ecology, environmental monitoring, and global change; topography, tectonics, and erosion; and glaciers and ice sheets. Indeed seminal advances in many of these fields have already been reported [26]. It is hard to think of another technology that has contributed so much toward the understanding of the many problems that confront us.

It is well known that the monetary cost of each earthquake occurring in densely populated regions runs into billions of dollars, and fatalities can by as high as 5000 (e.g. in the Kobe earthquake). The magnitude 7.4 earthquake of 17 August 1999 in north-east Turkey levelled thousands of buildings and resulted in over 15 000 deaths. Detection of minute surface changes over large areas will significantly aid our understanding of earthquakes and ultimately lead to the early evacuation of populations under threat.

A method of 'coherence synthesis', i.e. the calculation of the interferometric coherence for any possible combination of the polarimetric state of the transmitting and the receiving antennas between two passes, is being pursued [32].

1.3. RADIOASTRONOMY AIMS AND APPLICATIONS

Although radioastronomy is a pure science, the techniques developed by radioastronomers have been used in many radiometric systems of considerable applied value. Radiometers are passive systems that receive and measure the radiation emanating or reflected from objects of interest, e.g. celestial sources, ground, or sea.

From time immemorial people have wanted to extend their knowledge of the cosmos and define their place in it. Konstantin Tsiolkovsky (1857–1935), the Russian research scientist in aeronautics and astronautics, struck a resonance when he wrote: 'Man will not always stay on Earth; the pursuit of light and space will lead him to penetrate the bounds of the atmosphere, timidly at first but in the end to conquer the whole of solar system.' We have always had the urge to explore whatever is out there. We know our past successes but what are the modern challenges, apart from terra-forming the planets and their moons for future habitation [33]?

In section 1.7 we shall review the contributions of radioastronomy to key astronomical questions such as identifying the origins of emissions in the cosmos, the assembly of the alphabet of the chemical evolution in our galaxy, and searches for extraterrestrial intelligence; lastly we note here that radioastronomical measurements have placed cosmology on a firm scientific base.

Radioastronomy has uncovered a radio universe that turned out to be as interesting as the universe explored by optical astronomers since antiquity. The complementary knowledge derived from observations taken at different wavelengths, (radio, optical, X-ray, UV) has helped us to assemble the wondrous mosaic of the cosmos we live in.

1.3.1. The Aims

The aims of radioastronomy are many, but for a book dedicated to phased arrays we shall restrict our field of view to generic observations that in turn interact with current theories or generate new working hypotheses leading to new theories and paradigms.

Unlike radar systems, radiotelescopes are passive systems that measure emanations from celestial sources of interest. One of the major aims of radioastronomy is to deduce the mechanisms of emission taking place at or near to the celestial sources of interest. (As radiotelescopes can only 'see' along a line of sight, the phenomena observed take place at or near the celestial source.) For this purpose, high spatial resolution maps of sources taken at different wavelengths are required. By comparing the received energy emanated by a celestial source at different wavelengths with a knowledge of typical energy versus frequency relationships attributed to different emission mechanisms, we can deduce which emission mechanisms are at work in and around the celestial sources.

1.3.2. The Quest for High Spatial Resolution

High-resolution maps meet the requirement to have only one source supporting an emission mechanism in each high-resolution pixel; measurements of the angular extent of a source and of the energy emanating from it at many frequencies are required before we can make any definitive statements about the nature of the emission mechanism.

How fine should the spatial resolution be? The answer is clearly as high as possible, but what is a reasonable figure to aim for, derived from previous experience? The majority of radio sources outside our galaxy (extragalactic radio sources) have an angular extent of less than one arc minute, and conventional optical telescopes have a resolution of one to half an arc second, which is deemed as the minimum requirement. Spatial resolutions of the order of sub-milli arc seconds are required to study compact radio sources such as the cores of galaxies and interstellar masers—see section 1.7.1. The epic journey to the attainment of these spectacular spatial resolutions can be traced historically through the developments of:

- Compact phased arrays, sparsely populating a real estate of a few km^2.
- Compact phased arrays, sparsely populating a real estate of over 1000 km^2.
- Very long-baseline interferometry (VLBI) networks. These networks electrically connect various radiotelescopes located in different continents.
- Orbiting very long-baseline interferometry (OVLBI) networks. These networks electrically connect various radiotelescopes located in different continents to radiotelescopes orbiting the Earth.

We have outlined some of the important differences between radioastronomy and radar arrays. Here we note the following additional differences. The surveillance volume of a radar array is fixed in space and extends over ±45° or ±60° in azimuth and 45° in elevation. This volume of space is electronically scanned at high speeds. By contrast, the instantaneous field of view of a radioastronomy array is defined by the diameter of the antennas used and is of the order of a few degrees to a few arc minutes (depending on the frequency of operation). The array horizon is defined by the shadowing effects between adjacent antennas. The image of a weak celestial source is formed by pointing all antennas to the source of interest and by recording the observations while the antennas track the source for as long as it is necessary to achieve the required image quality. A very weak source is observed over several days and the derived observations are integrated.

1.3.3. Compact Arrays, VLBI and OVLBI Networks

The quest for spatial resolutions of the order of one arc second drove the efforts of radioastronomers for a long time; this target spatial resolution was based on cost considerations and the need to match the resolutions of radio and optical images for comparative studies. If we assume that the resolution of an aperture of diameter D operating at a wavelength λ is given by the relationship λ/D, a one arc second resolution can be attained by a radiotelescope operating at 1 or 10 GHz if its diameter is 61.88 km or 6.188 km, respectively. Considering that the largest (filled aperture) radiotelescope has a diameter of 305 m (hole-in-the-ground variety, located in Arecibo, Puerto Rico [34]), these diameters are very large indeed. Conventional filled apertures cannot meet this minimum resolution requirement, but phased arrays can.

References [34–36] provide up-to-date descriptions of optical telescopes and radiotelescopes. A special issue on radio telescopes in reference [34] includes a detailed description of the Jet Propulsion Laboratory's (JPL) Deep Space Network (DSN) consisting of three systems, each having one 70 m antenna, located in Goldstone (USA), Madrid (Spain), and Canberra (Australia). The network links several space probes and spacecraft to JPL in Goldstone. The same reference includes descriptions of single dish antenna radiotelescopes such as the Arecibo and Haystack systems as well as detailed descriptions of several phased array-based radiotelescopes located in the United States, Europe, and Japan.

In reference [37] the Very Large Array (VLA) located in the Plains of San Agustin, west of Socorro, New Mexico, USA, is fully described; the 27 antennas of the radiotelescope are 25 m in diameter and are located at positions along a rail line that has a 'Y' geometry. Each arm of the Y is 21 km long and the antennas can be transported along the rail lines to form a radiotelescope having the required overall diameter. This configuration yields the resolution of an antenna 36 km across, with the sensitivity of a filled reflector 130 m in diameter. This premier radiotelescope is the largest compact radiotelescope in which the antenna elements are electrically connected by low-loss transmission lines, oversized waveguides in this case. Table 1.4 lists the pertinent characteristic of the receivers at VLA. As can be seen, the array is designed to operate mainly at centimeter wavelengths.

Table 1.4 The receivers of the Very Large Array, VLA [38]

	P-band	L-band	C-band	X-band	Ku-band	K-band	Q-band
Frequency (GHz)	0.3–0.34	1.34–1.73	4.5–5	8–8.8	14.4–15.4	22–24	40–50
Primary beam (arcmin)	150	30	9	5.4	3	2	1
Finest resolution (arcsec)	6	1.4	0.4	0.24	0.14	0.08	0.005
System temperature (K)	150–180	37–75	44	34	110	160–190	90–140

The Nobeyama Millimeter Array (NMA) operates at millimeter wavelengths. The array, located in Nobeyama, Japan, consists of six transportable 10 m antennas that can be located at stations along two tracks, one extending 560 m in the E–W direction and the other 520 m inclined 33° from the N–S direction. Plans exist to connect the existing 45 m fully steerable antenna, capable of operating at millimeter wavelengths, with the NMA. Table 1.5 shows the essential characteristics of the NMA; as can be seen, its spatial resolution approaches the one arc second. A view of the NMA is shown in Figure 1.7.

Table 1.5 The essential characteristics of the Nobeyama Millimeter Array, NMA [39]

Frequency band (GHz)	Bandwidth (GHz)	T_{SYS} DSB[a] (K)	Maximum resolution (arcsec)
80–112	2	150–300	3–4
112–115	2	200–500	2–3
140–155	2	200–500	1–2
230	Not available for open use		

[a] Total system temperature, double sideband

The next step in attaining spatial resolutions significantly less than one arc second involved the 'connection' of radiotelescopes located in different continents with the aid of highly accurate clocks synchronized by hydrogen masers. The observations are taken by pointing all radiotelescopes to the same source at a given time and by recording the data at each site together with accurate timing information. The data derived from all

Figure 1.7 A view of the Nobeyama Millimeter Array, NMA. (Courtesy Prof. Masato Ishiguro.)

antenna elements are then processed at one site after the observations are taken [40–42]. The systems known as very long-baseline interferometer (VLBI) networks operate infrequently by the cooperation of several national radioastronomy observatories located in Europe, North America, Asia, and Australia.

A dedicated VLBI network, the Very Long-Baseline Array (VLBA) [43–45] consists of ten 25 m radiotelescopes located in St Croix (Virgin Islands), Hawaii, Washington, California, Arizona, New Mexico (2), Texas, Iowa, and Massachusetts. The system can image celestial sources at frequencies ranging from 300 MHz to 45 GHz with a capability for extension to 86 GHz, and its spatial resolution ranges from 20 to 0.2 milli-arc seconds. The VLBA can operate in conjunction with the VLA and other VLBI networks established in Canada and Europe either to increase the observation time for certain sources or to attain a higher sensitivity and image quality.

The control of the VLBA operations and the processing of all the observations taken by the ten array elements is undertaken at the Array Operations Center located at Socorro, New Mexico, USA. The wideband high-density recording system allows unattended operation at a sustained data rate of 128 Mb/s for a day, and peak rates of up to 512 Mb/s.

With the passage of time arrays having spatial resolutions below one milli-arc second were realized by 'electrically' connecting Earth-bound and spaceborne radiotelescopes [46–49]. Observations of many radio sources have been taken with antennas on board the geostationary Tracking and Data Relay Satellite System (TDRSS) satellite [46] and antennas located in two continents (Australia and Asia (Japan)) at 2.3 GHz [47,48] and at 15 GHz [49]. The TDRSS was originally designed to relay data between ground stations and satellites in low Earth orbit via satellites in the geosynchronous orbit.

Radiations emanated by celestial sources at 2.3 or 15 GHz are received, amplified, coherently translated to 14 GHz, and relayed to Earth. A tone signal from a frequency standard located on the ground is transmitted to the satellite antenna, where it is used to phase-lock all the on-board oscillators.

The OVLBI technique yields resolutions of the order of sub-milli-arc second and the attained resolution allows observers to better probe energy generation and motions at the heart of exotic objects like quasars. More accurate measurements of the relative positions and motions can be made of features exhibiting water vapor maser emission within star-forming regions. These measurements aid in the establishment of the fundamental distance scale of the universe [47].

While the technique of image formation described above is particularly suitable for radioastronomy and geodesy, it could find applications in other fields.

1.3.4. Applied Science Radiometric Systems

Meteorologists, oceanographers, environmentalists, climatologists, and the military require measurements of rain rates, soil moisture content, land and sea surface temperatures, and atmospheric water vapor content under all weather conditions and over the entire Earth at sufficient spatial resolution to understand how the weather engine works and to predict future conditions [50]. Techniques used to derive the required measurements are outlined in reference [51].

Passive microwave remote sensing of the Earth's surface has evolved over time.

Surface parameters of fundamental geophysical interest that have been measured and monitored by spaceborne/airborne microwave radiometers include sea surface temperature, salinity, wind speed, soil moisture content, and arctic sea ice concentration [52,53, and references therein].

In the early 1980s, airborne radiometry at millimeter wavelengths was pioneered by using one radiometer to scan the scene as the aircraft was moving [54,55]. The radiometer, fitted with a horn, illuminated a flat reflector that was in turn mechanically scanned in such a way that its successive footprints on the ground formed lines perpendicular to the direction of flight. As the airplane followed its line of flight, a complete radiometric image of the Earth's surface was recorded. The sensitivity of these systems was limited by the radiometer's short dwell time on each pixel, while the spatial resolutions attained by the systems were limited by the size of the reflector used. If an airborne radiometer has adequate spatial resolution and sensitivity, it can be used to perform all-weather ground surveillance and reconnaissance functions of large areas on the Earth's surface. These passive systems can complement the active radar imaging systems that we have already considered. Approaches to increase the sensitivity and spatial resolution of airborne/spaceborne radiometers utilizing phased arrays will be considered in section 1.7.5.2.

1.3.5. Searches for Extraterrestrial Intelligence

The searches for radio signals due to non-physical phenomena, such as signals that might be used for broadcasting on other planets, are relatively inexpensive and have the potential for a 'non-threatening contact' with other 'civilizations'.

In recent years several radiotelescopes have been used in the search for extraterrestrial intelligence (SETI). These searches continue, supported by the public at large. The public support of SETI and the popularity of several television serials related to the central question 'Are we alone?' are testimony to the public's interest in the most fundamental question that remains unanswered despite so many advances in science and technology spanning several centuries.

As we shall see in section 2.7.3.4, proposals have been put forward for second-generation radiotelescopes that can perform the required astronomical observations and searches for extraterrestrial intelligence.

1.4. PLANETARY RADAR ASTRONOMY

The exploration of the solar system with the aid of radar techniques dates back to 1946, when echoes from the Moon were first detected, and reference [56] outlines subsequent developments in this field. Radar observations employ the largest antennas, powerful transmitters, low-noise receivers, and high-speed data acquisition computers. Some observations are taken with one radiotelescope used to transmit high powers and receive the scattered radiations from the Moon, asteroids, comets, planets and their satellites (monostatic mode); other observations are taken using bistatic radars. The radiotelescopes commonly used are the Arecibo, Haystack, VLA, and the JPL system located at Goldstone [34].

An impressive list of achievements can be attributed to planetary radar astronomy [56]:

- The assembly of a wealth of information about the geological and dynamical properties of asteroids, comets, the inner planets and their satellites.
- The establishment of the scale of the solar system.
- A significant increase in the accuracy of planetary ephemeredes.
- Constraints on theories of gravitation.

1.4.1. Recent developments and discoveries

The JPL radiotelescope in Goldstone has been used as the transmitter and the VLA as the receiver of a bistatic radar to make a series of ground-breaking discoveries:

- Titan, Saturn's largest moon, is not covered with a deep, global ocean of ethane as previously thought [57].
- Mars has a region, aptly named Stealth, that extends over 2000 km in east-west extent, displaying no echo to the very low level of the radar noise [58].
- The cap image of Mars is interpreted as arising from nearly pure CO_2 or H_2O ice with a small amount of Martian dust and a depth greater than 2–5 m [58].
- Radar imaging supports evidence that Mercury has polar ice [59].
- Similarities between the radar scattering properties of Mars' and Mercury's polar anomalies and those of the icy Galilean satellites (Jupiter's satellites Europa, Ganymede, and Callisto) support the hypothesis that Mercury's polar features are deposits of water ice inside high-latitude craters and other concavities perpetually shaded from sunlight [56].

1.4.2. The Clementine Experiment

On the 25 January 1994 the spacecraft Clementine [60] was launched; it circled the Moon on a 400 km by 2940 km, 5-hour orbit. A transmitter on board the spacecraft illuminated the lunar surface and the scattered power was received by one of JPL's Deep Space Network radiotelescopes. A plethora of data was obtained, including observations of the north and south poles of the Moon. A probable explanation of the observations taken of the permanently shadowed regions of the south pole is the presence of low-loss volume scatterers such as water ice [61]. Later observations corroborated the observations taken by Clementine.

The discovery of water ice on the Galilean satellites and Mars and Mercury and the detection of water ice on the permanently shadowed regions of the Moon's south pole rank as major discoveries of our times. The task of terraforming some planets [33] becomes easier if water ice is already formed on other planetary satellites and planets. More importantly, these discoveries taken in conjunction with other discoveries of water vapor sources (see section 1.7.1) contribute to our knowledge of the chemical evolution of galaxies.

1.5. THE SUBTLE PHASED ARRAY ATTRACTORS

Having considered the defining characteristics of phased arrays and what they offer, we shall now examine how phased arrays meet the challenges imposed by radar requirements and the aims of radioastronomy.

1.5.1. The Nexus between Surveillance and Tracking

Radars utilizing large reflectors rotating in azimuth are a common sight in civilian and military airports. The apertures rotate mechanically over 360° in azimuth to monitor the surveillance volume centered around the airport, and a full rotation usually takes a few seconds. As the radar rotates, it detects and tracks objects of interest such as planes and rain fronts. Given that the radar operates on a track-while-scan (TWS) mode, the surveillance function is inextricably coupled to the tracking function. This radar is thereafter referred to as the conventional system or as conventional radar.

By contrast, a phased array radar yields an inertialess antenna beam, which in effect decouples these two main radar functions, and operates on a step-scan mode. The following example serves the purposes of illustrating this important distinction. Let us assume that we have two radars that have the same sensitivity but that one is the conventional airport radar and the other is a phased array radar; let us also assume that a target is approaching both radars. The radars record a marginal detection that is slightly above the receiver's threshold toward the direction defined by the angles θ and ϕ of the approaching target. The operators of both radars wish to increase the dwell or integration time toward this direction, so that a decision can be made whether the recorded marginal detection is due to a target or to noise. Longer integration times improve the sensitivity of most sensors.

In the example we have considered, the conventional radar cannot stop the surveillance function and dwell toward the direction θ,ϕ, but the phased array radar can interrupt its surveillance and dwell toward the same direction until either the recorded marginal detection is dismissed as noise or a target is declared. The conventional system will have to go around four times before it can declare three detections of the target (or after it had four 'looks' at the target) [62]. This is the conventional way in which a target is declared when the radar operates in the TWS mode [62]. The criterion used to declare a target for TWS radars is therefore 3/4, while the corresponding criterion for step-scan radars utilizing phased arrays is 1/1.

The target is eventually detected by the conventional system only because its range has decreased considerably, a situation which could be life-threatening at times. In a typical example, Billam [62] assumed a set of reasonably realistic conditions and deduced that the same target is detected at approximately 58 and 92 km by a conventional and a phased array radar, respectively.

More generally, the proportions of time spent performing the tracking and surveillance functions for phased array-based radars are negotiable and depend on the operational and environmental requirements. This is but one aspect of the flexibility the phased array radars offer to the user; other aspects are explored as we progressively erect the required framework.

The breaking of the nexus between the surveillance and tracking functions is partial, however, because one inertialess beam cannot dwell toward the directions of too many marginal detections. The assumption is made here that the sum of the time spent on surveillance, t_s, and that spent on tracking, t_t, remains constant in the interest of updating data regularly; update times of 2–10 s are typical. In section 2.7.4.7 we shall explore approaches that allow the user to break this nexus completely.

1.5.2. Radar Redefined

Let us reconsider the nose radar used in fighter aircraft such as the F-15 (see section 1.2.2) and compare its performance with that attainable by phased array radars. Both radars can operate in a step-scan mode but there are significant differences between them.

Let us assume that both radars are to visit targets A, K, and Z. The conventional nose radar wastes time traveling the angular distance from A to K and from K to Z. By comparison, the phased array takes 50 ns or less to direct its beam from one position to another [2] and can track up to 1000 targets on a time-sharing basis. Additionally, distributed logic can be used to speed up the beam-steering function of the array. The ability of phased arrays to perform high-speed update rates, and burst modes approaching 100 000 beam positions per second, offer new possibilities in multimode radars [2]. If we couple this capability with the multibeam capability of phased arrays, a new era of applications limited only by the designer's imagination is ushered in.

As the speed of steering of the beam of the nose radar increases to cope with the ever-increasing number of targets, its meantime between failures (MTBF) decreases while that of the phased array remains unaffected. If we accept that the MTBF of a conventional radar is at best 1000 h, the corresponding figure for a phased array is 100 000 h, a very significant improvement that impacts on the operational strategies of aircraft and ships and on the radar's life cycle costs (LCCs).

As the mechanically scanned antenna tracks a target, its RCS significantly increases the platform's overall RCS as seen by the target. The enhanced platform RCS in turn endangers the platform's survival. By comparison, the specular reflection of a phased array is stationary and can be directed toward a point in space where it is less likely to be intercepted.

Conformal phased arrays do not affect the aerodynamic properties of aircraft. This is a considerable advantage in comparison with conventional systems using reflectors in conjunction with radomes that preserve the aerodynamic properties of the platform.

Overall, the synergy of all these subtle phased array attractors form the basis for new systems that not only redefine the capabilities of radar but also meet future challenges.

1.5.3. The Nexus between Spatial Resolution and Costs

Phased arrays break the conventional nexus between the aperture real estate and spatial resolution. Given that aperture real estate is usually directly related to cost, trade-offs can be made between the required resolution and cost. If spatial resolution is of prime importance, the designer can attain the required resolution by realizing a thinned array and the extent of the thinning will be defined by the budget available for the array.

The first radiotelescopes operating at frequencies in the vicinity of a few hundreds of MHz were not too costly. Naturally the latest phased array telescopes are more expensive to construct but are well within the budgets of many nations. The cost of the VLA, for instance, was $78M in 1972 terms.

Let us examine a couple of extreme cases that will add value to the comparisons between phased arrays and conventional apertures. Even if one could realize a conventional reflecting aperture having a diameter of several kilometers, there are no known methods to mechanically move it toward the desired directions. The conventional aperture could be a hole in the ground, in which case the telescope's horizon is defined by the Earth's rotation and/or the limited scanning properties of reflectors. Even then the radiotelescope's horizon would be limited. The case for phased arrays becomes stronger if spatial resolutions of the order of arc seconds or milli-arc second are required.

The question that naturally arises is, given that phased array-based radiotelescopes and single dish radiotelescopes are still widely used, what are the relative domains of applicability for these diverse systems? We shall explore this question in section 1.7.2.2 and 1.7.3, after the required theoretical framework is established.

1.6. RADAR SYSTEMS

1.6.1. Introduction

In this section we consider the fundamentals of radar systems synoptically. If the radar's aperture has a geometric area, A, its gain, G, is given by the well-known equation

$$G = \frac{4\pi}{\lambda^2} A\eta \qquad (1.3)$$

where $A\eta$ is the aperture's effective area and η is the aperture efficiency. For conventional reflectors, η is a measure of the losses due to aperture illumination, spillover, and blocking. In the context of active phased arrays, η can be a measure of the losses associated with the components connecting the antenna elements and the LNAs/FPAs. For passive arrays, the feed systems and phase-shifters introduce significant losses. Another important measure of antenna efficiency is the beam efficiency η_b [63]; for single reflectors, η_b is defined as the ratio of the power received through the main beam to the power received through the main beam and the sidelobes. More explicitly,

$$\eta_b = \frac{\displaystyle\iint_{\text{main beam}} F(\theta, \phi)\, d\Omega}{\displaystyle\iint_{\text{main beam}} F(\theta, \phi)\, d\Omega + \iint_{\text{sidelobes}} F(\theta, \phi)\, d\Omega}$$

where θ and ϕ are the conventional polar angles.

For phased arrays, η_b is the ratio of the power received through the main beam to the power received through the main beam plus the powers received through the sidelobes and grating lobes:

$$\eta_b = \frac{\displaystyle\iint_{\text{main beam}} F(\theta, \phi)\, d\Omega}{\displaystyle\iint_{\text{main beam}} F(\theta, \phi)\, d\Omega + \iint_{\text{sidelobes}} F(\theta, \phi)\, d\Omega + \iint_{\text{grating lobes}} F(\theta, \phi)\, d\Omega}$$

1.6.2. The Radar Equation

The output signal-to-noise ratio (SNR) of a monostatic radar system is simply the ratio of the power P_r received at the input terminals of the receiver to the noise power of the system:

$$\text{SNR} = \frac{P_r}{kTB} = \left(\frac{P_T G}{4\pi R^2}\right)\left(\frac{A_e \sigma}{4\pi R^2}\right)\left(\frac{1}{kTB}\right)(10^{-0.2\alpha R}) \tag{1.4}$$

where

R	= the radar range or the distance between the radar and its target;
kTB	= total receiver noise power;
k	= Boltzmann's constant; 1.38×10^{-23} joules per kelvin;
T	= equivalent noise temperature of the system;
B	= system bandwidth;
P_T	= total transmitted power;
G	= aperture gain;
A_e	= $A\eta$, the effective antenna area;
σ	= average target RCS; and
α	= one-way atmospheric loss attenuation coefficient, expressed in dB/km.

The first bracketed term in the rhs of equation (1.4) is the power density available on the target. The product of the first and second bracketed terms of the same equation is the reflected power density available at the receiving antenna. The third term is the reciprocal of the receiver's noise power, and the fourth term is the atmospheric attenuation factor. It is here assumed that all spectral components of the received signal are within the receiver bandwidth, B. Here we have assumed that other losses, e.g. transmission losses and losses due to mismatches between subsystems, are negligible.

Inserting equation (1.3) into (1.4), we obtain

$$\text{SNR} = \frac{P_T G^2 \sigma \lambda^2 10^{-0.2R\alpha}}{(4\pi)^3 \, R^4} \frac{1}{kTB} \tag{1.5}$$

For a pulsed radar, P_T, the average power transmitted over a pulse period, is related to the peak transmitted power, P_{pk} by $P_T = P_{pk} d_t$, where d_t is the duty factor equated to the ratio of the pulse duration τ and the pulse period T. Although the radar equation looks simple, there are many hidden subtleties, some of which we shall explore here.

1.6.2.1. The RCS of Targets

Since the RCS of simple targets such as spheres and corner reflectors are well known [64], these standard reflectors are frequently used as calibrators of radar systems. The RCS of a sphere of radius r, for instance is equal to πr^2 at all look-out angles in the optical scattering region, where the target dimensions are much greater than the wavelength of operation.

By contrast, the RCS of a complex target, such as a ship or aircraft, depends on the look angle of the radar. As the geometry of the target's scattering points changes aspect with motion, the target's RCS varies, or scintillates. The same phenomenon takes place when the radar frequency changes because the relative position of the target scatterers, measured at different wavelengths, changes. Lastly, σ is polarization dependent (see section 1.6.12). Given that we have no *a priori* knowledge of the shape and polarization properties of the target, the possession of a full polarization capability of a system ensures that maximum information of the scene containing targets and clutter is attained. σ is therefore time, frequency, polarization, and look-angle dependent.

Moving targets can be discriminated from some clutter because the returns from clutter have zero Doppler shift, and the RCS of a target scintillates whereas that of clutter remains constant.

For the purposes of calculating the range of a target, some designers assume that σ is the time-average RCS of the target at one polarization. Measurements of σ of complex targets at different polarizations and frequencies, as well as approaches to minimize it, form an established field of research [64].

Table 1.6 Commonly accepted RCSs for different targets [65]

	RCS (m^2)
Jumbo airliner	100
Large bomber or airliner	40–100
Medium bomber or airliner	20
Large fighter	6–10
Small fighter	2
Small single engine aircraft	1
Cruise-type missile	1
Human body	1
Conventional unmanned winged missile	0.5
Bird	0.01

As might be expected, the RCSs of combat aircraft as a function of look angle, polarization, and frequency are not available in the open literature. In the recent past, designers have accepted the figures of 1, 10, and 100 m^2 for the RCSSs corresponding

Table 1.7 Estimated RCSs of some well-known aircraft [66]

Aircraft type	Estimated RCS (m^2)
B-52	100
Blackjack (Tu-160)	15
FB-111	7
F-4	6
Mig-21	4
Su-27	3
Rafale-D	2
B-1B	0.75
B-2	0.1
F-117A	0.025

to cruise-type missiles, fighter planes, and bombers, respectively [9]. Table 1.6 lists the commonly accepted RCS values of different targets, while estimates of the RCSs of some well-known aircraft are collated in Table 1.7.

It is worth noting here the prodigious efforts undertaken to minimize the RCS of some well-known aircraft, listed in Table 1.7; this development is one of the driving forces for the evolution of radars capable of meeting the challenges imposed by modern, low-RCS aircraft.

1.6.2.2. Atmospheric Attenuation

The attenuation experienced by EM waves from atmospheric gases, rain, and fog as a function of frequency is documented in many references (e.g. [67]). For a given site, the attenuation experienced by EM waves at the resonant frequencies of O_2, and H_2O is significant and depends on the height of the site above sea level.

The approximate atmospheric attenuation experienced by EM waves in clear weather and during light and heavy rain conditions at 10, 35, 94 and 140, GHz is shown in Table 1.8 [68]. Designers have to accept the free-space attenuation of the EM waves as well as the attenuation caused by scattering during periods of rainfall. The attenuation of EM waves caused by depolarization is recoverable, however, and we shall further consider the effects of rain on polarimetric radar and satellite communication systems in section 1.6.12.2.

Table 1.8 Attenuation factor α (dB/km) as a function of weather conditions and frequency

Conditions	Frequency [GHz]			
	10	35	94	140
Clear weather	0.05	0.11	0.44	1.8
Light rain, 10 mm/h	0.133	2.6	8.7	5.9
Heavy rain, 25 mm/h	0.385	6.50	12.55	

While communication and radar systems usually operate at frequency bands for which the atmospheric attenuation is minimal, e.g. 30–40 and 80–100 GHz, short-range secure communications and radar systems operate at frequency bands for which the atmospheric attenuation is high, e.g. 22 and 60 GHz. Knowledge of the magnitude of attenuation due to rain and the atmosphere at the designated frequency of operation of a system serves the purpose of setting safe operational margins.

More detailed studies of the many influences of rain on the propagation of EM waves have been undertaken by the millimeter-wave radar and satellite communications communities [68,69]. Estimates and measurements of the same attenuation under conditions of snowfall and fog have also been collated in reference [68].

1.6.2.3. SNR Considerations

Consider that we are monitoring the output, SNR, of the radar's receiver when the radar tracks an approaching target. At a range R_0 the $\text{SNR}|_{R_0}$ is equal to 1, and as the target approaches the radar the SNR increases. At a range R ($<R_0$) for instance, $\text{SNR}|_R = (R_0/R)^4$ and the detected output of the receiver is compared to a threshold bias voltage that can be set by the operator to an appropriate value; if the detected signal exceeds the threshold voltage, a target is registered.

Given that the RCSs of complex targets scintillate and that the receiver noise fluctuates, the setting of the bias voltage is crucial. A low threshold bias value, for instance, will result in the declaration of too many false targets, or in a high probability of false alarms, P_N. Similarly, low-RCS targets will be missed if a high bias setting is selected. While one requires a low P_N and a high probability of detection, P_S, both quantities are negotiable. If the P_N is known, the threshold bias is implicitly defined. Typical values for P_N and P_S range between 10^{-6} and 10^{-8} and 50% and 90%, respectively.

To calculate an adequate SNR for the detection of a target after a 'single look' we need to know whether the target fluctuates and the required P_S and P_N. The problem of detecting non-fluctuating and fluctuating targets has been treated by Marcum [70] and Swerling [71], respectively.

It is very seldom that a radar detection decision is based on a single pulse detection. N pulses usually illuminate the target and the returns are integrated coherently before one can declare a target detection. Thus, an effective value of SNR is obtained equal to the SNR attributed to a single pulse multiplied by N. With coherent integration, some clutter attenuation is attained [72] and acceptable detections can be declared even when the single pulse SNR is near to or below unity. However, the integration process requires that (i) the phase and frequency of all oscillators used in the radar system are highly stable; (ii) low noise is associated with the radar's transmitter chain and the local oscillators used in the receiving section; and (iii) the Doppler shift of the target returns is taken into account. As the stability of modern solid-state oscillators is more than adequate and signal processing techniques are highly developed, coherent radars are now widely used [73]. The recent development of cooled, ultrahigh-Q sapphire dielectric resonators [74] will contribute to the widespread use of ultralow-noise microwave signal generators in many systems—see section 4.3.2.

Some radars utilize non-coherent integration because it is easier to implement. In a typical example where 50 pulses are integrated, a 17 dB improvement in the derived

SNR is attained by a coherent radar, while the improvement attained by a non-coherent radar is only 13 dB [75].

Swerling [71] and Meyer and Mayer [76] have explored the cases when the targets are fluctuating and when the integration of several pulses takes place respectively. The latter reference contains a plethora of graphs to aid the designer of radar systems.

In the absence of a detailed knowledge of a radar system, false alarm probabilities of 10^{-6} to 10^{-8} and a SNR of 10–20 dB are commonly used.

1.6.2.4. Overview of the Radar Equation

The power available at the input of the radar's receiver, P_r, is

$$P_r = \frac{1}{4\pi}\left(\frac{P_T A_e^2}{\lambda^2}\right)\left(\frac{\sigma}{R^4}\right)(10^{-0.2aR}) \tag{1.6}$$

The first bracketed term of equation (1.6), is the power–aperture product scaled to the wavelength of operation squared. The designer of a polarimetric radar therefore maximizes P_r (and the system's SNR) by maximizing the power–aperture product of the system. As σ is polarization dependent, a radar system having full polarimetric capabilities measures the highest target RCS.

If we consider operation below 3 GHz, the last bracketed term of equation (1.6) can safely be neglected. If operation above 3 GHz is considered, P_r decreases because α increases as the frequency of operation increases (see Table 1.8) and the power derived from solid-state oscillators, P_T, decreases as the frequency of operation increases (see Chapter 4).

The maximization of a radar's range is only one of the many requirements a systems engineer has to meet. For secondary radars, other factors such as the radar's ability to handle a large number of targets in all weather conditions with no downtimes for maintenance are high-priority requirements.

1.6.3. Other Radar Characterizations

Moving-target indication (MTI) radars utilize the Doppler phenomenon and/or the fact that the RCS of moving targets scintillates to discriminate between moving targets and stationary targets.

The block diagram of a radar is shown in Figure 1.8; range gates 1, 2 , . . . and G ensure that returns from different ranges end up at different range bins and the filterbank connected to every range gate accommodates the returns from targets having different Doppler shifts. With this arrangement, returns from high-speed targets receding from the radar end up in filters 1, while high-speed targets approaching the radar end up in filter F and returns from stationary targets end up in the mid-filter of each filterbank. The output of each filter is connected to a threshold detector.

Most MTI radars utilize delay line cancelers instead of Doppler filters to distinguish moving targets from stationary ones. The basic principle used here is that the returns

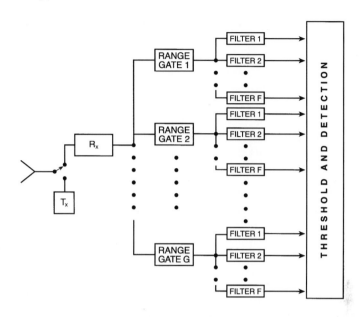

Figure 1.8 The block diagram of a radar. Range gates 1, 2, ... and G ensure that returns from different ranges end up at different range bins. The filterbank connected to each range gate accommodates
the returns from targets having different Doppler shifts.

from moving targets fluctuate in amplitude while returns from stationary targets have
a constant amplitude.

For a monostatic pulsed radar system, the requirement that the return from a target
is received before the next pulse is transmitted defines the radar's unambiguous range:

$$R\big|_{unam} = \frac{c}{2}\frac{1}{PRF}$$

Table 1.9 Maximum unambiguous ranges for a range of pulse repetition frequencies (PRF)

Radar PRF classification	PRF (kHz)	Unambiguous range (km)
Low	0.2	750
	1	150
	2	75
	3	50
	4	37.5
Medium	10	15
	15	10
	20	7.5
High	100	1.5
	200	0.75
	300	0.5

where PRF is the pulse repetition frequency. Table 1.9 lists the unambiguous ranges when the PRF is low, medium, and high. Doppler ambiguities result if the spread of Doppler shifts expected from targets is greater than the radar's PRF. The condition for unambiguous Doppler shift measurements is therefore

$$\text{Spread of Doppler shifts} \leq \text{PRF}$$

For airborne radars, the mainlobe ground clutter has a maximum Doppler shift of $2V_R/\lambda$, where V_R is the velocity of the platform. The maximum unambiguous Doppler shift is therefore given by

$$\text{Maximum unambiguous Doppler Shift} = \text{PRF} - \frac{2V_R}{\lambda}$$

It is recalled here that the target's range rate is often calculated from measurements of its Doppler shift.

Both kinds of ambiguities (range and Doppler) can be resolved by employing a variety of techniques, including changing the PRF of the radar and performing the appropriate signal processing operations [13]. Table 1.10 lists the commonly accepted frequency ranges of radars operating at X-band for low-, medium- and high-PRF radars together with an indication of whether measurements of range or Doppler shifts are expected to be ambiguous or unambiguous.

Table 1.10 Ambiguities related to X-band radars having different PRFs

$\text{PRF}|_{\text{X-BAND}}$	Frequency	Range	Doppler shift measurement
Low	250–4000 Hz	Unambiguous	Ambiguous
Medium	10–20 kHz	Ambiguous	Ambiguous
High	100–300 kHz	Ambiguous	Unambiguous

Medium- and high-PRF pulsed Doppler (PD) radars are mainly used in airborne applications, while low-PRF PD radars are becoming increasingly important in surface-based radars.

1.6.4. Large Time–Bandwidth Product Radars

For certain applications the ideal radar has a long detection range and fine range resolution. These two requirements are satisfied if high power is transmitted over extremely narrow pulses. Let us assume that a PD radar illuminates a target over a time interval τ_1, when its PRF is equal to $1/T_1$. The average power P_{ave}, is given by

$$P_{\text{ave}} = P_{\text{max}} \frac{\tau_1}{T_1}$$

where P_{max} is the maximum power. When the PRF and P_{max} are fixed, P_{ave} is increased when τ_1, is increased, a condition that does not meet the original requirements. Pulse compression methods allow us to meet the requirements by maximizing P_{ave} and minimizing τ_1. While there are several schemes to achieve this objective [13], we shall consider only the incremental frequency modulation approach.

N pulses of duration τ_1 are generated and each pulse gates different RFs ranging from $f_{highest}$ to f_{lowest}. The pulses are transmitted and, upon reception, they are passed through a special filter, usually a surface acoustic wave (SAW)-based filter that allocates different delays to different RFs so that the energy of all pulses is accumulated into one pulse of duration τ_1. With this arrangement the ideal conditions are approximated and the system has a compression ratio/processing gain or a time–bandwidth product given by $N\tau_1\,\Delta f$, where $\Delta f = f_{highest} - f_{lowest}$. Typically, time–bandwidth products of the order of 10 000 are common. These systems have a low probability of intercept (LPI) capability because they transmit low instantaneous powers.

We have already seen how the total bandwidth of an ISAR can be increased, and pulse compression is widely used in SARs and space radars. Systems utilizing just one cycle of the carrier form a particular type of ultra-wide-bandwidth (UWB) radar and can, theoretically, attain very high range resolutions (HRR) [77]. These systems are better known as monocycle radars and can be used to perform the identification function.

The same philosophy of spreading a signal in bandwidth before transmission and reversing the process upon reception is used in modern spread-spectrum communication systems [78] where the concept of processing gain (the ratio of the total bandwidth to the instantaneous bandwidth) is used. Modern communication systems operating in the frequency hopping (FH) mode can have processing gains of 7000 [79]. Communication systems having processing gains of 19 000 and 58 000 have been considered [80].

The spreading of the system's bandwidth results in an additional benefit. If a third party attempts to jam a spread-spectrum system, the jamming power has to be spread over a wide bandwidth; assuming that the power–bandwidth product of the EW system is finite, the jammer's effectiveness is decreased because the power emanated per unit bandwidth is decreased. The comparison is made here with respect to a conventional system having a narrow bandwidth centered at a known center frequency.

1.6.5. LPI Radars

We have already seen that PD radars utilizing pulse compression techniques, monocycle radars, and spread-spectrum systems can have the LPI characteristic in the previous section. Ideally, designers would like to transmit as much power as is necessary to attain a certain radar range requirement without being detected by spectrum surveillance or electronic support measures (ESM) systems. The availability of anti-radiation missiles (ARMs) emphasized the need for LPI radars. The Gulf War certainly proved the usefulness of both the ESM systems and the ARMs on the one hand and the vulnerability of conventional radars on the other.

In this section we shall consider the following three types of LPIRs:

- Frequency modulated continuous wave (FMCW) radars
- FM interrupted continuous wave (FMICW) radars
- Staring radars

In sections 1.6.6 and 2.5.7 we shall consider bistatic/multistatic radars, some of which have exceptional LPI capabilities.

1.6.5.1. FMCW and FMICW radars

The most straightforward LPI radar is the continuous wave (CW) radar, which has a 100% duty cycle. The radar can, in theory, perform the prime radar objectives of surveillance and detection and has LPI capability because its instantaneous power is low.

The most easily realizable CW radar is the frequency modulated (FMCW) radar where the transmitter signal is frequency modulated by a linear waveform [81,82]. The received signal has the same modulation but is delayed relative to the transmitted signal. If Δf is the peak-to-peak frequency deviation, SL is the slope of the ramp, and δf_1 is the frequency difference between the transmit and receive ramps, the target range can be measured with a range resolution of δR given by

$$ R = \frac{\delta f_1 c}{2(\text{SL})} \quad \text{and} \quad \delta R = \frac{c}{2\Delta f} $$

An excellent account of the developments related to FMCW radars is given in reference [83].

Acceptance of FMCW radar has been delayed because of the technological problems associated with its realization. More explicitly, the difficulties related to the realization of monostatic FMCW radars have centered around the isolation required between the transmit and receive EM waves. High isolation is required between the transmit and receive signals when the typical corresponding power levels are of the order of watts and picowatts, respectively. A satisfactory solution to this problem has been proposed and implemented [84].

Frequency modulated interrupted continuous wave (FMICW) radars resemble monostatic radars where the antenna is switched between the receiver and the transmitter. The system overcomes the difficulty of isolating the receiver from the transmitter and, provided certain conditions on the switching speed and the FMCW parameters apply, the essential characteristics of the FMCW are preserved [85].

1.6.5.2. Staring Radars

So far we have considered scanning radar systems in which one inertialess beam scans a given surveillance volume; all the power of the system is focused onto one pixel, which in turn is scanned through the surveillance volume. Scanning systems are therefore easily detected by radar warning receivers (RWRs) that work in conjunction with jammers capable of disabling the radar.

An LPI system is realized if the transmitted power is distributed evenly throughout the surveillance volume and contiguous staring beams receive the scattered radiation. As the power toward any direction is low, detection by a RWR is more challenging.

If P_T, M, and T_2 are the power transmitted, the number of beams covering the whole surveillance volume, and the time it takes to complete the surveillance volume, respectively, the energy per pixel for the scanning system is equal to $P_T(T_2/M)$, and for the staring system is $(P_T/M)T_2$. The two systems therefore provide the same energy per pixel. The overall effect upon detection, however, is primarily dependent upon the performance of the integration process employed and the target and noise statistics. While one can use coherent integration to ensure maximum efficiency, the target and noise statistics are application dependent.

The following comparisons between the scanning and staring systems can be made:

- The information update rate for the staring system is higher than that corresponding to the scanning system.
- The staring system uses parallel processing of the incoming information, while serial processing is used in scanning systems.

A small experimental radar that flood-illuminates the surveillance volume and utilizes staring receive beams has been reported [86].

1.6.6. Bistatic/Multistatic Radars

As the operation of radar systems does not depend on specular reflections, the transmit function can be performed at one site and the receive function at another; the separation between the two sites can be several hundreds of kilometers. Such an arrangement, the bistatic radar illustrated in Figure 1.9, has the following attractors:

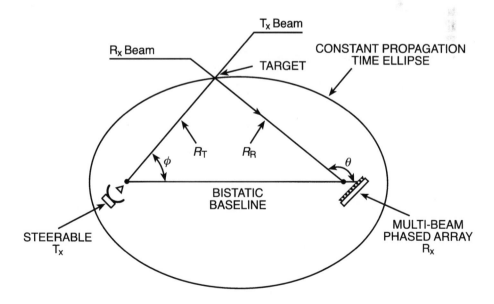

Figure 1.9 A simplified diagram of a bistatic radar.

- The receivers are immune to jamming directed toward the transmitters.
- Illuminators of opportunity can be used as transmitters to ensure covert operation. In section 2.5.7 we consider illuminators of opportunity, e.g. TV/FM stations or transmissions from the satellites of the GPS/GLONASS systems used to detect several targets [87]. Such LPI systems are ideal for air and space defense.

Multistatic radars have several transmitters/receivers. Transcontinental airlines, for instance, can be tracked by several illuminators/receivers encountered along their flight paths.

Bistatic radars are more complex than their monostatic counterparts and a communication link between the transmitter and the receiver is usually needed. The link relays the complete transmitter waveform and the transmitter's location and orientation. Line-of-sight, land line, satellite, or troposcatter systems can be used as links.

The benefits afforded by bistatic/multistatic radars are significant even though the receive system is required to incorporate a multibeam phased array—see section 1.6.6.2.

1.6.6.1. Target Measurements

As with monostatic radars, the target range can be deduced from the time delay incurred between the transmission and reception of the radar pulse. For the bistatic radar illustrated in Figure 1.9, the transmitter and receiver are the foci of an ellipse and the equal time-delay contours form ellipses with the same foci. For the general case, the ellipses become prolate spheroids. The elevation and azimuth angles of the transmitter then define the position of the target on the surface of the spheroid.

For monostatic radars the contours of constant detection range, or equivalently of constant received power, are spheres. For bistatic radars the contours of constant detection range are ovals of Cassini [88].

1.6.6.2. Receive Station Scanning Rate

If a bistatic radar is to have a data rate comparable to that of its monostatic counterpart, the receiving antenna has to be able to follow the directions from which energy is scattered toward it by any targets. For the bistatic radar shown in Figure 1.9, the maximum scan rate has been deduced to be [88]

$$\left[\frac{d\theta}{dt}\right]_{max} = \frac{c}{L}\frac{1+\cos\phi}{\sin\phi}$$

Given that in many applications the maximum scan rate can exceed a million degrees per second (as $\phi \rightarrow 0$), a multibeam phased array is required at the receive site.

1.6.7. Distributed Array Radar

The proposals for spaceborne distributed array radars (DARs) call for a constellation of independent radar systems set in medium/low Earth orbit (MEO/LEO). The antenna elements of the DAR are followed by transceivers and DARs are self-cohering.

Although the first DARs were ground-based [89] proposals for airborne and space-borne versions and recent experimental demonstrations of a DAR have been reported [90,91].

The attractive features of spaceborne DARs are enhanced survivability or graceful degradation of performance, and the detection of small-RCS airborne targets in severe clutter environments. The attractors of DARs performing radar surveillance and tracking functions are outlined in reference [90].

The costs of manufacturing, launching, and maintaining the constellations of MEO and LEO satellites used to usher in the wireless revolution are expected to decrease with the passage of time; the experience and knowhow thus gained can be used to decrease the cost of realizing spaceborne DARs.

1.6.8. Limitations of Ground-based Radars

The fundamental limitations of ground-based radars provide the raison d'être for a suite of airborne radars that essentially complement them. Additionally, considerations of these physical limitations will provide an historical background for the precursors of the early array-based radiotelescopes.

Most physics textbooks cover Lloyd's mirror diffraction experiment. A point source above a mirror produces a classic diffraction pattern on a screen at some distance away from the source because the direct ray interferes with the reflected ray from the mirror. A radio analog of Lloyd's mirror, shown in Figure 1.10a, was used by early radio-astronomers in Australia in 1947 to increase the spatial resolution of an antenna that was sited on a seaside cliff [92]. This important experiment was the precursor of the early phased array-based radiotelescopes.

With reference to Figure 1.10b, a target might be a sea-skimming missile aimed at a ship that uses its shipboard radar to track the missile. Alternatively, the target might be an unfriendly aircraft entering the surveillance volume of a radar by flying at a low altitude. The reflection coefficient of the surface, Γ, is here assumed to be -1.

The propagation factor, F, due to the multipath interference at the target can be defined by [65]

$$F|_{\text{TARGET}} = \frac{\text{Electric field strength in multipath field}}{\text{Electric field strength in free space}}$$

By reciprocity, the analogous power ratio at the radar's receiving antenna is given by [68]

$$F^4 = 16 \sin^4\left(\frac{2\pi h_1 h_2}{\lambda R}\right) \tag{1.7}$$

where λ is the wavelength of operation. Equation (1.7) represents the classic interference pattern at the radar's receiver. As the range decreases, the radar receives interference lobes that in turn prevent it from measuring the target's position. Thus, a low-flying target cannot be tracked by a ground-based radar that is used to defend a high-value platform.

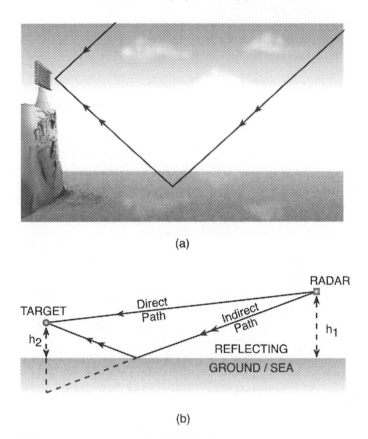

(a)

(b)

Figure 1.10 (a) The sea cliff interferometer or the radio analog of Lloyd's mirror, used to take the first high-spatial resolution radioastronomy observations in Sydney, Australia, 1947. (Courtesy CSIRO.) (b) A sea-skimming missile aimed at a ship. The ship's radar receives an interferogram caused by the direct and indirect paths of the radiation. A similar problem exists for ground-based platforms threatened by low-flying missiles.

Often Γ is not equal to unity and is polarization dependent, so it is convenient to separate the reflected energy into a coherent and an incoherent component.

1.6.9. Meeting the Challenges of Low-Flying Threats

The first line of defense against low-flying targets is to increase the height h_1 and/or the aperture of the radar's antenna, but there are limits as to how high a large aperture can be erected on a ship. An alternative approach would be to use a millimeter-wave radar that yields a narrow beam even when its aperture has manageable dimensions. This approach, however, will decrease the radar's range because the power of solid-state transmitters is low at millimeter wavelengths.

From the foregoing considerations it is clear that this limitation of ground-based radars is far from trivial and warrants considerable attention. Several systems have been developed to counter the threat of low-flying targets and here we shall consider only some of them. In an effort to narrow the elevation beamwidth of a ship's radar operating at X-band, an interferometer was used in the approach taken in reference [93]. High powers can be generated at X-band and four similar subsystems provide 360° protection for the ship against low observable (LO) low-flying threats out to the horizon. This approach, effective as it is, represents the limit for conventional radars utilizing the interferometric technique.

1.6.9.1. Over-the-Horizon Radars

Over-the-horizon radars (OTHRs) operate in the high-frequency (HF) band (3–30 MHz) and their radar ranges vary from a few hundreds to a few thousands of kilometers depending on the mode of propagation. The two types of radar use either the skywave or surface wave and typical ranges are 100–3500 km and up to 500 km, respectively. Thus, OTHRs are often used for the strategic surveillance of large areas.

When the skywave is used, power is radiated toward the ionosphere and illuminates large areas on the Earth after downward refraction caused by the stratified ionosphere; the scattered radiation is in turn received by the radar's receiving system. As the frequency of operation is low, the transmitting and receiving apertures need to be phased arrays, extending over at least one kilometer. Apart from performing the surveillance function over very large regions of land or sea with spatial resolutions of the order of tens of kilometers, OTHRs are used for the monitoring of a variety of environmental phenomena including wind patterns over vast stretches of the ocean [94].

The surface wave OTHR utilizes propagation along the sea–air interface and the resulting systems are less expensive and independent of the state of the ionosphere except for clutter that results from the following mechanism. Some transmitted power is leaked to the ionosphere, refracted downward, reflected from the Earth's surface, and received by the surface radar after a similar downward refraction from the ionosphere. Large powers have to be generated to achieve the maximum range of 500 km.

OTHR systems measure the range, azimuth, signal amplitude, and Doppler characteristics of observed objects. The evolution of digital signal processing (DSP) and the availability of low-phase-noise transmitters enables OTHRs to reach their full potential [94]. We shall consider OTHRs in some detail in section 2.7.1.

1.6.9.2. Airborne Early Warning Radars

Airborne radar systems look down at low-flying targets, so multipath propagation is not an issue. With these systems a low-flying target is located and tracked by the airborne system and intercepted by ship-launched missiles. These systems effectively extend the ship's horizon so that their defense becomes viable. The AWACS perform this function effectively.

While fully airborne early warning (AEW) systems are expensive but can be used anywhere at short notice [8], airship-borne systems are more affordable but slow moving; typical speeds of 70 knots enable the airships to keep pace with naval vessels in virtually all weather conditions [95]. Their horizon is typically 240 km for very small-RCS airborne targets and 650 km for conventional targets. The evolution of AEW systems together with an outline of the technical problems solved with the passage of time are given in reference [96].

A demonstration utilizing a drone launched from a ship to increase its horizon was successfully completed at Kauai, Hawaii during January and February 1996. The demonstration (the cruise missile defense advanced concept technology) is also known as Mountain Top [97].

As air threats having ever-decreasing RCS are developed and fly closer to the land or sea, a defensive missile faces problems similar to those confronting surveillance radars, i.e. how to discriminate the threat from background clutter. Detailed studies of the seeker's perspective of the radar environment are time consuming and expensive. A series of experiments in which a seeker is 'captive' to an aircraft have been reported [98]; with this arrangement valuable data are collected when the low-flying threat is illuminated by a ship and the scattered radiation is received by the captive seeker. The assembled data are valuable for the evolution of the next generation seekers.

1.6.9.3. Eternal Airplanes/Pseudo-satellites

While AWACS are highly mobile and effective, the acquisition cost and life-cycle costs associated with the equipment and the maintenance/operational crews are high. The costs associated with airship-borne systems are lower, but the mobility and horizons of these systems are limited; drone-borne systems share similar limitations.

Pilotless aircraft powered by electric energy derived from solar cells and/or by ground-based transmitters can support radar and comunication functions. In the literature the terms eternal airplanes or pseudo-satellites have been coined for these platforms. Nikola Tesla predicted in 1928 that 'some day' airplanes would be 'powered by wireless' [99] and several attempts to realize the eternal plane have been summarized [99,100]. A solar-powered airplane has been reported [100] and a detailed account of the Stationary High Altitude Relay Platform, SHARP-5, is found in reference [99].

The SHARP-5 was powered by batteries during takeoff and by microwave power once the airplane reached the height of 300 ft (92 m). The transmitter's frequency was 2.45 GHz, and the airplane was sustained by a 10 kW microwave beam, derived from a 15 ft (4.5 m) paraboloid. A lightweight rectifying antenna or rectenna, 3.5 ft (1.07 m) in diameter consisted of 400 dipole antennas that received the transmitted electric energy and rectified it to power the plane's high-performance electric motor using samarium–cobalt magnet technology.

While the SHARP-5 was only a 4.5 m featherweight aircraft, it is seen as a forerunner of a full-size craft that could some day stay afloat for as long as a year, circling at an altitude of 21 000 m [99]. Several aspects related to the realization of these pseudo-satellites are the subjects of ongoing research and development outlined in sections 1.8.2 and 2.5.7. The pseudo-satellites are expected to be cost-effective platforms for lightweight, miniaturized radar, radar-related and communications systems over an area approximately 1000 km in diameter.

In section 1.8.2 we shall explore the opportunities offered by electrically powered airplanes when the required power is derived from a densely populated phased array of transmitters. The array collimates the available power from all the transmitters onto one rectenna, which in turn powers the electric motors of an aircraft. The same collimated power can be used to disable the electronic systems of unprotected aircraft. These contemporary threats are far more serious than the threats imposed by conventional EW systems.

1.6.9.4. Two Novel Approaches

The use of a high-peak-power transmitter obviates the need for pulse compression systems and their attendant limitations when used to detect and track low-flying targets [101]. However the generation of short-duration pulses in space of sufficient energy to provide useful detection was not practical until the advent of a suitable transmitter in the form of a microwave relativistic backward-wave oscillator [101].

In a collaborative experiment between the United Kingdom and Russia, a 10 GHz radar system has been used to collect data from clutter, static targets, and low-flying aircraft. The transmitter generates short-duration pulses of 5 ns of high peak power, 500 MW, when the pulse repetition frequency is 150 Hz. Apart from the collection of useful data, the system was used to define future transmitter and system enhancements. While the current system has a fixed beam, it is desirable that future systems have steerable beams.

It is recalled that in a pulsed Doppler or moving-target indicator system, targets are discriminated from clutter by their Doppler shift. High range resolution radars offer the user an alternative approach to performing the same task. More explicitly, clutter rejection is accomplished by detecting the change in location from pulse to pulse produced by a moving target. This approach is viable only because of the extremely high range resolution of an ultra-wide-bandwidth (UWB) radar. For a radar operating from 8.5 to 11.5 GHz, 30% bandwidth, the range resolution is 5 cm [102]. A conceptual design of an UWB radar is outlined in the same reference.

While there has been some controversy associated with early claims of the capabilities of the UWB radar, reference [103] put UWB radar in the proper perspective by showing that some of these claims were not valid.

1.6.10. Optimal Frequency Bands for Radar Functions

In this section we shall derive the optimal frequency bands for the two major radar functions by taking into account theoretical and real-world considerations.

1.6.10.1. The Surveillance Function

Let us assume that a radar searches a volume of Ω steradians in time t_s and that the radar's antenna beam Ω_b dwells for a time t_0 in each direction subtended by the beam. The total scan time is $t_s = t_0 \, \Omega/\Omega_b$ and, if P_{av} is the average power during t_0, $P_T = t_0 P_{av}$. By inserting these equations into equation (1.5) and recalling that $\Omega_b = \lambda^2/A_e$ we obtain

$$R^4 = \frac{P_{ave}A_e}{T}\left(\frac{\sigma}{\text{SNR}}\frac{1}{4\pi\Omega}\frac{t_s}{kB}\right)10^{-0.2aR} \tag{1.8}$$

The parameters in the bracket of equation (1.8) are either constants or given. At first sight the range attainable is not dependent on the wavelength of operation. However α, P_{av}, and T are frequency dependent. At frequencies below 3 GHz, α can be safely ignored and the lower the frequency the longer the range. The assumption is made that as the frequency of operation decreases, the transmitter power available increases and the noise temperature of receivers decreases (see Chapter 4). Below L-band, however, the sky noise increases significantly and low-noise systems are not viable. It is therefore preferable for surveillance radars to operate at L-band, where the radar range can be maximized.

While maximizing the range is appropriate for long-range searches, other criteria have to be considered for horizon searches. For a given aperture, the radar's beamwidth should be as narrow as possible to avoid the problems associated with multiple paths considered in section 1.6.8. If we adopt this criterion, higher frequencies are favored. The frequency band, however, cannot be too high because the atmospheric attenuation and noise temperature of receivers increase as the frequency of operation increases and the transmitter power generated by solid-state transmitters decreases. Considering these real-world aspects, X-band operation is preferred for horizon surveillance.

In principle, therefore, low-frequency bands (e.g. L-band) are preferred for long-range and medium-range surveillance and the X-band is preferred for horizon surveillance.

1.6.10.2. The Tracking Function

If the radar tracks the target continuously for an interval of t_0, equation (1.5) can be written as

$$R^4 = \frac{P_{ave}A_e^2}{\lambda^2 T}\left(\frac{t_0\sigma}{4\pi kB(\text{SNR})}\right)10^{-0.2\alpha R} \tag{1.9}$$

As can be seen, the range is inversely proportional to the wavelength of operation. However, we have already noted that the frequency band cannot be too high; real-world considerations therefore dictate that the X-band is preferred for the tracking function.

Apart from detectability, the tracking radar should also yield good angular accuracy, which in turn can be attained by radars operating at the higher frequencies. The K-band (27–40 GHz), between centimeter and millimeter wavelengths, offers reasonable ranges and spatial resolutions. There are clearly many trade-offs to be considered when a radar is designed to perform a given set of functions. Ultimately, the choice will depend on detailed studies of the powers available at the different bands and the costs at the time the radar is designed. Here we have outlined the issues a designer will normally consider.

1.6.10.3. A Golden Mean Band for Multifunction Radars?

Given that it is too expensive to have one radar operating at L-band and another at X-band to perform the surveillance and tracking functions, respectively, radars operating at frequency bands between the two bands have been realized. The AEGIS (AN/SPY-1)

[104], EMPAR (European multifunction phased array radar) [105] and MESAR (multifunction electronic scanning adaptive array) [106] operate at S-, C-, and S-band, respectively. These are multifunction shipboard phased array radars or are designated for shipboard operation. From the foregoing considerations it is clear that neither function is performed efficiently at bands between the L- and X-bands.

Systems operating over 2–3 octaves have many attractions since they can perform the surveillance and tracking functions at optimal frequencies as well as other radar-related or unrelated functions that contribute to the platform's survival. The communications function, for instance, can be performed at the designated frequencies as well as the EW functions. If these systems can operate over large bandwidths, some protection from jammers is also inherent; these systems therefore have an electronic counter-countermeasure (ECCM) capability. We have assumed that third parties have no *a priori* knowledge of the frequency bands the wideband system uses to perform the different radar/non-radar functions, so that the jamming power has to be spread over 2–3 octaves.

Within these broad requirements, many realizations are possible and a plethora of systems and technological issues are explored in sections 1.6.14 and 2.7.4.

Many authors have proposed wideband multifunction radars [e.g. 107–110] and plans for dual-frequency multifunction radars have been reported [111]. More specifically, it is now recognized that the surveillance and tracking functions can no longer be undertaken at one compromise frequency band and the frequencies of 1 and 10 GHz have been accepted for the surveillance and tracking functions [111].

Experimental work has also been reported on advanced shared aperture arrays (ASAAs) capable of operating over several bands and performing a suite of functions [112]. As some multifunction arrays are inherently narrowband, we shall use the term SAAs to describe arrays that operate over 2–3 octaves and perform a variety of functions at optimal frequencies. In a similar vein, a wideband radar is capable of performing a set of civilian functions related to safe air traffic control in airports [14]— see section 2.7.4.

In summary, SAA radars are capable of performing a multitude of interdependent and interrelated tasks at optimal frequencies and with unprecedented efficiency. We shall address the technological and systems challenges imposed by the realization of these systems in sections 1.6.14 and 2.7.4.

1.6.11. The Identification Function

We have already included identification as an important radar function. Without a reliable identification the most advanced weapon systems are rendered impotent. As the spatial resolution of an image is dependent on the wavelength of the illuminator, an optical image of a target has a finer resolution than an image obtained by illuminating the same target by radiation at centimeter or millimeter wavelengths. Optical systems, however, have short ranges and are usually passive. One can therefore propose sensor complementarity in the sense that a radar acquires and tracks the targets and the imaging of some targets of interest is undertaken at infrared (IR) or at optical wavelengths. Alternatively, the imaging function is performed with the aid of SAR or ISAR techniques that offer reasonable spatial resolutions that are independent of the target ranges in all weather conditions.

As these approaches to target identification are expensive to implement, other more affordable approaches to the identification function have been sought. We have already mentioned that HRR radars are often used to identify targets. For approaching targets, neither ISARs nor SARs are effective in identifying aircraft and HRR radars are used [113,114]. The difficulties of identifying targets using HRR range profiles are addressed in reference [115].

In this section we shall outline some other inexpensive approaches to target identification. Polarization target information is often helpful in the identification process and a movement toward automatic radar target identification is evident [116].

An active identification-friend-or-foe (IFF) interrogator is often used to trigger coded responses from a transponder aboard an aircraft of interest as a means of identifying it. This approach is inexpensive and effective in times of peace. The concern is that existing IFF techniques can be spoofed, jammed, or otherwise rendered ineffective in times of hostile engagement.

It is known that airplanes are literally 'glowing' in the electromagnetic spectrum and that the transmissions are highly characteristic of the particular airplane. Spectrum surveillance or electronic support measures (ESM) systems can therefore be used for aircraft identification. However, there are two problems with this approach:

(i)　Range information cannot be usually derived by ESM systems, which provide ID and azimuth information only.
(ii)　The airplane might switch off its jammers to deny the detection of its presence by the ESM user.

Given that a radar can provide range and azimuth information, an ESM system coupled to a radar forms a powerful complementary sensor combination capable of performing the identification function [117]. Furthermore, the radar is switched on for relatively short periods of time, so that ESM systems do not readily detect its presence. The generic requirements for the ESM system and possible approaches to realization are outlined in reference [117].

Considerable effort has been expended in attaining target location by using two or more passive sensors with the aid of triangulation. In realistic scenarios, however, ghost noise jammers can reduce the system's effectiveness. The Passive Jammer Locator is described in reference [118] as well as some experimental results supporting the feasibility of the system. As can be expected, the accuracy of the positional information of different jammers has to increase as the number of jammers increases.

Jet engine modulation (JEM) of radar returns is a commonly observed phenomenon that occurs when a radar illuminates a jet airplane at an aspect angle that allows the electromagnetic radiation to be backscattered from the moving parts of a jet engine's compressor and blade assembly. This has been observed at angles as large as 60° from a nose-on aspect between the radar and the aircraft. Since these engine parts are in rotational periodic motion with respect to the aircraft's airframe, they impart a periodic modulation to the signal scattered from the engine structure, which in turn generates a radar signature that can be useful for target identification. The emphasis of current research in this area is on reliable identification when the observation time on the aircraft is minimized [119].

1.6.11.1. VHF Multifrequency/Multipolarization Radar

A useful approach to target identification is based on the inherent relationship between an object's geometrical shape or composition and its EM scattering characteristics. While one can derive the exact shape of a target by observing it over an unlimited range of frequencies and aspect angles, this approach is impractical.

It has been shown that the essential information related to the overall dimensions, approximate shape, and material composition of a target can be deduced from observations taken over a range of wavelengths [120,121]. The longest and shortest wavelengths can be one-half and ten times the target's maximum dimension,

Table 1.11 Dimensions of some representative aircraft

Aircraft type	Designation	Dimensions (m)		
		Length	Wingspan	Height
Combat	Sea Harrier F/A Mk2	14.17	7.7	3.71
	Hawk 200	11.33	9.94	4.16
	Rafale	15.30	10.9	5.34
	Mirage 2000C	14.36	9.13	5.2
	Mirage III	15.03	8.22	4.5
	F-4	17.76	11.7	4.96
	F-14 Tomcat	19.10	19.54/11.65[a]	4.88
	F-16 Fighting Falcon	15.03	9.45	5.09
	F-111 Aardvark	22.40	19.2/9.74[a]	5.22
	F-117A	20.08	13.20	3.78
	F/A-18C	17.07	11.43	4.66
	B-2 Spirit	21.03	52.43	5.18
	MiG-17	11.36	9.63	3.8
	MiG-21	15.76	7.15	4.1
	MiG-23	15.6	13.96/7.78[a]	4.82
	MiG-25	23.82	14.01	6.1
	MiG-31	22.7	13.464	5.15
Reconnaissance/Patrol	An-72P Coaler	28.1	31.9	8.6
	E-767 AWACS	48.5	47.5	15.8
	E-8 Joint STARS	46.6	44.4	12.95
	E-2C Hawkeye	17.5	24.6	5.6
	E-4B	70.5	59.6	19.3
	P-3 Orion	35.61	30.37	10.2
Transport	An-225 Mriya	84	88.4	18.2
	C-5B Galaxy	75.5	67.8	19.8
	C-130 Hercules	29.8	40.4	11.7
Civilian	A340 Airbus	59.39	60.3	16.74
	Concorde	62.10	25.56	11.4
	Boeing 707	46.61	44.42	12.93
	Boeing 747	70.7	59.6	19.3
	Boeing 777	63.7	60.9	18.5
	DC-10	55.5	50.4	17.7
	SJ30	12.9	11.1	4.2
Light	PA-32-301	8.4	11	2.5
	Beachcraft Skipper	7.3	9.1	2.3

[a] Wingspan swept.

respectively [120,121]. Observations taken at shorter wavelengths are useful only when they are accompanied by observations taken at longer wavelengths.

Lin and Kriensky [122] found experimentally that a high reliability of classification is achieved with the aid of systems using only two frequencies and that even a single frequency can provide over 95% reliability when both the phase and amplitude of the two orthogonal polarizations are used. The data set involved eight military aircraft whose sizes varied over a 3:1 range.

Table 1.11 lists the dimensions of several aircraft [123]. The dimensions of different sets of aircraft vary considerably, so classification is rendered easier; when the dimensions of aircraft within one set exhibit only small variations, classification by the above method is generic.

Preliminary results from a VHF multifrequency and multipolarization radar dedicated to the identification function have been reported [124]. The Maquette Orientee pour un Systeme d'Analyse de Resonances (MOSAR) has the following characteristics [124]:

- The system is a coherent, pulsed, quasi-monochromatic radar operating in the frequency range from 20 to 100 MHz.
- Initially two frequencies are transmitted simultaneously, but ultimately four frequencies are to be used.
- Two orthogonal polarizations are used.
- The system records the amplitude, phase, and Doppler information of the received signals.

As the efficiency of radar-absorbing materials is reduced with decreasing frequency, the use of low frequencies enhances the probability of detection and identification of stealthy targets in the VHF range.

1.6.12. Polarimetric Radars

The requirement to derive maximum information from systems led to polarimetric systems. Dual-polarization communication systems, for instance, carry twice the number of communication channels compared to single-polarization systems; commercial returns are therefore increased. Similarly, dual-polarization radioastronomy systems derive maximum information related to the observed celestial sources.

Given that the polarimetric properties of complex targets are not known *a priori*, polarimetric radars measure the σ of a target at all possible polarization combinations. Thus polarization is used to maximize the SNR of a scene and for the purposes of discriminating targets from clutter. The underlying assumption made here is that the polarimetric properties of targets (human-made objects) are fundamentally different from the polarimetric properties of clutter such as sea, foliage, and rough terrain. Armored targets, for instance, have a significant amount of even-bounce returns, whereas clutter tends to have mostly odd-bounce returns [68]. Apart from detection, polarimetric radars are used for target identification also.

Work on polarimetric radars began at the Georgia Institute of Technology and the Georgia Tech Research Institute in the 1960s [125] and a plethora of research papers originated from research carried out at these establishments [68]. An excellent handbook [126] summarizes work on polarimetric radars.

Without loss of generality, let us consider the four principal polarizations, i.e. vertical (V) horizontal (H) left-handed circular (LHC) and right-handed circular (RHC). The two linear polarizations are orthogonal while the two circular polarizations are opposite.

The reflected power from any target or clutter can be completely described by the radar scattering matrix (often referred to as the polarization scattering matrix, PSM). The matrix describes the radar reflectivity characteristics in terms of two orthogonal transmitted polarizations and two orthogonal received polarizations. Thus for vertical and horizontal polarizations the received signals from a target can be expressed by the following matrix:

$$\begin{bmatrix} E_{rH} \\ E_{rV} \end{bmatrix} \propto \begin{bmatrix} \alpha_{HH} & \alpha_{VH} \\ \alpha_{HV} & \alpha_{VV} \end{bmatrix} \begin{bmatrix} E_{tH} \\ E_{tV} \end{bmatrix} \tag{1.10}$$

Here α_{HH} is a coefficient corresponding to the case where the horizontal polarization is transmitted and received. Similarly, α_{HV} is a coefficient that corresponds to the case where a horizontal polarization is transmitted and a vertical polarization is received. Subscripts r and t stand for receive and transmit modes.

Each of the α_{MN} terms of the matrix is complex, i.e. it has an amplitude and a phase term, and can be related to the RCS of the target by

$$\alpha_{MN} = \sqrt{\sigma_{MN}} \exp j\phi_{MN} \tag{1.11}$$

For the monostatic radar case, α_{VH} equals α_{HV}, so that there are only three complex, independent coefficients in the scattering matrix. This redundancy is often used to calibrate polarimetric radars.

While the two orthogonal linear polarizations have been considered here, two opposite circular polarizations are often used. In reference [68] the scattering components for circular polarization are related to those for linear polarization.

The phase between the horizontally and vertically polarized components of the received electric field is called the polarimetric phase, ψ, which can be derived by performing a dot-product vector operation between the horizontal and vertical received components of the electric field [68]. Thus,

$$\cos \psi = \frac{\alpha_{HH}\alpha_{HV} + \alpha_{VH}\alpha_{VV}}{(\alpha_{HH}^2 + \alpha_{VH}^2)^{1/2} (\alpha_{HV}^2 + \alpha_{VV}^2)^{1/2}} \tag{1.12}$$

The measurements of the quantities α_{MN}, which are related to σ_{MN}, are performed in two stages. For the case where the two linear orthogonal polarizations are used, the transmitter transmits one pulse of vertically polarized radiation and the scattered radiation is received in two separate channels accommodating the vertical and horizontal polarizations. From these measurements the coefficients a_{VH} and a_{VV} are deduced. The polarization of the next pulse changes to horizontal and the scattered radiation is received in the same two receivers. The assumption is here made that no changes occur between pulses. From these measurements the coefficients a_{HH} and a_{HV} are deduced. Similar measurements are performed when the polarizations used are opposite circular polarizations.

1.6.12.1. Hardware Requirements

While measurements of the full PSM yield the maximum benefits for a radar system, measurements of two elements of the PSM yield some benefits, which are usually application dependent. References [68] and [126] outline several approaches of maximizing the benefits of polarization radars using combinations of polarizations.

It is possible to derive an optimum transmitter–receiver polarization combination for target detection in the presence of background clutter [127]. Alternatively, the contrast of radar images can be maximized by the appropriate selection of polarization combinations [128]. In the latter approach an optimal polarimetric matched filter is sought; this filtering task can be performed adaptively [129–133] .

Brown and Wang have shown that an improved radar detection capability results from using adaptive multiband polarization processing [134,135].

Given the trend toward adaptive systems, the quest for procedures and techniques that maximize the benefits of polarimetric radars adaptively is of considerable import.

1.6.12.2. Depolarization of EM Waves

Rainfall introduces depolarization of EM waves, which in turn causes additional attenuation and a decrease in the isolation between the channels accommodating the two polarizations. The depolarization of EM waves is caused by the introduction of a differential phase and amplitude attenuation to the two channels accommodating the two polarizations of a dual-polarization system [136]. Methods of reducing the effects caused by the depolarization of EM waves adaptively in real-time have been proposed for dual-polarization satellite communication systems [137] and the application of these techniques to polarimetric radars has been proposed [138].

1.6.12.3. Instrumental Aspects

Apart from polarization agility, adaptability, and frequency diversity, the attainment of polarization isolation is significant for polarimetric radars for the following reasons:

- The minimization of errors in measuring the polarization coefficients.
- Protection of the radar from jammers. More explicitly, jammer power can render a radar that has poor cross-polarization isolation inoperative by transmitting considerable power having a polarization state orthogonal or opposite to the radar's polarization.

Detailed calculations [139] related to the position and shape of the cross-polarization lobes of conventional reflectors resulted in the following conclusions. To obtain an accuracy of 0.5 dB for the cross-polarized scattering coefficient lying x dB beneath the like-polarized scattering coefficient, the required one-way isolation P_{ISO} is given by $P_{ISO} = x + 16$ (dB). In reference [140] it is shown that the difference between the co-polarization and cross-polarization scattering coefficients can be as high as 25 dB, a figure that can be taken as the upper limit condition.

Designers therefore opt for cross-polarization isolation levels between 25 and 40 dB. Polarization isolation figures in excess of 40 dB are routinely obtained in satellite communication systems, even during rainy periods [137, 141]; it is noted here that communication systems are narrowband systems. Polarizers and their pertinent characteristics are considered in some detail in section 3.3.6.

1.6.12.4. Polarimetric Radars

For a long time polarimetric radars were seen as systems that increase the SNR of a scene. The capability of these systems to identify targets is a recent development. The benefits offered by polarimetric radars were appreciated as early as 1954, when an improvement of at least 10 dB in clutter reduction was claimed for a radar using circular polarization [142]. The radar transmitted and received power with the same circular polarization, so rain clutter returns that had the opposite circular polarization to that originally transmitted were considerably attenuated. Speckle reduction in SAR images was reported by the use of polarization information [143].

An improvement of a few decibels was reported in reference [128] in the contrast of scenes by the illumination of the scene with power having the optimum polarization and the reception of the scattered powers by a dual-polarization receiver having optimal polarizations.

The definition of the benefits that dual-polarization radars offer to the user is not a trivial task, for these benefits are application dependent and other techniques are usually used in conjunction with polarimetric information. Here we shall consider the case of the airborne detection of stationary objects [144], a typical application in which other conventional target discrimination techniques are not applicable.

When the spatial resolution of a polarimetric radar is low, compared to the target dimensions, the maximum increment in the probability of detection for a polarimetric radar ranges between 2 and 4 dB [144]; here the comparison is made with a conventional system utilizing one polarization only. If the spatial resolution of the polarimetric radar is comparable to the target dimensions, the resulting benefits are considerable and the probability of detection is independent of the intercell signal to clutter, S/C ratio [144]. By contrast, the probability of detection for a conventional radar depends on the S/C.

From these considerations it is clearly seen that the maximum benefits of polarimetric radars reside with high-resolution systems, e.g. SARs, ISARs, and high-resolution phased array radars.

1.6.13. EW Aims and Systems

We have already seen that EW systems can diminish the effectiveness of radars through the use of jamming. More explicitly, a typical EW system consists of a spectrum surveillance or electronic support measures (ESM) subsystem and an electronic countermeasures (ECM) subsystem. The ESM subsystem detects and measures as many parameters of operating radars as possible and hands over the derived information to the ESM subsystem, which in turn delivers jamming power to marginalize a selected radar's effectiveness; in extreme cases unprotected radars can be rendered useless.

An overview of surface navy ESM/ECM development has been reported [145].

Although it is notionally convenient to consider ESM and ECM subsystems separately, in practice there is a high degree of integration between the two subsystems.

As different radars operate over many different bands, EW systems are inherently wideband systems. Our interest in EW systems is therefore threefold:

1. Knowledge of the capabilities of EW systems is necessary for the evolution of radars that have electronic counter-countermeasures capabilities. From a radar point of view it is a simple case of knowing the capabilities of your enemy.
2. It is impractical to adapt and/or modify radars designed to operate in benign environments so that they can cope with the challenges imposed by realistic EW environments.
3. Given that EW systems are wideband and the general trend toward wideband systems, the radar and communications designer can derive a plethora of generic design guidelines.

1.6.13.1. ESM Functions/Systems

The functions of ESM systems include the interception, identification, analysis, and location of hostile sources of EM radiation. ESM systems operate in a receive-only mode but, unlike the receiving part of conventional radars, the systems operate over very wide bands of frequency, i.e. over a few octaves. ESM systems or spectrum management systems are often used for the purposes of managing the frequency spectrum, a valuable commodity.

Spectrum surveillance is undertaken by elaborate, and complex ESM systems continuously over a long period of time. The continuous surveillance of the spectrum is mandatory if important intermittent emissions are not missed. These systems, usually referred to as electronic intelligence (ELINT) and communications intelligence (COMINT) systems, are at some distance from the transmitters of interest.

Inter alia these ESM systems have a high probability of interception (POI) in the spatial and frequency domains and perform the following functions:

- The de-interleaving of the radar pulses. Even a 10° field of view in azimuth can encompass millions of pulses per second [117].
- Assigning the derived emanations to radars.
- The prioritization of threat radars.
- The definition of as many parameters of the threat radars as possible.

Radar warning receivers (RWRs), on the other hand, are used onboard aircraft, ships, and submarines and by ground-forces personnel for their protection from surface-to-air and air-to-air missiles and anti-aircraft gun systems. Let us consider a victim radar that is at a distance R from the RWR. If the radar transmits power P through an aperture that has a gain G_R, the SNR of the RWR, SNR_{RWR}, is given by

$$SNR_{RWR} = \left(\frac{P}{4\pi R^2} G_R\right)\left(\frac{A_e}{kTB}\right)\frac{1}{L}$$

where A_e and B are the RWR's effective aperture and bandwidth, kTB is the noise power of the RWR, and L is the system losses at the RWR.

As can be seen, the SNR_{RWR} is inversely proportional to the square of the radar range; the jammer therefore has an inherent advantage over a radar, the SNR of which is inversely proportional to R^4.

Ideally, the requirement to maximize the POI of an ESM system in the frequency and spatial domains can be met by wideband receivers and staring phased arrays yielding enough independent and contiguous antenna beams to fill the required surveillance volume. This is a tall order that is hard to meet efficiently because of the following real-world limitation:

- The noise temperature of receivers operating over several octaves is poor when compared to receivers operating over one octave.
- Jammers can drive the wideband receiver into a non-linear operation.
- Passive components have limited bandwidths.
- The formation of a very high number of staring beams is not trivial—see Chapter 5.

The above concerns can be succinctly summarized by the words of H. A. Weaver 'broadband design means equally bad all over'. The solutions derived by the EW community are of a generic importance. The input bandwidth is often split into many bands by a low-loss frequency multiplexer. Each band, extending over one octave only, is then accommodated by a low-noise amplifier, to maximize the sensitivity of the system, and after amplification each band is further processed separately. In one realization of an ESM system, banks of contiguous filters are used to examine the spectrum of each band. Jammers are excised in the spatial domain by nulls generated by the array and tunable band rejection filters in the frequency domain. Here we have assumed that efficient frequency-independent antennas are used.

1.6.13.2. Active ECM Systems

ECM systems are usually divided into passive and active systems, often used to disable or at least marginalize the effectiveness of victim radars. Intentional jamming, for instance, is used to protect aircraft from ground-based radars, while blinking jammers deny ARMs their guidance signal during their off-times. The requirements for active ECM systems are therefore wideband operation; the maximization of the jamming power; and the delivery of the jamming power to the victim radar.

As phased arrays can direct considerable powers toward any given direction within their surveillance volume, phased array-based ECM systems are often used. Typically, phased array-based ECM systems can provide high effective radiated power (ERP) in the vicinity of 1 MW, for airborne platforms [146].

We have already seen that the jammer has a considerable advantage over the radar; here we shall further quantify this advantage by deriving the jam to radar ratio, J/S, which can be derived from the ratio of the powers received by the radar from the jammer and from a target having an RCS of σ. It can be shown that the J/S ratio is given by

$$J/S = 4\pi \left(\frac{P_j}{P_r}\right)\left(\frac{G_j}{G_r}\right)\left(\frac{R^2}{\sigma}\right)10^{0.1\alpha R} \qquad (1.13)$$

where G_j is the gain of the antenna used by the jammer. The first two bracketed terms in equation (1.13) can be significantly lower than unity, while the third is significantly higher than unity.

The derived J/S corresponds to the case in which the boresight axes of both the radar and the jammer are coincident, which is at best valid for only a small fraction of the time as the radar antenna moves in azimuth to cover the surveillance volume. When the two axes are not aligned a similar calculation can be made, taking into account the radar–jammer geometry, the polarization state of the jammer, the sidelobes, and the cross-polarization lobes of the radar's aperture.

1.6.13.3. Passive ECM Approaches

Passive ECM approaches include:

* Chaff and decoys/flares.
* The evolution and usage of radar-absorbing materials and radar target signature modification.
* Foliage/natural cover; camouflage screens.
* False targets to produce confusion.

Use of chaff involves dispensing a large number of 'reflectors' from aircraft or ships for the purpose of marginalizing the radar's effectiveness. The 'reflectors' take mainly two forms: chaff dipoles and non-resonant streamers of conducting material, also called 'rope'. Typically the chaff dipoles consist of aluminized-glass fibers cut to roughly half the wavelength of the target radar frequency; the streamers consist of aluminum foil 30 to 300 m long. There are other realizations of chaff and details on chaff systems used for the defense of ships are outlined in reference [147].

1.6.13.4. RCS Minimization Approaches

The minimization of the RCS of high-value platforms is the most significant driver for the development of modern radars. Let us consider that an important platform is detected by a radar at a range R_{min} when its RCS is σ_0; using the radar equation we can calculate the reduction in R_{min} as σ_0 is decreased by 10, 20, and 30 dB to be $0.56R_{min}$, $0.32R_{min}$ and $0.18R_{min}$, respectively. It is clear that further reductions in the platform's RCS will render the radar useless. If the radar increases its transmitting power to detect the platform at ranges longer than R_{min}, the radar's detectability by ESM systems is increased.

Two other reasons why the reduction of a platform's RCS can improve its survivability are that the enemy may misjudge the size and shape of the platform; and the amount of chaff needed to protect the platform is reduced.

While new platforms can be designed to have geometric shapes that minimize the platform's RCS, the RCS of platforms already in operation is minimized by the application of radar-absorbing materials (RAMs). Early work in the development of RAMs is described in references [148,149]. The proceedings of the IEE-sponsored colloquium on 'Low Profile Absorbers and Scatterers' (see, for example, [150,151] contain a plethora of papers that describe theoretical and experimental work related to many approaches for the minimization of the RCSs of platforms. The application and effectiveness of RAMs to armored fighting vehicles and aircraft structures are described in references [150] and [151], respectively. Design tools of broadband RAMs for large angles of incidence have been developed in reference [152].

1.6.13.5. Approaches to ECCM

From the foregoing it is not hard to appreciate that radars need considerable protection from ECM systems. It is also appropriate to emphasize here the often-forgotten dictum that it is more economic to incorporate electronic counter-countermeasures (ECCM) in the initial radar or communication system design rather that modify a system to protect it from the threats of ECM systems at a later stage.

From our considerations we can easily draw the requirements for radar apertures to have low sidelobes and cross-polarization lobes. The other requirements are:

- Agility/flexibility of as many of the radar's pertinent parameters as possible, e.g. central frequency, pulse repetition frequency, polarization, pulse width and intrapulse modulation on a pulse-to-pulse basis.
- Ability to use pulse compression, spread-spectrum techniques, and pulse integration techniques against ECM. Pulse compression can typically add a 30 dB advantage to the radar, while pulse integration can add a 10–17 dB advantage (the integration of 10 or 50 pulses is assumed) [75].
- The other LPI capabilities that were explored in section 1.6.5.

The coherent integration of a radar's pulses operating in a frequency-hopping mode [75] deserves further research. Frequency hopping is here assumed to occur on a pulse-to-pulse basis.

1.6.14. Broadband Arrays: Definitions and Applications

We have already made a case for broadband phased array systems performing sets of interrelated and interdependent functions in the military and civilian arenas. Corroborative support for truly broadband phased array systems comes from the systems that have large time–bandwidth products and are capable of performing the identification function.

Let us define what we mean by broadband/wideband, since there is considerable confusion in this area. We have already seen that most operational systems operate over a fractional bandwidth of 10% or less; at the design stage these systems met most of the requirements drawn at the time.

If we restrict attention to military systems for the time being, we can discern two types of broadband phased array systems:

Type 1   The system operates over a bandwidth of at least 2–3 octaves. It performs all radar, EW, radiometric and communications functions necessary for the survival of a platform at the optimum/assigned frequencies and has LPI and ECCM capabilities. The system is ideal but expensive. The technological challenges of realizing and operating the system are:

(i)   Issues centered around subsystems such as antennas, multiplexers, LNAs, FPAs, and transmitters that can operate efficiently over the required broad bandwidth.

(ii) Scheduling problems related to a system that performs a wide range of functions when the threats vary dynamically and the prime power available is finite.

(iii) The requirement arising from the need to excise jammers in the spatial and frequency domains.

Type 2 The system operates at widely separated frequencies $f_1, f_2, \ldots, f_N$, but the fractional bandwidth centered at each frequency is 10% or less. It performs all radar, radiometric, and communications functions necessary for the survival of a platform at the optimum/assigned frequencies and has some LPI and ECCM capabilities. Additionally, it can perform the EW functions over a limited frequency range.

The two systems can be complementary in the sense that they are suited to the strategic and tactical environments, respectively. Civilian systems are likely to be type 2 systems. It has been claimed that type 2 systems are likely to be more popular than type 1 systems [153]; from a theoretical standpoint, the complementarity of the two systems is attractive.

While there is consensus that these systems are needed, there are divergent views on their realization and a detailed discussion of the realization issues of these systems is taken up in section 2.7.4. The relative costs of wideband systems compared to several systems performing the many required functions for the survivability of a valuable platform is another issue considered in the same section.

1.6.15. Radar Arrays: A Record of Progress

Detailed descriptions of different radar phased array systems can be found in [8]; here we shall thematically outline the record of progress with the aid of Table 1.12; themes 1–17 are dedicated to different aspects of radar phased arrays. Non-phased array systems and precursors of phased arrays are also considered so that continuity is maintained and comparisons are made. In a similar vein, some of the proposed systems are included for the purposes of tracing future trends and minimizing surprises.

Theme 1 The early ground-based radars met military and aviation requirements. Airborne/spaceborne SARs (flying phased arrays) have military roles such as surveillance and reconnaissance and also non-military roles. Maps derived from airborne and spaceborne SARs contribute significantly to the management of the Earth's resources.

It is expected that the exploration of the capabilities of SAR systems is followed by the exploitation of SAR systems with the aid of proposals for LightSARs. The latter systems are high-performance, lightweight, and cost-effective systems that meet the demonstrated need for Earth science, commercial and civil applications—see section 1.11.2.

Seminal advances in earthquake studies, volcano monitoring, hydrology, topography, tectonics, and monitoring of movements of glaciers and ice sheets are based on data derived from spaceborne interferometric SARs. It is impossible to overemphasize the applied science value of these contributions.

Table 1.12 Radar phased arrays: A record of progress

1

Early ground-based arrays met military and aviation requirements.

⇓

Airborne and spaceborne SARs have military roles and contribute significantly
to the management of the Earth's resources.
Exploration of SAR capabilities.

⇓

LightSARs are high-performance, lightweight and cost effective systems that
meet the demonstrated need for Earth science commercial and civil applications.
Proposal stage—exploitation of SAR capabilities

⇓

Seminal advances in earthquake studies, volcano monitoring,
hydrology, topography, tectonics, glacier and ice sheet movements
are based on data derived from spaceborne interferometric SARs.

⇓

Radars on board constellations of LEO/MEO satellites can perform
military roles at short notice anywhere on the globe.

2

Early military radars reflected cold-war realities and doctrine
(very wide surveillance areas; high RCS targets; long re-visit times)

⇓

Military radars reflect post-cold-war realities and doctrine
(theater battle support in low-intensity conflicts; system mobility;
small surveillance area; low RCS targets; short re-visit times).

3

One beam derived from a parabolic reflector; mechanical scanning in two dimensions.

⇓

One beam derived from a 'flat phased array'; mechanical scanning in two dimensions.

⇓

Mechanical scanning of a phased array in one dimension and electronic scanning in the other.

⇓

One inertialess beam derived from a phased array (phase scanning); the beam is used in a benign
environment where a few targets exist.

⇓

Many inertialess beams are derived from a phased array.
Multiple beams are required in a realistic multitarget
environment characterized by clutter, rain, and chaff. Proposal stage.

Table 1.12 cont.

4

Airborne non-aerodynamic aperture protected by a radome.

⇓

The evolution of 'smart-skin', aerodynamic, conformal arrays.
Radome attenuation is eliminated.
Potential for lower costs attributed to the array antenna elements—see section 3.6.

5

Mechanically rotating radars have invariable
data rate, dwell time per pixel and transmit equal power to every pixel.

⇓

Phased arrays: Data rate, dwell time per pixel, and energy for each
pixel are variable and scenario dependent.

6

Mechanical scanning: Nexus between the surveillance and tracking functions.

⇓

Narrowband multifunction phased arrays: Partial breakage of the nexus between the
above radar functions.

⇓

Shared aperture phased arrays: Complete breakage of the nexus between
the above functions; adequate dwell times per pixel, higher
data update rates and a more robust tracking capability.
Proposal stage.

7

Cold-war environment.
Array cost is an independent variable
(not mentioned in polite conversations, not treated in books).

⇓

Array cost is equated to acquisition cost.

⇓

Post cold-war environment.
Array cost is *a priori* constrained—impetus for innovation.

⇓

Array cost is the sum of acquisition and life-cycle costs.

⇓

From affordable to low-cost, 'intelligent' arrays: self-scanning, Radant arrays and lenses.

Table 1.12 cont.

8

Passive arrays: One tube transmitter; high voltage and high power at the array face; transmission losses and losses due to phase-shifters degrade the system's performance; bulk; low MTBF and no graceful degradation of the system's performance.

⇓

MMIC-based active arrays: The effect of the above losses on the system's performance is marginalized; compact and low-voltage; high MTBF and graceful degradation of the array's performance.
MMIC-based beamformers augment the array's capabilities.
Active arrays using MMIC-based T/R modules are compact and exhibit controlled uniformity ⇒ lower sidelobes/sharper nulls.

⇓

Use of cryogenically cooled HTS-based subsystems, e.g. antennas, filterbanks, low-phase-noise transmitters, multiplexers and delay lines.

⇓

Renaissance of vacuum technology.

⇓

Marriage of vacuum and solid-state technologies.

⇓

Consideration of new-look passive arrays.

9

Canonical phased arrays perform radar functions over a narrow band.

⇓   ⇓

Separate arrays perform a variety of functions necessary for the survival of a high-value platform.	Separate arrays perform a variety of functions necessary for airport traffic management/control.
⇓	⇓
Shared aperture array: One wideband phased array performs a variety of military functions. Interrelated/interdependent functions. Proposal stage.	Shared aperture array: One wideband phased array performs a variety of civilian functions. Interrelated/interdependent functions. Proposal stage.

10

Phased arrays: Transmit and receive one polarization.

⇓

Phased arrays: Transmit and receive one polarization only but the polarization can be either of the linear orthogonal polarizations or one of the opposite circular polarizations.

⇓

Full polarization capability phased arrays have significant advantages over their single polarization counterparts.

Table 1.12 cont.

⇓

Adaptive polarization capability phased arrays.
Proposal stage.

11

Phased arrays with no ECCM capabilities have high sidelobes, narrow bandwidth,
poor polarization isolation, and invariable waveforms.

⇓

Phased arrays with excellent ECCM capabilities have LPI capability, low sidelobes,
high polarization isolation, variable waveforms on a pulse-to-pulse basis and their bandwidth
is centered at a variable frequency or is spread.

12

Phased arrays with no LPI capabilities: Transmission of maximum power
directed at one pixel at the shortest time.

⇓

A plethora of phased arrays with excellent LPI.

⇓

Bistatic/multistatic arrays.

⇓

Bistatic/multistatic arrays using illuminators of opportunity.
Excellent LPI capabilities.

13

Basic phased array radar functions: Surveillance and target tracking.
A few targets in a benign environment ⇒ multitarget environment in the
presence of clutter, rain, and chaff.
RCS of targets decreases significantly
+
Weapons delivery (fire control)
+
Ground following/avoidance capability
+
High resolution mapping/imaging/reconnaissance resulting
in the identification of moving/stationary targets from airborne platforms.

⇓

Inexpensive target ID systems.

14

Monostatic radar.

⇓

Bistatic/multistatic radars.

⇓

Table 1.12 cont.

Multistatic radar utilizing illuminators of convenience: transmissions of FM/TV stations or of constellations of satellites such as NAVSTAR GPS and GLONASS. Excellent LPI capabilities.

⇓

Spaceborne distributed array radar.
Proposal stage/some experiments.

15

Many transmitters, forming a compact phased array, illuminate the same target.

⇓   ⇓

Threats imposed to the unprotected electronic systems of aircraft by these arrays is far more serious than that imposed by conventional EW systems.
Proposal stage.

Power pseudo-satellites that perform radar, radar-related, and communication functions over an area of 1000 km in diameter.
Proposal stage.

16

Infrequent array calibrations.

⇓

Frequent array calibrations during operations.

⇓

Array self-healing during operations.

17

Active phased arrays: Power addition on target; number of antenna elements equals the number of modules.

⇓

One antenna and many T/R modules: Power addition at the array.

⇓

Many antenna elements and each antenna element is fed by many modules (some power addition at the array and some on target).

Radars on board constellations of LEO satellites can perform military roles at short notice anywhere on the globe—see the DARPA proposal in section 1.11.5.1.

Theme 2   The early military radars reflected the cold-war realities and doctrines and were characterized by very wide surveillance areas, relatively high RCS targets, and long re-visit times. By contrast, the post-cold-war doctrine is characterized by systems suitable for theater battle support in low-intensity conflicts. The surveillance area is smaller, the targets have low RCS and the re-visit times are shorter. The mobility of these systems is imperative—see section 1.11.5.1.

Theme 3   We start with a conventional radar system utilizing a parabolic reflector mechanically scanning the surveillance volume. The parabolic reflector is then substituted by a 'flat phased array' that produces a stationary beam that can be mechanically scanned; the inertia of the latter system is decreased and all the elements of the 'flat phased array' are connected to a summing point via equal lengths of transmission lines or waveguides. The 'flat phased array' is substituted by a phased array that is scanned electronically in one dimension and mechanically in the other (e.g. AWACS). The latter antenna is again substituted by a phased array that yields one inertialess beam and the system is appropriate for operation in a benign environment where a few targets exist. An array yielding many inertialess beams is required for operations in a realistic multitarget environment characterized by rain and chaff—proposal stage see section 2.7.4.

Theme 4   We begin with an airborne radar utilizing a mechanically scanned aperture that is protected by a radome to preserve the aerodynamic properties of the platform. The aperture, a reflector, 'flat phased array', or a fully steerable phased array is substituted by a conformal array distributed on the aircraft's skin. The 'smart skin' array preserves the aerodynamic properties of the platform and radome losses are eliminated; there is potential for lower costs attributed to array antenna elements—see section 3.6.

Theme 5   Mechanically rotating radars have invariable data rate and dwell time, and transmit the same power to every pixel within the surveillance volume of the radar. For a phased array system the data rate and dwell time on a pixel are variable; therefore the energy transmitted to any pixel is also variable and scenario dependent.

Theme 6   A nexus between the surveillance and tracking functions exists for conventional radars. For conventional narrow band multifunction radar arrays this nexus is partially broken and is completely broken by proposed shared aperture arrays having a multitude of beams. The latter arrays therefore have adequate dwell times per pixel, higher update rates, and a more robust tracking capability—proposal stage, see section 2.7.4.7.

Theme 7   Array cost was an independent variable for the early military radars, and the total array cost was equated to the acquisition cost. In the post-cold-war environment the array cost is *a priori* constrained, like many electrical radar parameters, and the total array cost is equated to the sum of the acquisition and life-cycle costs. At first sight it would appear that cost constraints impact on the radar's quality; cost constraints, however, often promote innovations in many areas of array design—see section 1.11.5.2. Lastly, the widespread application of phased arrays is based on the premise that affordable arrays and low-cost 'intelligent arrays' can be realized—see Chapter 4 and section 2.6.

Theme 8   A passive phased array utilizes a tube transmitter, which in turn brings the following suite of unattractive characteristics to the system: high voltage and high power at the array face; short life, bulk and the losses due to transmission lines and phase-shifters affect the system's performance; lastly there is no graceful degradation of the array's performance.

   Active arrays using solid-state devices, on the other hand, exhibit graceful degradation of the array's performance, long life, and the effect of the above losses on the system's performance is marginalized (see sections 2.5.3 and 4.7). Low voltages and powers at the array face and adequate power density on the target characterize the active phased arrays.

MMIC-based beamformers augment the capabilities of an active array by deriving several independent and simultaneous beams. The controlled uniformity of the MMIC-based T/R modules results in lower array sidelobes or sharper nulls (see sections 2.5.6 and 4.3).

Cryogenically cooled high-temperature superconductor (HTS)-based subsystems offer very attractive features to all systems including radar and EW systems. It is expected that the trend to use of HTS-based subsystems such as antennas, filterbanks, multiplexers, and low-phase-noise transmitters will accelerate as the cost of cryodynes is lowered—see section 1.11.3.

The renaissance of vacuum technology results in many important devices, while the marriage of vacuum and solid-state technologies results in high-power transmitters operating at centimeter and millimeter wavelengths. Attractive passive array variants can now be considered. See sections 4.4 and 4.5, respectively.

Theme 9   Canonical phased arrays perform radar functions over a narrow frequency band; the other functions necessary for the survival of a high value platform, e.g. EW and communications, are performed by other phased array systems. Similarly, many separate radars are used in airports to implement airport traffic control. One shared aperture array operating over a wide band can perform all the interrelated and interdependent functions necessary for the survival of a high-value platform or to implement airport traffic control in the civilian arena. These are but two applications for shared aperture arrays—see section 2.7.

Theme 10   The simplest radar systems transmit power and receive the scattered radiation in one polarization. Significant system advantages are gained if the array polarizations on receive and transmit can be selected. Arrays having full polarization capability on a pulse-to-pulse basis are more expensive but yield significant system benefits. Lastly, radar systems have been proposed that adjust the receive and transmit polarizations adaptively to maximize the derived SNR of a target or a scene—section 1.6.12.1 and 1.6.12.2.

Theme 11   A phased array with no ECCM capabilities has the following attributes: high sidelobes, poor polarization isolation, a narrow band width centered at a fixed frequency, and a waveform defined by its invariable PRF and duty cycle.

By contrast, an array that has excellent ECCM capabilities has the following attributes: LPI capability, low sidelobes, and high polarization isolation; and its waveform is variable on a pulse-to-pulse basis. Its bandwidth is centered at a variable frequency or is spread.

Theme 12   Phased arrays that transmit maximum power toward one pixel at the shortest time and have no LPI capability. By contrast, LPI phased arrays can be as effective as the above arrays, but the power transmitted at any time is low, so the system is hard to detect. A plethora of LPI radars is considered in sections 1.6.5 and 1.6.6. Bistatic/multistatic radars using illuminators of opportunity have excellent LPI capabilities.

Theme 13   The radar functions performed by phased arrays have increased with the passage of time. The surveillance and tracking functions are the most basic; with the passage of time the number of targets that the array is capable of tracking in a realistic environment has increased and target RCS has decreased significantly. The following additional functions have been added: (i) weapon delivery (fire control); (ii) ground following/avoidance capability; and (iii) high-resolution imaging, mapping and reconnaissance, resulting in the identification of targets (moving/stationary) from airborne platforms. The integration of all these functions into working arrays presents considerable challenges. Inexpensive target ID systems have evolved—see section 1.6.11.

Theme 14   While monostatic arrays are the norm, bistatic/multistatic arrays are gaining popularity; for the latter systems, illuminators of opportunity can be used, such as FM/TV transmissions or those from NAVSTAR GPS and GLONASS. As more LEO and MEO constellations of satellites are in the planning stage, bistatic radars will become more attractive—see sections 1.6.6 and 1.9. These systems have excellent LPI capabilities. Spaceborne, self-cohering DARs have been proposed and some experimental work has been reported (see section 1.6.7).

Theme 15   Several transmitters of a compact array can illuminate the same target, e.g. a flying aircraft, to increase the power density on the target. The collimated power of the same array can at least disable an aircraft's unprotected electronic systems. The threats offered by the proposed arrays is far more serious than that offered by conventional EW systems—see sections 1.6.9.3 and 1.8.2. The same array can be used to power a pseudo-satellite to perform radar, radar-related, and communication functions over an area of 1000 km in diameter. Both systems are at the proposal stages.

Theme 16   For an array to operate efficiently, calibrations are necessary and the frequency of these calibrations has increased with the passage of time. Self-healing arrays have evolved—see section 4.2.2.

Theme 17   Conventional active phased arrays have many antenna elements and as many T/R modules. A wider definition of phased arrays includes systems of one antenna and several T/R modules (power addition at the array), and arrays where each antenna element is fed to a number of T/R modules (power addition at the array and on target)—see section 1.8.1.

1.7. RADIOASTRONOMY SYSTEMS

In section 1.3 we defined the aims of radioastronomy and alluded in general terms to some of the contributions made by this young branch of science. In this section we shall outline some of the key discoveries made and provide the connecting links between observations taken by radiotelescopes and the derivation of the physical processes taking place in and around celestial sources. Lastly, the theoretical framework of imaging is considered before applications are outlined.

1.7.1. Discoveries and More Discoveries

Nowadays we take it for granted that celestial sources emanate radiation at radio frequencies and that some of them are compact. As late as 1950 some astronomers held the view that radio observations would play no important role in the study of the cosmos.

They assumed that thermal radiation was the only significant source of radio waves and calculations indicated that the radiation from thermal sources is so faint that no useful radio measurements could be made of any sources except perhaps the Sun and planets.

The early discoveries of Karl Guthe Jansky (1905–1950) [154] established radioastronomy as a new science that raised a degree of curiosity. Noise emanating from the galaxy was confirmed by Grote Reber, who measured radiations from the Sun and from other sources [155]; an account of these pioneering observations is found in the many references included in [156]. The radio intensities observed by Jansky and Reber were, however, about 10 million times higher than those expected if thermal emission is assumed. Obviously other emission mechanisms are at work.

These and later observations, taken with conventional apertures connected to sensitive radiometers, resulted in blurry maps of sources. Radiotelescopes having higher spatial resolutions were required to derive maps having resolutions comparable to those attained by optical telescopes. The moon's occultation with the Sun [157] was used to increase the resolution of conventional apertures. With the aid of this technique the correlation of intense solar burst activity with sunspots on the solar disk was established.

The radio analog of Lloyd's mirror [92] was used to form a two-element interferometer that increased the spatial resolution of conventional antennas. The interferometer was used to determine the accurate position of sunspots and of several strong radio sources, some extragalactic [158,159]. Optical identification of the newly discovered radio sources was made possible by the relatively high resolution afforded by the interferometric system, which can be considered as the precursor of the first phased array-based radiotelescopes [160–164]. With the passage of time, large array-based radiotelescopes achieved the spatial resolutions we have already considered, and radioastronomy became a respectable branch of science.

The Sun, the planets, our galaxy and other galaxies, and a variety of sources emanate radiations in the radiofrequency bands and the emission mechanisms at work in many sources are not due to black body radiation.

The following significant discoveries were made chronologically:

1932   Radio waves of extraterrestrial origin [154]
1940   First radio maps of radio sources [155]
1951   Detection of the 1420 MHz line radiation from the ground state of atomic hydrogen [164–166]
1960–3 Detection of quasi-stellar objects [167]
1964   Detection of the hydroxyl radical, OH [168] in many sources
1965   2.7 K cosmic background radiation [169]
1967   Discovery of pulsating stars, pulsars [170], detection of hundreds of pulsars
1968–   Detection of the first interstellar molecules: ammonia (NH_3) and water vapor (H_2O) [171,172]; detection of a variety of interstellar molecules [173]—see also Table 1.13
1971   Discovery of black holes [174]?
1974   Detection of the first binary pulsar PSR1913+16 [175] and an update on pulsar research [176]
1994   Discovery of ripples on the cosmic background radiation with the aid of the cosmic background explorer, COBE [177].

Table 1.13 A representative sample of the 90-plus interstellar molecules

Formula	Name	Derivatives		
Masers	OH	Hydroxyl radical		
	H_2O	Water vapor	Deuterated H_2O:	HDO
	SiO	Silicon monoxide	Silicon sulfide:	SiS
Simple molecules	CO	Carbon monoxide	CO isotopes:	^{13}CO, $C^{18}O$
	CS	Carbon monosulfide	Carbonyl sulfide:	OCS
	SO	Sulfur monoxide		
A hydrogenation series	CN	Cyanide radical		
	HCN	Hydrogen cyanide	Deuterated HCN:	DCN
	H_2CNH	Methanimine		
	H_3CNH_2	Methylamine		
Aldehydes	HCHO	Formaldehyde	Thioformaldehyde:	H_2CS
	CH_3CHO	Acetaldehyde	Formamide:	NH_2CHO
Alcohols	CH_3OH	Methyl alcohol	Methyl cyanide:	CH_3CN
	CH_3CH_2OH	Ethyl alcohol		
Acids	HCOOH	Formic acid	Isocyanic acid:	HNCO
	CH_3COOH	Methyl formate		
Various other molecules	NH_3	Ammonia		
	CH	Methylidene		
	HNCO	Isocyanic acid		
	CH_3CN	Methyl cyanide		
	HC_3N	Cyanoacetylene		
	$(CH_3)_2O$	Dimethyl ether		
	CH_2CHCN	Vinyl cyanide		
	CH_3C_2H	Methyl acetylene		
	HC_9N	Cyano-octatetrayne		
	$HC_{11}N$	Cyanotetraacetylene		

Spectral-line emissions from atoms, radicals, and molecules occur at unique frequencies, transition frequencies, defined by their molecular constants; the term 'line radiation' is often used as a contraction of the term 'spectral-line radiation'. In the interstellar medium, the bandwidth of line emission is usually determined by the thermal, turbulent, and galactic motions of the molecular clouds. Most of the transitions detected in the interstellar medium at microwave frequencies are rotational transitions associated with end-over-end rotation of the entire molecule.

Emanations from hydrogen, hydroxyl radical, and simple and complex molecules have been detected. Clouds of OH, H_2O, and SiO in the galaxy emanate strong radiations due to maser transitions. Table 1.13 gives a representative sample of the 118 interstellar molecules [173a&b] discovered so far. Even this short list illustrates the range of the molecules and to some extent the chemical evolution of the galaxy. In addition to masers, simple molecules, a hydrogenation series, the aldehydes, alcohols, acids, and a list of various other molecules, some of the derivative molecules from those listed are also shown. The alphabet of chemical evolution is being assembled with every new discovery of yet another interstellar molecule. Most of the interstellar molecules have been detected toward the center of the galaxy and/or the Orion nebula, but some molecules, such as carbon monoxide, are widely distributed.

Carbon monoxide is considered to be associated with sites where star formation is taking place. Low-resolution maps of molecular clouds define the extent of these clouds and high-resolution maps of carbon monoxide molecular clouds are required to detect the first condensations of material taking place.

As each molecular species requires a certain range of densities and temperatures before one of its transitions can be observed, different species of molecules act as density and temperature gauges of the regions where molecular clouds exist. High-resolution and high-sensitivity maps are therefore required for many clouds of interstellar species in several galactic sites, so that the physical conditions throughout the galaxy can be estimated with some certainty.

Quasars are the most luminous known objects in the cosmos, some of them having luminosities more that thousands of times greater than that of the galaxy. The rapid change of their intensity indicates that the main source of their energy is at most only a few light-months in size. By comparison, the galaxy extends over 10^5 light years in diameter. A phenomenal outpouring of energy has therefore been observed. Could they be the most distant galaxies known? Is there a black hole at the center of every quasar ? Are quasars nearby objects that 'appear' to be receding from us with high speeds? The jury is still out [167].

The discovery of the 2.7 K cosmic background radiation [169] was attributed to the remnant radiation of the Big Bang. This important observation placed cosmology on a firm scientific basis. The cosmic background experiment, COBE, revealed ripples in the cosmic background radiation [177] and this discovery will provide badly needed measurements for promising theories related to the origin of the background radiation.

The discoveries of several hundreds of pulsars provided observational evidence for neutron stars rapidly rotating to their eventual death. More importantly, however, the discovery of pulsars added a missing species in the stellar evolution chain from gas and dust to nebulae, protostars, main sequence stars, white/black dwarfs, rotating neutron stars, and eventually black holes. Theoretical work supporting the existence of black holes has been formulated over many years and observational evidence for their existence is mounting [174]. Black holes pose challenging and fundamental problems for modern physics.

The binary pulsar first detected in 1974 provided a laboratory in space for yet another verification of the theory of relativity [175]; an update on the ever-evolving pulsar research is outlined in reference [176].

Radioastronomical observations can be broadly divided into three categories:

1. High-spatial-resolution observations that establish the spectrum of sources also known as 'continuum observations'. Here the intensity of the source is measured at spot frequencies extending over two or more octaves. At each spot frequency, the instantaneous band of the radiometer has a fractional bandwidth of 10% .
2. High- and low-spatial-resolution spectral-line observations (or simply line observations); the line radiation emanates from clouds of atoms, radicals, or molecules located in the interstellar medium.
3. Variable continuum or line observations: observations taken at many frequencies simultaneously by several radiotelescopes located in different continents. Usually

these observations are complemented by observations taken at other wavelengths (e.g. optical and X-ray).

We have already noted that high-spatial-resolution observations are usually taken by compact phased arrays, VLBI, or OVLBI networks. Low-spatial-resolution observations on the other hand are taken by single aperture antennas. The discoveries of interstellar molecules, for instance, were made with the aid of single aperture antennas.

Let us derive some fundamental quantities often used in radioastronomy before we consider phased array-based radiotelescopes.

1.7.2. Measurements and Fundamental Quantities

In this section we shall consider the case where a typical radiotelescope is used to measure the fundamental quantities of a radio source. As is depicted in Figure 1.11, the radiotelescope consists of a total power radiometer that is connected either to the antenna or to matched resistors at physical temperatures T_1, T_2, and T_3. The antenna can be pointed to any source and is characterized by its beamwidth, Ω_A and effective area A_e, while the total power radiometer has a system noise temperature T_s, bandwidth B, and integration time τ.

The aim is to measure the fundamental quantities of a source, e.g. its flux density, S, and brightness, B_1, by measurements afforded by the radiotelescope. The flux density S of a source is defined as the power received by a radiotelescope per square meter of its collecting area per hertz. From a knowledge of the flux density of the source, as a function of frequency and the source's angular extent, one can usually deduce the physical processes at work near or at the source. Here we have assumed that the spectrum of various emission processes is known. The observed or apparent brightness of a source denoted by B_0 is defined as the received power per unit effective collecting area per unit solid angle per unit bandwidth.

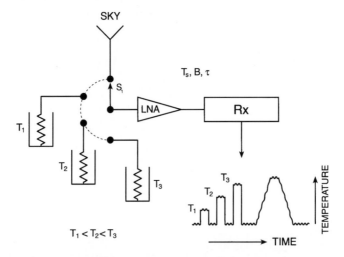

Figure 1.11 A total power radiometer used to measure the equivalent antenna temperature of a source. At the lower right of the figure, the record obtained is shown when the radiometer is switched consecutively to matched resistors at temperatures T_1, T_2, and T_3 before the celestial source is scanned.

1.7.2.1. Thermal Radiation

All bodies at a temperature above zero kelvin emanate thermal or black body radiation. We have already seen that the entire universe appears to be within a black body 'enclosure' where the temperature is about 2.7 K.

Some notable sources supporting black body radiation are the quiet Sun, the Moon, and Mars. Let us examine how the brightness of celestial bodies varies as a function of frequency for thermal sources. The brightness, B_1, of the radiation emanated by the black body at a temperature, T, is given by Planck's law,

$$B_1 = \frac{2h\nu^2}{c^2} \frac{1}{\exp(h\nu/kT) - 1} \tag{1.14}$$

where ν, h, and k are the frequency, Planck's constant (6.62×10^{-34} J s), and Boltzmann's constant (1.3804×10^{-23} J K^{-1}), respectively. If operation at radio wavelengths is considered, the photon energy $h\nu \ll kT$, hence the Rayleigh–Jeans approximation to Planck's law can be applied to transform equation (1.14) into

$$B_1 = \frac{2kT}{\lambda^2} \tag{1.15}$$

As can be seen, B_1 decreases as the wavelength of operation increases. Given that the wavelengths corresponding to the radio spectrum are much longer than the wavelengths corresponding to lightwaves, B_1 is expected to be negligibly low at radio frequencies for thermal sources other than the Sun and the planets.

If a radiotelescope that has an antenna pattern $P_0(\theta, \phi)$ is pointed toward a source of $B_1(\theta, \phi)$ brightness distribution, the received power P, per unit bandwidth from a solid angle Ω, of the sky is given by

$$P = \frac{A_e}{2} \iint_\Omega B_1(\theta,\phi)\, P_0(\theta,\phi)\, \mathrm{d}\Omega \tag{1.16}$$

The assumptions made here are that the radiation received is of an incoherent, unpolarized nature and that one polarization is received. If the antenna is placed inside a black body enclosure at a temperature T, and $B_1(\theta, \phi)$ is constant, uniform, and equal to

$$B_c = \frac{2kT}{\lambda^2}$$

equation (1.16) can be written as

$$P = \frac{kTA_e\Omega_A}{\lambda^2} \tag{1.17}$$

If we introduce the well-known relation $\Omega_A A_e = \lambda^2$ into equation (1.17), we can deduce that $P = kT$, which is exactly equal to the noise power per unit bandwidth available at the terminals of a resistor at temperature T.

As the power received by the antenna in the frequency interval dv, is equal to kT dv, T is the equivalent antenna noise temperature, T_A, attributed to the source and the power received by the antenna is equal to kT_Adv. A straightforward calibration procedure for T_A is therefore available, since there is a one-to-one correspondence between T_A and T. Furthermore, T_A can be produced by any emission mechanism at work in the direction of the celestial source under consideration. An antenna connected to a radiometer therefore measures the equivalent antenna temperatures, or simply the antenna temperature of radio sources that are within its field of view.

With reference to Figure 1.11, the radiometer is first connected to the matched resistors at temperatures T_1, T_2, and T_3 sequentially before it is connected to the antenna. The source is then scanned by the antenna and the output powers of the radiometer is measured on a chart recorder. As $T_1 < T_2 < T_3$, the output power of the radiometer measured by the deflections on the chart record are higher in the same proportions. The linearity of the radiometer is verified by comparing the recorded deflections on the chart with the known temperatures. The source's T_A is then measured by comparing the maximum deflection on the chart corresponding to the source with the temperature scale established by the deflections recorded when the radiometer is connected to resistors at T_1, T_2, and T_3. These measurements are now automated and the antenna temperatures of sources are measured with the aid of a calibrated cold-load and comparisons with the antenna temperatures of standard non-variable sources used as calibrators.

1.7.2.2. Flux Density and Antenna Temperature

When an isolated source is much smaller in extent than the beam area, the following relationships have been deduced [178]:

$$P = kT_A \, dv = \frac{1}{2} \, SA_e \, dv$$

or

$$S = 2k\left(\frac{T_A}{A_e}\right)$$

Given that T_A is measurable, S can be deduced; more importantly, for a given flux density corresponding to a source, T_A is proportional to A_e.

In section 1.5.3 we raised a question related to the fields of application of phased arrays and conventional apertures. While phased arrays can provide the spatial resolution required, large antennas provide affordable and significant effective areas. The ratio of the geometric area of the Arecibo antenna to that of the VLA, is for instance about 5.5. For point sources, therefore, it makes sense to use single antennas of very large diameters. The binary pulsar, to cite one observation only, was discovered with the aid of the Arecibo antenna.

If the size of the source is Ω_s, its flux density S can be defined as the integral of the radiation brightness, b_1, over the source, or

$$S = \int_{s} b_1 \, d\Omega = \frac{2k}{\eta_b \lambda^2} T_A \Omega_s \qquad (1.18)$$

where η_b is the main beam efficiency. In deriving equation (1.18) the assumption is made that T_A is uniform over the source. As T_A, η_b, and Ω_s are measurable, S can be deduced.

Table 1.14 lists the flux densities of some of the strongest sources at 1, 3, and 10 GHz in flux units (one flux unit, or fu, is equal to 10^{-26} W m^{-2} Hz^{-1}). By comparison, the flux densities of the quiet and active Sun at 3 GHz are ~10^{-20} and 8×10^{-18} W m^{-2} Hz^{-1}, respectively. As can be seen, the Sun is not only intense but also variable. Its emanations

Table 1.14 Flux density of some of the strongest celestial sources at 1, 3, and 10 GHz

Source	*Flux density (fu)*[a]		
	1 GHz	3 GHz	10 GHz
Cassiopeia A, CasA	3185	1340	520
Cygnus A, CygA	2270	690	162
Orion A	330	420	420
Virgo A, VirA	285	112	40
Hydra A	60	22	7
DR21	4	20	21

[a] 1 fu = 10^{-26} W m^{-2} Hz^{-1}

are several orders of magnitude more intense than those attributed to the strongest sources. Burst radiation usually associated with dark spots on the Sun is extremely intense and variable in timescales ranging from one second to hours.

In general, $S \propto \lambda^n$, where, n is an index corresponding to the emission mechanism related to the source. If the source is a thermal source (black body), $n = -2$ as per equation (1.15). For sources where the emission mechanism is predominantly synchrotron (one of the important non-thermal emission mechanisms), S is proportional to $\lambda^{0.7}$. The first sources from which synchrotron radiation was observed were the Crab Nebula and a large galaxy near the constellation of Virgo [179–181]. In reference [182] other emission mechanisms and their spectra are considered.

1.7.2.3. Radiometers

Radiometers are essentially low-noise, wideband receivers exhibiting gains that are stable over long periods of time. The LNA of the radiometer, shown in Figure 1.11, is usually cryogenically cooled to lower the total system equivalent temperature. The fractional bandwidth of radiometers working in conjunction with one antenna is usually at least 10%.

In communication theory terminology, the measured antenna temperature, T_A, when the antenna is pointed toward a source, can be considered as the 'signal'. The 'noise' of the total power radiometer determined by the rms of its minimum detectable signal, $\Delta t_{min}|_{TP}$, is given by

$$\Delta t_{min}\big|_{TP} = (T_A + T_s)\left[\frac{1}{B\tau} + \left(\frac{\Delta G}{G}\right)^{1/2}\right]^{1/2} \tag{1.19}$$

where the ratio $\Delta G/G$ is a measure of the gain variations of the radiometer and is typically equal to 0.01% for modern radiometers [183]. If the gain variations of a radiometer are substantial over the integration period, τ, a Dicke radiometer [184], minimizes the gain variations ($\Delta G/G \rightarrow 0$) by switching its input to a calibrator periodically and using synchronous detection. The $\Delta t_{min}\big|_{Dicke}$, however, is equal to $2\Delta t_{min}\big|_{TP}$. The conditions for using either system are [183]

$$\text{Total power system} \Leftarrow \frac{3}{B(\Delta G/G)^2} > \tau > \frac{3}{B(\Delta G/G)^2} \Rightarrow \text{Dicke switched system}$$

Typical values for bandwidth of radiometers connected to a conventional radiotelescope range from 1 to 2 GHz.

1.7.3. The Complementarity of Radiotelescopes

Let us consider the complementarity of the different radiotelescopes so that phased array-based systems are seen in the appropriate perspective.

- Single antennas having very large diameters are ideally suited to observing point sources; the Arecibo hole-in-the-ground radiotelescope is an archetype.
- Other fully steerable single antenna radiotelescopes are used:
  - (i)   in conjunction with other radiotelescopes to form VLBI or OVLBI systems for the purposes of deriving high resolution maps of celestial sources; or
  - (ii) to search for new interstellar molecules or continuum sources; or
  - (iii) to derive low-resolution maps of molecular clouds or continuum sources.
- The VLBA is dedicated to the derivation of high-resolution maps of continuum sources or molecular clouds of interest.
- Future dedicated OVLBI systems will be assigned to the derivation of continuum/molecular maps having the highest spatial resolution available.

Basically it is easy and inexpensive to explore the spatial and frequency domains with low-resolution systems and to augment the acquired knowledge with the aid of high-resolution systems. In essence it is less expensive to have one large fully steerable radiotelescope equipped with a suite of radiometers covering many octaves in frequency than to equip the many antenna elements of a phased array with several sets of radiometers to cover a comparable frequency range.

The fundamental relations we have derived here support the approaches outlined.

1.7.4. Image Formation Theory

We have assumed all along that a notional filled aperture has the same spatial resolution as a thinned array having the same overall dimensions and in this section we shall

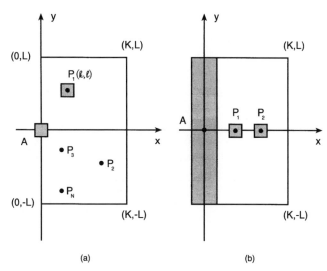

Figure 1.12 The principle of the aperture synthesis technique used to measure the brightness distribution $B(\theta,\phi)$ of celestial sources. (a) When two similar antennas are used. Antenna A is stationary while antenna P is moved to occupy the positions P_1, P_2, ... P_N sequentially. (b) When two dissimilar antennas are used. Antenna A is again stationary.

explore how an image of a celestial source is formed by a thinned array. So far we have considered thinned arrays having many antenna elements, but in this section we shall consider the simplest phased array having only two antennas.

With reference to Figure 1.12a one antenna, A, is stationary and the other, P, occupies the positions P_1, P_2, ... P_N, defined by the x–y coordinates successively. At all times the two antennas are pointed toward the source of interest and at each successive position the resulting amplitude and phase are recorded. The movable antenna can occupy the positions that the different elements of a multi-antenna element array occupy. The sum of all the resulting measurements is further processed to yield the required map. The assumption is made here that the statistics of the recorded emanations are not time-variable.

The waveform of a two-element interferometer, at a given spacing, is a sinusoid and its phase is measured by recording its sine, S, and cosine, C, terms. Let us assume that both antennas are similar, and that the stationary antenna is conveniently placed at the origin of the coordinate system and the other at positions P_2, P_3 ... P_N or more generally at position P_1 (k, ℓ) as shown in Figure 1.12a. The output of the receivers may be written in a general way as

$$C|_{k,\ell} + jS|_{k,\ell} \propto \iint D(\theta, \phi)B(\theta, \phi) \exp\left[-2\pi j(k\theta + \ell\phi)\right]\mathrm{d}\theta\, \mathrm{d}\phi \qquad (1.20)$$

where $C|_{k\ell}$ and $S|_{k\ell}$ are the sine and cosine terms of the waveform, $D(\theta, \phi)$ is the radiation pattern of the antennas, $B(\theta, \phi)$ is the brightness of the source, and θ and ϕ are the conventional spherical angles. Inspection of equation (1.20) indicates that it is possible to determine the smoothed Fourier transform of $B(\theta, \phi)$ by moving one antenna from one location to the other, record the resulting outputs, and hence by Fourier inversion derive a map of $B(\theta, \phi)$. When this operation is repeated over many spacings, the brightness distribution of the sky can be obtained from [185–186].

$$B(\theta, \phi) \propto \frac{1}{D(\theta, \phi)} \sum_{-K}^{K} \sum_{-L}^{L} (C|_{k,\ell} + jS|_{k,\ell}) \exp[2\pi j(k\theta + \ell\phi)] \qquad (1.21)$$

Given that $D(\theta,\phi)$, is known, the brightness distribution, $B(\theta,\phi)$ can be derived. While the above method of deriving $B(\theta,\phi)$ serves the purpose of illustrating the principle of image formation, it is seldom used because it is time-consuming. Antenna A can take the form illustrated in Figure 1.12b and antenna P can take the successive positions shown in the figure. Usually $B(\theta,\phi)$ is derived either by recording the outputs of several pairs of interferometers simultaneously or by recording the outputs of several interferometers when the interferometers occupy two or more sets of positions.

Given that the observations can take several days to complete, the Earth's rotation is used to increase the Fourier components that an array yields and the system integration time. Put differently, the spacings between antenna elements change as the celestial source travels its course during the observations.

This method of attaining the brightness distribution of a source by recording as many Fourier components as possible is referred to as aperture synthesis and it is used routinely to map celestial sources. References [187,188] are two excellent review papers related to image formation and self-calibration in the radioastronomy context—see section 1.7.7.

As we have already seen, burst radiation from the Sun is extremely intense and variable on timescales ranging from one second to hours. The mapping of the Sun therefore calls for real-time imaging. Similarly, the mapping of the Earth's radiations at microwave wavelengths from an airborne or spaceborne platform calls for real-time image formation as the platform moves with respect to the Earth. This being the case, another method of mapping more familiar to radar researchers is used.

The method calls for the formation of many independent and simultaneous antenna beams separated by an angular distance that ensures no loss of information is incurred. For radioastronomical observations, the half-Rayleigh limit of resolution ($\lambda/2D$, where D is the aperture's diameter) is used. For convenience we shall use the term staring antenna beams as a shorthand for independent and simultaneous beams. The output of each beam yields the $B(\theta,\phi)$ of a pixel of the scene or source and this method of image formation on a picture-point by picture-point basis is referred to as image synthesis. Although the hardware employed by the two methods is different, both methods share the same theory.

There is one more issue of practical importance left to consider. While the resulting beam from a densely populated radar array can have low sidelobes or grating lobes, the resulting beam from a sparsely populated radioastronomy array is almost useless because it has high sidelobes or grating lobes. The sidelobes of the same beam, however, can be corrected and used to form a low-sidelobe image of a scene—see section 1.7.5.1. In section 2.4.2 we shall consider a highly thinned circular array used to image the active Sun and a method of correcting the resulting beam in real time. The array used for the Culgoora Radioheliograph derived two images of the active Sun every second and has been used for studies related to the burst radiation of the active Sun. Similarly, the early radiotelescopes that consisted of two crossed arrays operated in the image synthesis mode and the derived beams were used for mapping after correction—see next section. The same techniques of beam correction (or variants thereof) can be applied to non-radiometric, sparsely populated arrays.

From the foregoing considerations it is clear that there is a continuum between radar arrays having short integration times and radiometric or radioastronomy arrays having short and long integration times; short integration times are used by arrays operating in an image synthesis mode and long integration times are used by arrays operating in an aperture synthesis mode.

Given that the collecting area of phased arrays is thinned, the minimum detectable signal, $\Delta t|\,array$, has to be redefined. If $T_{sys} = T_s + T_A$, $\Delta t|_{array}$ for phased arrays is given by [52]

$$\Delta t\Big|_{array} = \frac{T_{sys}}{\sqrt{Bt}} \; \frac{A_{sys}}{nA_e} \tag{1.22}$$

where A_{sys} is the equivalent area of the synthesized antenna, and A_e the effective area of an actual antenna used for the measurement. Under a reasonable set of assumptions, n is equal to the number of antennas used in the measurement (or the total number of independent antenna positions in a system where the antennas might be moved to obtain the desired baselines). When n is equal to the number of antennas used in the array, nA_e is equal to the actual collecting area employed in the measurement. In the limit when the array is completely filled, $nA_e = A_{syn}$. Under these conditions the minimum detectable signal of a phased array is equal to that corresponding to a total power radiometer, when $\Delta G/G \rightarrow 0$.

1.7.5. Image Synthesis Applications

While the aperture synthesis technique is now routinely used by modern phased array-based radiotelescopes, the image synthesis technique was used by some of the early phased arrays and is used in a variety of applied science applications. In this section we shall outline radioastronomical and applied science applications where several approaches for performing image synthesis are used.

1.7.5.1. Radioastronomy applications

One of the early planar radiotelescopes was the Mills cross [162], which consisted of two perpendicular linear arrays, and many other crossed radiotelescopes have been built. Let us consider how one or several pencil beams were formed in the early realizations of the Mills cross [162].

The East–West and the North–South arms of the cross yield fan beams that extend from the North to South and East to West directions, respectively. If the two fan beams are multiplied, a pencil beam results, occupying the common area of the two beams. The method is known as the Mills periodic 0°–180° switching technique, a name that reflects the early realizations of the method used. It can be shown that the spatial resolution of the pencil beam resulting from a cross radiotelescope is the same as that resulting from a cross telescope when one of its four arms is removed, to form a T. The sensitivity of the former array, however, is higher. Multiple staring, pencil beams are formed when the unit collecting areas of the array are appropriately phased with the aid of resistive networks—see Section 5.2.1. The sidelobes of the resulting beams were lowered by the application of an amplitude taper that increased as the distance from the array center increased.

The Culgoora Radioheliograph [189], completed in 1967, was a radiotelescope that yielded two images per second of the two opposite circular polarizations of the Sun at 80 MHz. It consisted of 96 steerable paraboloids, 13 m in diameter, equally distributed along the perimeter of a circle 3 km in diameter. The two images were formed when 48 staring beams generated along a North–South line [190] were swept along an East–West direction by changing the phases of the local oscillators of the system to form a 48×64 pixel image. The staring beams were formed with the aid of a multitude of accurately calibrated low-loss transmission lines—see section 5.2.2. The maximum resolution attainable was 3.5 arc minutes and the observations taken contributed significantly to the understanding of short- and long-duration solar burst radiations. This unique radio-heliograph ceased operations in the early 1980s.

1.7.5.2. Applied Science Applications

We have already noted that passive microwave remote sensing of the Earth's surface has progressed considerably in recent years. With reference to Figure 1.13a, airborne scanning radiometers were used to generate the ground image as the airborne platform moved. Airborne/spaceborne phased arrays capable of generating several staring beams along a line perpendicular to the flight direction can meet the high sensitivity and spatial resolution requirement. We shall cite several realizations and proposals addressing these requirements.

In the next section, we shall consider an airborne system that generates a fan beam along a line perpendicular to the direction of flight. As the aircraft moves, the fan beam forms an image of the Earth's surface in a push-broom mode, as illustrated in figure 1.13b. The radiometric sensing of the sea surface temperature at 5 GHz with the aid of a satellite-borne cross-beam interferometer radiometer (CBIR) has been proposed [191]. According to the proposal the CBIR, in the form of a T, is on board a satellite in an 833 km orbit and the resulting ten colinear quasi-pencil beams measure the surface

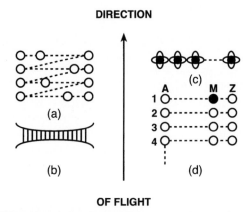

Figure 1.13 The footprints of several airborne radiometric imagers: (a) the early scanning radiometers; (b) the electronically scanned thinned array radiometer, ESTAR; (c) the cross-beam interferometer radiometer, CBIR; (d) the high-resolution and high-sensitivity radiometer, operating at millimeter wavelengths. The imager generates several rows of pixels to increase the dwell time on each pixel.

temperature to a sensitivity of 0.5 K as the satellite moves in a push-broom mode. The staring beams in the two perpendicular dimensions are formed with the Butler matrix and the Mills periodic 0°–180° phase-switching technique is used to form the instantaneous pencil beams. The footprints of the derived beams are shown in Figure 1.13c, where the pencil beams are shaded. The information obtained is of interest to the meteorological and oceanographic communities and a CBIR operating at 1.43 GHz may be used for sea surface salinity and moisture sensing.

In another proposal the concept has been extended to a crossed phased array radiometer operating at millimeter wavelengths to attain the maximum spatial resolution for a given airborne platform. The sensitivity of the system is increased by the generation of *N* rows of staring beams along directions perpendicular to the direction of flight, and Figure 1.13d illustrates the footprints corresponding to the proposed system. The proposed array is designated for high-resolution surveillance and reconnaissance applications [192].

1.7.5.3. Minimum-Redundancy Arrays and Applications

Minimum-redundancy arrays (MRAs) were first proposed and used by radioastronomers in the 1950s [193], see section 2.2.2.2. A MRA is capable of generating the maximum number of non-redundant spacings using the minimum number of antenna elements and yields the highest spatial resolution.

In reference [194] a theoretical framework is developed to support radiometric observations taken from airborne sensors pointed toward the earth. The measurement of soil moisture is optimally undertaken at 21 centimeters and requires spatial resolution of about 10 km to achieve meaningful understanding of the global hydrological cycle and its coupling to the atmosphere and energy cycle [52,53, and references therein]. A MRA has been used on board an aircraft to monitor the surface soil moisture [52,53,194–196]. The FOV of the electronically scanned thinned array radiometer, ESTAR, is a fan beam perpendicular to the direction of flight and a number of staring beams are generated within the array FOV as illustrated in Figure 1.13b. The beams generated by cross-correlating the outputs of the array elements scan the scene in a push-broom mode.

Thus the technique of image synthesis has been used to image galactic sources, the active Sun, and the Earth's surface in applied science applications. In section 2.2.2.2 we shall consider the underlying theory related to MRAs and the radar-related applications of the same arrays. *Inter alia* the applications include ground/sea sensing, surveillance/reconnaissance, direction finding, direction of arrival, interference cancellations, and adaptive beamforming.

1.7.6. Radioastronomy Polarimetric Systems

The requirement to derive maximum information from observations has led to systems that measure the polarization of EM waves emanated by celestial sources that are assumed to be partly polarized noise signals that can be uniquely resolved into polarized and unpolarized components. Here we shall assume that the two opposite circular polarizations are used; parallel formulations can be derived if two orthogonal linear polarizations are used.

Chandrasekhar [197] set out the theoretical framework for polarimetric measurements in 1955, and Cohen [198] related the four Stokes parameters I, Q, U, and V to parameters that can be measured by a polarimeter in a radioastronomy context. Suppose I_R and I_L are the intensities of the left- and right-handed circular polarization, respectively. If RL is the product of the polarized components and γ is the phase difference between them, the following four Stokes parameters have been derived [198]:

$$I = I_R + I_L$$
$$Q = 2RL \cos \gamma$$
$$U = 2RL \sin \gamma$$
$$V = I_L + I_R$$

The intensity, orientation, and axial ratio of the polarized and unpolarized components of the received radiation can in turn be derived from the four Stokes parameters. The unique definition of the incoming radiation from celestial sources can only be attained when both polarizations (either two orthogonal linear or two opposite circular polarizations) are measured simultaneously. Most radiotelescopes utilize dual-polarization radiometers and a typical polarizer operating at 18 centimeters intended for radio-astronomical observations is considered in detail in section 3.3.6.1.

1.7.7. Image Formation by Self-calibration

Data taken by an array are often corrupted by phase and amplitude errors due to instrumental and propagation effects. As the number of antenna elements in an array and/or the extent of the array in wavelengths increase, a self-calibrating scheme becomes mandatory. Put differently, high-resolution observations are useless if they are severely corrupted.

Mathematically, we are required to form an image from an incomplete knowledge of its Fourier coefficients and the main source of errors encountered in arrays is due to errors associated with individual antennas. These errors can be minimized or eliminated if data taken from three or more antennas forming a triangle or a polygon are suitably processed and the constraint that the measured brightness is positive is accepted.

References [187,188] are excellent review papers that outline the self-calibration schemes used in radioastronomy to overcome these real-world limitations associated with phased arrays.

1.7.8. Radioastronomy Arrays: A Record of Progress

In this section we shall outline the progress made in radioastronomy phased arrays in the same manner in which we outlined the progress made in radar arrays in section 1.6.15. With reference to Table 1.15 we can discern the following themes.

Theme 1 Radioastronomy is a new science that raises some curiosity. Phased arrays establish radioastronomy as a new science that is as respectable as optical astronomy. The radio universe is as interesting as the universe explored by optical astronomers

Table 1.15 Radioastronomy phased arrays: A record of progress

---

1

Radioastronomy is a new science that raises some curiosity.

⇓

Phased arrays establish radioastronomy as a respectable science.

⇓

The radio universe is as interesting as the universe explored
by optical astronomers since antiquity.

⇓

Observations taken at different wavelengths help us assemble
the wondrous mosaic of the cosmos we live in.

⇓

The techniques developed by radioastronomers are used
in several applied science applications.

2

Nexus between the cost of a conventional aperture and its spatial resolution.

⇓

The complete breaking of the nexus between the array cost and spatial resolution.

3

Arrays have the highest spatial resolution and adequate collecting area
OR
Arrays have adequate spatial resolution and maximum collecting area (sensitivity).
Trade-offs between these two extremes are possible.

4

Array spatial resolution of compact arrays: a few arc minutes ⇒ one arc second or less.

⇓

Array spatial resolution of VLBIs: milli-arc second.

⇓

Array spatial resolution of OVLBIs: sub-milli-arc seconds.

5

The antennas of a compact array are interconnected with
low-loss cables.

⇓

The antennas of a compact array are interconnected with
low-loss waveguides or optical fibers.

⇓

VLBI and OVLBI systems are synchronized with the
aid of highly stable reference clocks.

Table 1.15 cont.

6

Array antenna elements are fixed on the ground.

⇓

Array antenna elements can occupy different positions within a prescribed real estate ⇒ variable resolution and the maximization of interferometric spacings.

⇓

The Earth's rotation is used to further increase the interferometric spacings.

7

The spacings between array antennas have high redundancy.

⇓

The spacings between array antennas have minimum redundancy.

⇓

MRAs are eminently suitable for many applied science, radar, and radar-related applications

⇓

Arrays having a plethora of short spacings between array antennas have been proposed.

8

Array has one staring beam.

⇓

Array has many staring beams ⇒ image synthesis.

⇓

The sidelobe level of the resulting beams has to be lowered before the beams can be used.

⇓

Aperture synthesis and the Earth's rotation are used to obtain the image of a celestial source.

9

Array has one field of view.

⇓

Arrays can have several fields of view—Proposals. These arrays are expected to perform radioastronomy observations and SETI.

10

Array operates at one frequency band.

Table 1.15 cont.

Array operates at many bands but the change of frequency bands is not instantaneous.

⇓

Arrays that operate at many frequency bands simultaneously—Proposal.
These arrays are expected to perform observations of variable/invariable sources and SETI.

11

Front-ends of the receiving systems operate at room temperature.

⇓

Cryogenically cooled parametric amplifiers used as LNAs.

⇓

Cryogenically cooled solid-state LNAs, at cm $\Rightarrow$ mm wavelengths.

⇓

The use of cryogenically cooled HTS-based subsystems, e.g. filters, low-noise oscillators and
multiplexers in receiving subsystems of the second-generation radiotelescopes—Proposal.

12

Infrequent array calibrations.

⇓

Continuous array self-calibration.

13

Arrays operating at decimeter $\Rightarrow$ meter wavelengths.

⇓

Arrays operating at centimeter $\Rightarrow$ millimeter/submillimeter wavelengths.

⇓

Arrays operating at optical wavelengths.

14

Analog signal processing.

⇓

Two-bit quantization and digital signal processing.

15

Single-polarization arrays.

⇓

Dual-polarization arrays.

16

Linear arrays.

⇓

Table 1.15 cont.

Planar arrays.

⇓

Multiplicative arrays.

17

Most radiotelescopes are located in low industrial noise sites.

⇓

RFI due to the transmissions attributed to constellations of LEO/MEO satellites dedicated to navigation/communications affect all radiotelescopes.

⇓

Mitigation of the effects of RFI on astronomical observations is a major research field.

---

since antiquity and observations taken at different wavelengths (radio, optical, UV and X-ray) help us assemble the wondrous mosaic of the cosmos we live in.

The techniques developed by radioastronomers are used in several applied science applications.

Theme 2 Phased arrays break the nexus that exists between the cost of a conventional aperture, usually defined by the real estate it occupies (and the mechanism required to steer it), and its spatial resolution. The spatial resolution required defines the dimensions of the array and the funds available determine the extent of array thinning. The assumption is made here that the requirement for high spatial resolution is paramount.

Theme 3 If the array spatial resolution is not of the utmost importance and array sensitivity—defined by the total number of antennas used—is important, another array realization is possible. The number of antennas/radiometers is chosen to meet the requirement and the real-estate used by the array is defined by how closely the antennas can be placed.

These two requirements (Themes 2 and 3) illustrate the range of the trade-offs between spatial resolution and sensitivity that a designer usually considers.

Theme 4 The quest for arrays with ever-increasing spatial resolution defined progress in radioastronomy: the evolution of compact phased arrays that yield spatial resolutions ranging from a few arc minutes to one arc second. The VLBI and OVLBI networks evolved, yielding maximum spatial resolutions of the order of one milli-arc second and sub-milli-arc second, respectively.

Theme 5 When the dimensions of the compact arrays are a few kilometers, the antenna elements are interconnected via low-loss cables; as the array dimensions increase, the antenna elements are interconnected via low-loss waveguides or optical fibers.

Data received by all elements of a VLBI network are synchronized in phase and frequency. A highly stable hydrogen maser clock located at each receiving station forms the core of the synchronization subsystem. The synchronization scheme for OVLBI networks is similar to that used in the VLBI networks except that the synchronization of data derived by the orbiting antennas is slaved to that of the Earth-bound antennas.

Theme 6   The antennas of the early arrays were fixed on the ground, but in later realizations the antennas are transportable to different positions within a given real estate. This innovation enabled the arrays to have variable spatial resolution and allowed maximization of the available interferometric spacings of an array having a given number of antenna elements. The Earth's rotation is used to further maximize the available interferometric spacings.

Theme 7   The spacings between antenna elements of the early arrays had high redundancy. Arrays having minimum redundancy between antenna spacings evolved subsequently. The resulting arrays are used to make radioastronomical observations or to yield radiometric images in applied science applications. Additionally, MRAs are eminently suitable for a variety of radar and radar-related applications, see section 2.2.2.2.

Short spacings between antennas are hard to obtain owing to shadowing between adjacent antennas; arrays where a plethora of short spacings are available have been proposed—see section 3.5.

Theme 8   Image formation results from either image or aperture synthesis. Although the hardware corresponding to the two approaches differs, both approaches share the same theory. Image synthesis is implemented by generating one or more staring beams corresponding to one or more pixels of the image and by recording the received powers. The sidelobe level of the resulting beams has to be lowered before the beams can be used.

Using the aperture synthesis mode, an image is derived by recording the amplitude and phase corresponding to many interferometric spacings and suitably processing them. The Earth's rotation is used to increase the spacings available from a given number of antennas. The assumptions made here are that: (i) each observation takes several days and (ii) the array antennas occupy different positions during the observations.

Theme 9   Radioastronomy arrays yield high-spatial-resolution maps of celestial sources and the array FOV is defined by the half-power beamwidth of the array antenna elements. For extended sources many FOVs have to be taken, a time-consuming task. Arrays having many FOVs are required to accelerate the mapping of large regions of the cosmos; the same arrays can be used for searches for extraterrestrial intelligence (SETI)—see proposals for second-generation radiotelescopes, section 2.7.3.4.

Theme 10   Radioastronomy arrays yield high-spatial-resolution maps of celestial sources at one frequency band. The frequency band can be changed, but the change is not instantaneous. Arrays capable of operating at several frequencies simultaneously are required to (i) accelerate the conventional radioastronomy observations, (ii) observe variable sources, and (iii) search for extraterrestrial intelligence. See section 2.7.3.4.

Theme 11   The early radiotelescopes had front-ends that operated at room temperature; in an effort to decrease the system noise temperature, cryogenically cooled parametric amplifiers replaced the room-temperature front-ends. Solid-state low-noise amplifiers (LNAs) eventually replaced the parametric amplifiers in systems operating at centimeter wavelengths. For systems operating at millimeter wavelengths, the front-end consisted of a down-converter followed by a solid-state low-noise intermediate frequency (IF) amplifier. These front-ends are substituted by solid-state LNAs operating at millimeter wavelengths. As the cost of cryodynes decreases and subsystems utilizing high temperature superconductors (HTS) become more available, the trend toward HTS-based filters, low-noise oscillators and multiplexers will be hard to resist—see sections 1.11.3 and 2.7.3.3. These receiving subsystems can be elements of the second-generation radiotelescopes—see section 2.7.3.4.

Theme 12 Calibrations are necessary to ensure the quality of the observations taken. Initially only infrequent array calibrations were possible; with the passage of time, self-calibration techniques that ensure the quality of observations evolved—see section 1.7.7.

Theme 13 The frequency of operation of phased arrays increased with the passage of time. The discoveries of interstellar molecules provided the impetus for the exploration of the cosmos at centimeter, millimeter, and submillimeter wavelengths. The techniques developed for radioastronomical arrays have been used for the operation of phased arrays operating at optical wavelengths—see section 3.3.3.

Theme 14 Initially analog signal processing for the received signals was the norm. The received signals after low-noise amplification are now often digitized before they are processed further. A two-bit digitization is often used and some loss of information is tolerated to facilitate the digital signal processing [37].

Theme 15 While the early arrays received only one polarization, the reception of two orthogonal linear polarizations or opposite circular polarizations is now the norm.

Theme 16 The progression from linear arrays to planar arrays has been witnessed. Multiplicative arrays have some uses in compound interferometers.

Theme 17 Most existing radiotelescopes are located in low industrial noise sites. Radiofrequency interference (RFI) attributed to the transmissions from constellations of LEO/MEO satellites dedicated to navigation/communications affect all radiotelescopes. The mitigation of the effects of RFI on astronomical observations is a major research field—see section 3.10.

1.8. VARIANTS OF PHASED ARRAYS

So far we have explored passive and active phased arrays that consist of several antenna elements followed by T/R modules. The power generated by each module of an active array is minuscule and yet the power density on target is substantial and meets requirements. Here we shall explore important variants of phased arrays.

1.8.1. Many T/R Modules and One Aperture

Let us revisit the radar system that consists of a high-power tube transmitter, a conventional receiver, a parabolic reflector, and auxiliary subsystems. If the tube transmitter is substituted by an ensemble of MMIC-based transmitters all appropriately phased, a new species of phased array emerges. In the resulting system, power addition occurs at the aperture and the system exhibits graceful degradation of performance. Additional benefits attributed to the derived system include the following:

- Long life and transmitter purity are attributed to the use of solid-state transmitters.
- Faulty modules are substituted while the system is running: a hot maintenance concept that allows the system to operate continuously on a 24-hour, 365-day basis has evolved.

This phased array variant is a cost-effective step toward phased arrays and a representative system is described in reference [199].

The next variation on the theme is to have phased arrays where some power addition is implemented at the array face and some on target [200]. According to this proposal a power combiner that combines the power of a number of MMIC-based FPAs feeds each array antenna. This phased array variant possesses all the characteristics of phased arrays and one can easily implement an amplitude taper of the transmitted power across the array, so that decreased sidelobes are attained [200]—see section 2.7.4.7.

1.8.2. Some Wireless Power Transmission Applications

The early history of wireless power transmission (WPT) has been delineated in several publications authored by W. C. Brown included in reference [201]. The topic has gained importance during the last two decades as the pivotal subsystems of WPT are reaching ever-improving efficiency. In this section we shall consider a couple of applications of WPT and the realization aspects of these applications are outlined in section 2.5.7.

Conventional radar systems have one common attribute: a target is illuminated by one transmitter only. From the early days of radar the hope surfaced that a high-power beam directed toward an aircraft might not only illuminate the target but also destroy it. As a conventional radar beam derived from a filled aperture diverges too rapidly when the radar range increases, it has been proposed to illuminate a given target by several widely dispersed transmitters [202]. While the receiving systems of widely dispersed radioastronomy arrays are phase coherent, this proposal calls for phased coherent transmitters that are widely dispersed. According to the proposal, an array resolving 0.1 arc second could concentrate its power in a circle 10 centimeters across at a distance of 100 km. Focused to this intensity, 100 kW of microwave power would melt a hole in any aircraft or missile and wreck its electronics.

While reference [202] serves the purpose of flagging the idea of focusing microwave power onto airborne targets, further research on this topic is considered in section 2.5.7. The threat to electronic equipment onboard airplanes is confined to densely populated phased arrays transmitting substantial powers at millimeter wavelengths; this threat is far more serious than that posed by the diverging beams of conventional EW systems. In the same section, WPT is seen as having a variety of applications, but the challenges of realizing these systems are significant. The major challenges are to achieve phase coherence of the powers of all transmitters onto a moving target in real time and the generation of sufficient powers at millimeter wavelengths.

The same system, a densely populated phased array of transmitters, can be used for peaceful purposes. An aircraft utilizing electric motors can derive its power from such an array of transmitters and the derived energy would power the aircraft for a long time. The feasibility of the basic concept of powering an aircraft by microwave energy transmitted from the ground has already been demonstrated (e.g. [99]). What follows is a short account of the reported progress.

The proposals center on an airship at an altitude of about 20 km powered by microwave power beamed to the airship by a densely populated phased array [99,100,203–205]. To increase the transmitted power density, two polarizations are used [203,205]. From this height, the airship or 'pseudo-satellite' can perform the following tasks over an area 1000 km in diameter:

- Communications, remote sensing and large area surveillance.
- Reconnaissance, border patrol, and disaster management during forest fires or earthquakes.
- Severe storm tracking.

A self-steering phased array to power the stationary high-altitude relay platform (SHARP) was proposed in reference [206]. The proposal is based on the 'beam tagging' technique that focuses the microwave power of several ground-based antennas onto the SHARP's rectenna without knowledge of the accurate position of the SHARP—see section 5.7.4. Pseudo-satellites can be more affordable than systems performing similar tasks onboard constellations of LEO/MEO satellites. More importantly, the aircraft can be put into operation over an area selected on an *ad hoc* basis to meet unforeseen requirements.

In the long term this application of phased arrays could easily qualify as one of the significant applications of phased arrays.

1.9. COMMUNICATION SYSTEMS

There are many ways of looking at communication systems. One can, for instance, map the prodigious increase in transmissions as a function of time, data rate, and number of users. At one end of the scale we have low-data-rate communications between small groups of users through the use of semaphore-based systems; at the other end we have high-data-rate multimedia services, delivered to millions of users via optical fibers.

We have come a long way since the launch of Echo 1 in 1960, which marked the beginning of the satellite communications era. Communications via a LEO active satellite, the Telstar 1, followed before the GEO satellites ushered in communication systems capable of handling ever-increasing traffic. Despite their many differences, however, all systems share a linear architecture. Small traffic streams from local communities are routed into groups, supergroups, and master groups before being assigned to a trunk transmission route.

We are now poised to exploit many evolutionary architectures of non-geostationary earth orbit systems, based on constellations of LEO or MEO satellites.

1.9.1. Geostationary Earth Orbit (GEO) Satellite Systems

GEO satellites offer communications between widely separated sites on a 24-hour basis through the lifetime of a satellite, that is in excess of 5 years. Let us consider the transmission of signals between a satellite and an Earth station through free space when the gains of the transmit and receive antennas are G_t and G_r, respectively. The resulting carrier-to-noise ratio at the receiver is given by

$$\left. C/N \right|_{GEO} = P_t G_t \frac{1}{B} \left(\frac{G_r}{T_s} \right) \left[\left(\frac{\lambda}{4\pi R} \right)^2 \frac{1}{k} \right]$$

(1.23)

where
P_t = is the transmitted power;
R = 35 786 km;
λ = the wavelength of operation;
T_s = the equivalent system noise temperature;
B = the system's bandwidth;
k = Boltzmann's constant.

The term in parentheses, the G_r over T_s ratio, is a significant parameter that characterizes the receiving Earth station system. The terms in the square bracket are constants for a GEO satellite operating at a wavelength defined by spectrum availability and overall costs. Two orthogonal or opposite circular polarizations are used to double the system's capacity.

In the early days of GEO satellite communications, the diameter of the receiving antenna was 30 m and the cost of the Earth station exceeded one million dollars. With the passage of time, however, we have witnessed the availability of more directive transmit beams, an increase in P_t, and a decrease of T_s. These evolutionary advances were instrumental in the decrease of the size of the receive antenna diameter to about a meter, with an attendant decrease of costs. Now that space-qualified, affordable T/R modules are readily available, phased arrays are destined to substitute the offset antennas used by GEO satellites to enhance their operational versatility. In the longer term, communications systems using GEO satellites will continue to complement ground-based fiber systems.

The drivers for novel satellite communication systems having evolutionary achitectures are outlined in reference [207]. The realization of the global positioning system completed in 1993 [208] and LEO satellite systems used for weather monitoring, resource mapping and Earth sensing provided an invaluable knowledge base for the many future non-GEO systems.

1.9.2. LEO/MEO Satellite Systems

As LEO/MEO satellites orbit the Earth at heights ranging from 500 to 2000 km and at 1500 to 15 000 km respectively, their footprints on Earth are not as large as those corresponding to GEO satellites. Constellations of satellites with intersatellite links (ISLs) and on board processing (OBP) facilities are therefore required to attain global coverage. Table 1.16 lists the pertinent characteristics of some LEO/MEO constellations of satellites drawn from reference [209a]; descriptions of LEO satellite-based systems are given in reference [209b].

As can be seen in Table 1.16, the systems listed will usher in the communications revolution by providing substantial capacity on a worldwide basis. Large LEO satellite systems are designated to carry voice telephone and low-speed data. The Iridium

Table 1. 16 Existing and future navigation and communication systems

System	Up-link frequency (GHz)	Down-link frequency (GHz)	Cross-link frequency (GHz)	Capacity
GPS		1.237±0.020, 1.575 ±0.020		Navigation system
Odyssey	1.6, 29	2.5, 19		3000–9500 voice channels
ICO	2.2, 5	2.0, 7		4500 voice channels
Globalstar	1.6, 5	2.5, 7		2000–3000 voice channels
Iridium	1.6, 28	1.6, 19	23	1100 voice channels
Teledesic	K_a^a	K_a^{1a}	60	100 000 voice channels

[a] 27–40 GHz

constellation of 66 satellites (each weighing 689 kg) at a height of 780 km is representative of large LEO satellites.

The Teledesic system is representative of broadband LEO systems and it is estimated to be operational before the year 2002; as can be seen, its capacity is substantial. The complexity of these systems is considerable, but their attractive characteristics stem from the following considerations:

- Lower transmitted powers are required; the comparison is made here with respect to GEO-based systems.
- The time delay of 0.5 s corresponding to GEO-based systems is significantly reduced.

In the next five years the GEO and non-GEO systems and fiber-based systems will offer complementary services of unheard of capacity that matches the seemingly unsatiable demand for more multimedia services. The Iridium constellation of satellites utilizes multiple-beam phased arrays [210] and the phased arrays on board GEO and non-GEO satellites are destined to play a crucial role in ushering in the wireless revolution. While systems based on constellations of LEO/MEO satellites will revolutionize communications, similar radar systems will play a leading role in meeting the military requirements derived by the post-cold-war environment—see section 1.11.5.1.

From a radiastronomy perspective, the strong emissions from these systems are seen as unintentional jammers for all radiotelescopes regardless of their location on Earth. Here the comparison is made with respect to the very weak signals emanated from celestial sources. Work toward the mitigation of the effects of these signals on radioastronomical observations is outlined in section 3.10.

1.10. THE IMPORTANCE OF RADIOFREQUENCY (RF) SUBSYSTEMS

The RF subsystems of many systems often define significant system parameters such as system noise temperature and the directions from which reception is accepted and to which transmission is directed. For phased array systems, the RF subsystems set or influence an even greater number of significant system parameters. The RF subsystems consist of the array antenna elements/polarizers, the low-noise amplifiers (LNAs), the transmitter chains, final power amplifiers, and beamformer(s).

The following lists the important system parameters, determined by the RF sub-systems of an active array radar:

- The array antennas and geometry determine the array radiation pattern as a function of frequency and scan angle. The array radiation pattern is in turn defined in terms of its HPBW, sidelobe level, and grating and cross-polarization lobes.
- The antennas used in conjunction with the appropriate processing subsystems determine the extent to which the system can derive the full benefits that polarimetric systems offer.
- The antennas used contribute significantly to the array weight, the cost, and whether the array is conformal to the skin of the platform.
- The array system noise temperature and the intermodulation distortion are defined by the LNAs and the losses between the antenna elements and the LNAs.
- The antenna array and LNAs define the gain over temperature, G/T, ratio of the system.
- The transmitter chain defines the purity and the frequency/phase stability of the transmitted signals, a prerequisite for high-quality coherent radars.
- The FPAs available now have a significant bearing on the approaches used to minimize the array sidelobes—see Chapter 4.
- The uniformity with which T/R modules and beamformers are manufactured influences the array sidelobe levels.
- The architecture of beamformers used in conjunction with active arrays defines whether one or more inertialess beams are generated.
- Cost minimization schemes are focused on the array front-ends.
- The array operational bandwidth is determined by the antennas and its front-ends.

From the foregoing one can easily appreciate the importance of RF signal processing in phased array systems. Signal processing of the incoming radiation was for a long time equated with the processing of signals at some baseband and the processing of signals at RF was seldom considered. It is not an exaggeration to state that many engineers imbued in information technology (IT) systems consider that a coat-hanger followed by an ADC constitutes the front-end of a system.

These mindsets were based on the following real-world considerations:

- Digital signal processing ICs operating at low frequencies are affordable and have exceptional signal processing powers.
- Digital electronics have been successful in many systems.
- EM theory and RF techniques are hard to master when compared to digital circuits.
- The requirements of narrow band, unsophisticated systems are easily met by standard horns, dipoles, or monopoles and narrowband subsystems
- Active phased arrays were not affordable for a long time.

The following drivers opened the floodgates of innovation in the area of RF signal processing:

- It has been realized that unsophisticated narrowband systems can no longer meet the challenges imposed by the modern multitarget and multithreat environment.
- MMIC-based T/R modules are now affordable; furthermore, 'intelligent' or self-focusing/self-cohering arrays hold the promise for the realization of even more affordable or cheaper arrays.

- Lightweight, low-cost, high-quality conformal microstrip antennas that are easily integrated with MMIC-based modules have been developed.
- Lightweight, low-cost beamformers of unprecedented versatility are now available.

Our aim here is not to overemphasize the role of RF signal processing at the expense of signal processing at basebands but to stake a claim for the importance of signal processing at RF. In the near term, substantial benefits to systems are expected from signal processing at RF. Ultimately, the convergence of digital, photonic, and RF signal processing is bound to yield further benefits to future array systems.

1.11. THE WIDER THRUSTS

Phased arrays evolve by efforts specifically focused on intrinsic issues and problems and by the leverage of developments that take place in allied fields. While we shall consider specific phased array research throughout the remainder of the book, we briefly survey developments taking place in other, allied areas in this section.

We have already stressed the importance of MMIC-based T/R modules for the evolution of affordable phased arrays and because of its importance this issue will be thoroughly explored in Chapter 4. The evolution of affordable MMIC modules operating at RF was the natural progression of long and continuous research and technological thrusts toward low frequency ICs and digital signal processing ICs.

Photonics is another field that evolved mainly within the communications engineering community and is now used to realize wideband beamformers operating in conjunction with active phased arrays; it is also used for the distribution of digital and RF signals in phased arrays. Again we shall explore these applications in Chapter 5 because of their importance.

We have already seen that the US and Russian global navigation systems can be used as illuminators of opportunity for bistatic radars. Other constellations of satellites are planned for the near future and their transmissions can be used for the same purpose. In the remainder of this section we shall briefly outline developments in microelectromechanical systems (MEMS), LightSARs, and subsystems utilizing key high-temperature superconducting (HTS) components, so that the impact of developments in these allied areas on phased arrays is appreciated. The issues raised by electromagnetic missiles (EMMs) are also considered because of their potential to disable the electronics of many high-value platforms.

1.11.1. Microelectromechanical Systems

MEMS are minuscule, light, and mobile information systems that intimately couple information systems with the physical world and people. The defining characteristics of MEMS are:

- Miniaturization, which implies minuscule components measured in micrometers.
- Multiplicity or batch processing, which make it possible to produce millions of components concurrently as easily as one component.

- Microelectronics that provides the intelligence to MEMS and allows the monolithic merger of sensors, actuators, and logic to build closed-loop feedback components and systems.

Important MEMS are described in many references of which [211–213] are but a few representative samples. Roadmaps for future applications of MEMS are outlined in references [214,215]. The emerging importance of MEMS can be assessed by the following graphic projections [215]:

- The majority of MEMS in current use have about 10 mechanical components and 10^3 transistors.
- Future MEMS will have about 10^7 mechanical components and 10^8 transistors.

What follows are samples of promising applications that have a bearing on array systems. A single chip, for instance, can perform sensing, signal processing, and data processing to achieve a total system solution to a given problem with dramatically reduced size and cost. Micromechanical signal processors that exhibit active Q-control and temperature stabilization have been reported [216]. Other possibilities include electromagnetic/optical beamsteering and distributed networks, sensors, actuators, and low-loss RF switches.

Cooling of high-power chips presents considerable problems. A device that produces microjets of air to cool high-power chips has been reported [217] and consists of a box with one flexible wall and a hole or several holes in the opposite wall. Vibrating the flexible wall causes cooling jets of vortices to emerge from the holes. A device with a hole 1/16 of an inch (1.5 mm) in diameter allowed the researchers to boost the power of an array of chips by 150% with no increase in temperature. The development of such devices constitutes a significant breakthrough for high-power chips used in phased arrays, which are currently cooled by forced or liquid air.

At a systems level, MEMS systems are ideal candidates for low-cost, lightweight, and volumetrically attractive airborne or spaceborne systems. While the exploration of the capabilities and potentials of airborne or spaceborne systems was undertaken with conventional full-size subsystems, the exploitation of the derived capabilities on a commercial basis can be aided by MEMS.

1.11.2. LightSARs

LightSAR is a spaceborne lightweight SAR system intended for commercial and civil applications. As the payload for the SAR system is only 250 kg, a lower-cost launch vehicle can be used and the total cost per mission is projected to be less than US$100M [218–1219]. Unlike other spaceborne SAR systems, the lightSAR has a useful life of five years, a feature that allows users to fully exploit the SAR capabilities over an extended period.

Table 1.17 lists the essential parameters of a lightSAR that is designated to operate in the following six modes: high-resolution spotlight, strip map, quad polarization, dual polarization, repeat pass interferometer, and scan SAR. Applications envisaged for lightSARs include:

- Mapping and charting—topographic maps

- Resource monitoring and management—agricultural/forest mapping
- Pollution and waste threats—geological interpretation
- Natural hazards—coastal monitoring (for shipping)
- Oceans and ice—ice mapping (for shipping)
- Enforcement and surveillance—disaster monitoring (news services)

As can be seen, the applications have pronounced civil and commercial thrusts.

Table 1.17 Essential parameters of a LightSAR

Frequency (GHz)	1.2
Polarization options	HH or VV
	VV, VH
	VV +VH
Spatial resolution options (m)	3, 6–10, 50, 25 or 100
Peak power (kW)	2
Average power (W)	64
Payload mass	<250 kg
Orbit altitude	529 km
Exact repeat time	8 days
Revisit time	3 days for 50% of the Earth's surface

1.11.3. HTS-based Subsystems

It is well-known that some materials conduct electricity without resistance when cooled to temperatures as high as 35 K [220,221]. Electronic devices made of these materials therefore exhibit low losses and high Qs. The other important development that contributed to the widespread use of HTS components and subsystems was the lowering of the costs of cryodynes [222]. While the use of cooled front-ends for LNAs in radioastronomy systems is widespread, other systems are rapidly adopting cooled HTS-based subsystems, because of the diverse range of benefits these systems offer. These attractive features include:

- Ultra-stable, low-phase-noise oscillators—see section 4.3.2.
- Low-loss multiplexers, filters/filterbanks, and delay lines.
- High-efficiency antennas (see section 3.7) and frequency-selective surfaces (FSSs).

The utilization of HTS-based components and subsystems in EW and radar systems [223], airborne platforms [224], and communication systems [225] has been outlined. Indeed, it is challenging to imagine systems that cannot benefit from HTS components and systems. Examples of work related to antennas and FSSs using HTSs are given in [226–228] and [229], respectively.

Coming to components, the following references constitute representative samples of work in the indicated areas:

- Multiplexers for satellite applications [230]
- Space-qualified multiplexers [231] and delay lines [231,232]
- Filters [233,234]
- Filterbanks [235]

The generation of ultra-stable, low-noise oscillators constitute the basic building blocks of any high-quality coherent system. Resonator unloaded Qs of the order of 500 000 can be obtained in a sapphire dielectric resonator (DR) operating on a low-order mode at 77 K and employing HTS films installed in the DR enclosure covers [236].

While it is important to survey the breadth of promising applications for HTS-based components and subsystems, here we shall revisit applications of particular importance to phased array systems in the coming chapters of the book.

1.11.4. Electromagnetic missiles

We have already seen that coherent microwave power can be aimed toward a target from a compact array of transmitters. The microwave power can be used either to disable the electronics of an aircraft or for the purposes of propelling electric-powered aircraft. This is just one approach that focuses EM energy onto a target and other approaches are also valid. Our interest in these areas of research stems from the fact that all electronic systems, including phased array systems, are vulnerable against high-power microwave energy. The research areas of interest are categorized as RF directed-energy weapons (DEWs), which include high-power microwave and non-nuclear EM pulse (NNEMP) technologies; and electromagnetic missiles (EMMs).

The protagonists of theoretical papers related to EMMs are Tai Tsun Wu , Hao-Ming Shen, Changua Wan, Chengli Ruan and Weigan Lin (note, for reference, that the surnames are given last) and a succinct summary of theoretical work is given in reference [237].

Apart from generating and transmitting high-power electromagnetic pulses, methods of assessing the effects of these threats on electronic systems and approaches taken to harden the front-ends of these systems are currently subjects of significant research interest. Some solutions to these problems are outlined in sections 4.3.3 and 4.4.3.

1.11.5. Contemporary Trends

The post-cold-war environment is markedly different from the era when an all-out war between the two superpowers was a distinct possibility. The system requirements therefore had to change to reflect the new realities. In this section we shall review the changes and map the new technological opportunities the post-cold-war era has ushered in. Without a knowledge of the new landscape it is hard, if not impossible, to appreciate current or future developments in phased array technology.

1.11.5.1. Post-Cold-War Environment and Technology

The political changes in Africa and Asia since the Second World War were followed by major changes that occurred after the dissolution of the former Soviet Union and the creation of new states. These changes include the drawing of new borders between the new nations and some political realignments. These changes reduced the threat of an all-out nuclear war, but generated regional instability on a global scale. The shrinking of the defense budgets of many countries followed and we shall pursue the impact of this trend in the next section.

Cold-war era surveillance requirements were based on large targets (ships and aircraft) and long revisit times of very wide areas. Post-cold-war surveillance requirements have shifted to theater battle support and low-intensity conflicts. The surveillance area is smaller, revisit times are of the order of 10 s and targets of less than 1 m^2 travel at speeds as low as 2 m/s. Radar systems onboard constellations of LEO or MEO satellites can provide near-continuous surveillance and reconnaissance of small, slow-moving targets in focused areas.

DARPA's proposed Surveillance, Targeting and Reconnaissance Satellite (Starllite) system [238] meets this new set of requirements. The proposal calls for a constellation of satellites orbiting the Earth at a height of 756 km supporting SAR and MTI systems. The response time of the system as a function of the number of satellites used is shown in Table 1.18. Here the response time of the system is calculated as the time required to deliver images to the user with a probability of 90% over areas of the Earth between 65 degrees North and South latitude. Commanders should be able to see moving vehicles moving as slowly as 5–10 km/h. Starllite's SAR/MTI sensors would enable theater commanders to see vehicle formations some 500–1000 km behind the front line. Currently airborne radars cannot look much beyond 300 km into a foreign territory. A pronounced decrease in the response time is seen in Table 1.18 as the number of satellites increases.

Table 1.18 The response time of the Starllite system as a function of the number of satellites used [238]

Number of Satellites	Response time (min)
8	94
12	60
18	30
24	15
37	8
48	5

The following additional capabilities for the Starllite system are claimed [138]:

- Spot observations would examine grids 4 km × 4 km (2.5 miles × 2.5 miles) with a resolution of 25 cm (1 foot).
- Backscan would allow each satellite to look at 20% of a 102 000 km^2 (40 000 square mile) theater each hour in 10 km (6-mile) wide strips with a 1 m (3 foot) resolution. For example, strip observations could scan all of Korea each hour with a 3 m resolution.

The Starllite system illustrates how the emerging post-cold-war requirements can be met and the attractors offered by constellations of LEO satellites.

1.11.5.2. Cost Minimization/Sharing Approaches

During the cold-war era the cost of military systems was treated as an independent variable; now cost is constrained and treated like any other system parameter. In the commercial world, systems have to be produced at a cost that is *a priori* constrained. It is not uncommon, for instance, for a CEO of a large organization to decree that a new product should be twice as good as a current product but costing half as much. However, cost constraints are not necessarily associated with low quality but with the fostering of product innovation that is used to derive affordable high-quality products.

Given that the threat of an all-out nuclear war has diminished decidedly, the defense budgets of many countries have decreased. This being the case, the military is using commercial off-the-shelf (COTS)-based systems that exhibit high reliability and are affordable. Most of us are aware of the high reliability of our computers, microwave ovens, and television sets. Commercial radar, communication, and navigation systems are just as reliable.

In another era the cost of a system was equated to the acquisition cost. Now the total cost of a system is seen in more holistic terms and is given by

$$\text{Cost}\big|_{\text{total}} = \text{acquisition cost} + \sum_{T_2}^{T_1} \text{Development cost} + \sum_{T_3}^{T_2} \text{LCC}$$

where
T_1 to T_2 = the time interval over which the development of a system takes place;
LCC is the life-cycle cost that extends from T_2 to T_3;
T_2 = the system launch time;
T_3 = the system substitution time.

Development costs for a system can be minimized if precompetitive R&D consortia are formed from representative interested parties [239]. More than 200 consortia had been formed by 1993 in the United States and the R&D fields include, *inter alia*, microelectronics and computer technology, software engineering, manufacturing sciences, semiconductor technology, biotechnology, transportation, and superconductivity. Parallel developments and support for precompetitive R&D is evident in Japan and Europe. Often government institutions support the commercial enterprises in the funding of precompetitive R&D.

Development costs for small subsystems are often shared between interested organizations. R&D papers related to cryogenically cooled MMIC-based LNAs, for instance, have authors drawn from the radioastronomy and radar/communications communities. Basic chips can be used in turn used in radar, radar-related, and communications systems.

Economies are also achieved by the use of many high-value systems to perform tasks other than those originally planned for the systems. The following examples will illustrate this trend:

- The VLA has been used in bistatic radar planetary experiments in conjunction with the Arecibo telescope.
- The JPL's DSN antennas and the Parkes (Australia) radiotelescope have been used for searches for extraterrestial intelligence.
- The JPL's DSN antennas have been used for many VLBI experiments.

- Earth-orbiting satellites designated for communications links have been used for OVLBI experiments.
- Interferometric SARs are used in a plethora of civilian and military applications.

The skills available in many diverse disciplines are often used to derive key components and subsystems that are in turn used in many systems. Affordable MMIC-based T/R modules, for instance, have evolved because the skills of solid-state physicists, production engineers, and systems engineers have been harnessed to solve a set of related problems over a long period of time. The knowledge base built over the years can now be used to derive affordable subsystems for the radar, EW, communications, and radioastronomy communities.

1.12. CONCLUDING REMARKS

In this chapter we have demonstrated that all phased arrays share the same theory and that there is a continuum between radar imaging arrays having short integration times and radiometric/radioastronomy arrays having short or long integration times. While the instantaneous beam derived from a densely populated radar array can have low sidelobes and the corresponding beam of a sparsely populated radioastronomy array is almost useless because it has high sidelobes, the latter sidelobes can be lowered to an acceptable level in quasi-real time. Similarly, the gain of a thinned array is very much lower than that of a filled aperture having the same extent. In the next chapter we shall consider in detail an important example where the sidelobe level of a highly thinned array was lowered and derive an equation for the gain of thinned arrays.

The linkages between arrays performing a variety of functions not only enrich and deepen one's understanding of array systems but also allow a designer imbued in one discipline to utilize the techniques and approaches developed by another discipline. Lastly, the defense of high-value platforms or the efficient operation of airports depend on array systems that derive interdependent and interrelated data. At a technological level, it is impossible to design a robust radar without knowledge of the capabilities of EW systems designed to disable it.

Typical or notional radar and radioastronomy arrays are delineated before the many fundamental attractors of phased arrays are considered. Graceful degradation of performance, radiation patterns on demand and multiple agile or staring beams and nulls are now the hallmarks of phased arrays. Even at this early stage two unattractive array characteristics—the presence of grating lobes and variable spatial resolution—have been considered because a substantial portion of this book is devoted to approaches developed to marginalize their effect on performance of the array. Overall, however, it is fair to state that the array attractors far outweigh these two limitations.

Some archetypical radar and radioastronomy systems are described and the reader is referred to references where complete descriptions of other operational systems are available.

The theoretical infrastructure necessary for an appreciation of radar fundamentals and radioastronomy aims is progressively developed, while the applications considered constitute a remarkable set of the services and products on offer to the user. A seemingly endless list of applied sciences benefit from data derived from flying phased arrays or SARs and of interferometric SARs. It is now impossible to overemphasize the contributions of these systems to a variety of applied sciences. More importantly, we are now witnessing a transition from an era when SAR capabilities were explored to an era in which the same capabilities are exploited on a commercial basis. In the military arena, we are now witnessing the influence of the post-cold-war doctrine on systems that are markedly different from the systems based on the cold-war doctrine. Records of progress for radar and radioastronomy arrays have been delineated in sections 1.6.15 and 1.7.8, respectively.

It is not an exaggeration to assert that phased arrays have redefined radar and established radioastronomy as a respectable new science. Radar and EW systems are considered together, while radioastronomy/radiometric and communication systems are treated separately. The discoveries in radioastronomy are intrinsically important since they complement the knowledge of the cosmos accumulated at other wavelengths. Radiometric systems derived from radioastronomy systems are now routinely used in several applied science applications.

The major drivers for future system developments are that the user wants to know more and to know it sooner; and that system cost is no longer an independent variable.

Developments toward wireless power transmission used for military and civilian aims will continue. Affordable pseudo-satellites can perform national defense and communications roles over locations chosen on an *ad hoc* basis. If surveillance over enemy territory is required, constellations of LEO satellites are a clear option, again on an *ad hoc* basis.

It is appropriate to end this section by enlisting promising areas of R&D:

- Work related to truly wideband multifunction systems capable of deriving interrelated and interdependent data in the civilian and military sectors has just begun (see also Chapter 2).
- The emergence of bistatic radars using illuminators of opportunity provided by TV/FM broadcast stations and the transmitters onboard the many constellations of LEO/MEO satellites. These systems have excellent LPI capabilities.
- Self-cohering distributed radars.
- The development of large self-cohering arrays (LSAs) to power eternal planes or pseudo-satellites.
- The evolution of measures and techniques to protect high-value assets from the harmful effects of LSAs and/or EMMs.
- The realization of second-generation radioastronomy arrays (see also Chapter 2).

The practicing engineer or scientist can derive help to meet future challenges from developments related to the areas of MEMS, HTS-based systems, photonics, MMIC T/R modules, vacuum technology, and vacuum microelecronics devices. We shall revisit the last five areas in Chapter 4.

It is fairly clear that we are witnessing a revival of interest in the RF subsystems that often define the significant system parameters. The new opportunities for signal processing afforded by active phased arrays at RF/IF and at baseband augment the capabilities of phased arrays.

REFERENCES

[1] Friis H. T. and Feldman C. B. A multiple unit steerable antenna for short-wave reception. *Proc. IRE* **25**, 841 (1937).

[2] Longuemare R. N. Advanced phased arrays redefine radar. *Defense Electronics* (Apr.), 63 (1990).

[3] Rao J. B. L., Trunk G. V. and Patel D. P. Two low-cost phased arrays. *1996 IEEE Int. Symp. on Phased Array Systems and Technology*, Boston, USA, pp. 119–124 (1996).

[4] Chow Y. L. Comparisons of some correlation array configurations for radio astronomy. *IEEE Trans. Ant. Propag.* **AP-18** (4), 567–569 (July 1970).

[5] Swenson G. W. Jr and Mathur N. C. The circular array in the correlator mode. *Proc. IREE (Aust.)* **28** (9), 370–374 (Sep. 1967).

[6] Keizer W. P. M. N. New active phased array configurations. *Conf. Proc. Mil. Microwaves 1990*, p. 564, (1990).

[7] Jaszka P. R. Aegis system still gets high marks. *Defense Electronics*, (Oct.), 48 (1988).

[8] Brookner E. (ed.). *Aspects of Modem Radar*. Artech House, Boston and London 1988.

[9] Parrish A. J. Electronically steerable arrays: current and future technology and applications. MSc Project Report, Loughborough University of Technology and Royal Air Force College, Cranwell, UK (1989).

[10] Coriolis K. The true cost aircraft in service. *Jane's Defence Wkly* (27 Sept.), 680 (1986).

[11] Petty R. T. (ed.). *Jane's Weapon Systems, 1987–1988*. Jane's Publishing, London (1987).

[12] Brookner E. Phased array radars. *Sci. Am.* **252**(2), 94 (1985).

[13] Stimson G. W. *Introduction to Airborne Radar*. Hughes Aircraft Company (1983).

[14] Brukiewa T. F. Active array radar systems applied to air-traffic control. *IEEE Natl. Telesyst. Conf.*, pp. 27–32 (1994).

[15] Michelson M., Shrader W. W. and Wieler J. G. Terminal Doppler weather radar. *Microwave J.* **33** (Feb.) 139–148 (1990).

[16] Wurman J., Randall M., Frush C. L., Loew E. and Holloway L. Design of a bistatic dual-Doppler radar for retrieving vector winds using one transmitter and a remote low-gain passive receiver. *IEEE Proc.* **32**(12), 1861–1871 (Dec. 1994).

[17] Kim Y., Lou J., van Zyl J., Maldonado L., Miller T., Sato T. and Skotnicki W. NASA/JPL airborne three-frequency polarimetric/interferometric SAR system. *IGARSS 96, Remote Sensing for a Sustainable Future*, pp. 1612–1614.

[18] Evans D. L. (ed.). Spaceborne synthetic aperture radar: Current status and future directions. *NASA Technical Memorandum 4679* (1995).

[19] Stofan E. R., Evans D. L., Schmullius C., Holt B., Plaut J. J., van Zyl J., Wall S. D. and Way J. Overview of results of spaceborne imaging radar-C, X-band synthetic aperture radar (SIR-C/X-SAR). *IEEE Trans. Geoscience Remote Sensing* **33** (4), 817–838 (July 1995).

[20] Johansson F., Rexberg L., and Petersson N. O. Theoretical and experimental investigation of large microstrip array antenna. *IEEE Colloq. Recent Developments in Microstrip Antennas*, **4**(1) (1993).

[21] Wahl M., Otti H. and Velten E. H. Programme aspects for future multifrequency/multipolarization SAR facilities. *IGARSS, 1995. Quantitive Remote Sensing Symposium*, vol 3, pp. 1834–1835 (1995).

[22] Winokur R. S. Operational use of civil space-based synthetic aperture radar (SAR). *JPL Publication 96-16* (1996).

[23] Hovanessian S. A., Jocic L. B. and Lopez J. M. Spaceborne radar design equations and concepts. *1997 IEEE Aerospace Conference*, Snowmass at Aspen Co., USA, vol. 1, pp. 125–136 (1997).

[24] Wehner D. R. *High Resolution Radar*. Artech House, Dedham, MA (1987).

[25] Chen C. C. and Andrews H. C. Target-motion-induced radar imaging. *IEEE Trans. Aerospace Electronic Syst.* **AES-16**(1), 2–14 (Jan. 1990).

[26] Rosenstiel School of Marine and Atmospheric Science. Technical Report TR 95-003, *SAR Interferometry and Surface Change Detection*. Report of a workshop held in Boulder, CO. (readily available from NASA/JPL) (1994).

[27] Goldstein R. M., Zebker H. A. and Werner C. Satellite radar interferometry: two-dimensional phase unwrapping. *Radio Sci.*, **23**, 713–720 (1998).

[28] Massonnet D. and Rabaute T. Radar interferometry: limits and potential. *IEEE Trans. Geoscience Remote Sensing* **31**(2), 455–464 (Mar. 1993).

[29] Adams G. F., Ausherman A., Crippen S. L., Sos G. T. and Williams P. The ERIM interferometric SAR: IFSARE. *Proc. 1996 IEEE Nat. Radar Conf.*, pp. 249–254 (1996).

[30] Zebker H. A., Farr T. G., Salazar R P. and Dixon T. H. Mapping the world's topography using radar interferometry: The TOPSAT mission. *IEEE Proc.* **82**(12), 1774–1786 (Dec. 1994).

[31] Massonnet D., Rossi C., Carmona C., Adragna F., Peltzer G., Feigl K. and Rabaute T. The displacement field of the Landers earthquake mapped by radar interferometry. *Nature* **369**, 227–230 (1994).

[32] Papathanasiou K. P. and Morelka J. R. Interferometric analysis of multifrequency SAR data. *IGARSS 96, Remote Sensing for a sustainable future*, vol. II, pp. 1227–1229 (1996).

[33] O'Neil G. K. *The High Frontier*. Corgi Books (1978).

[34] Radio Telescopes Special Issue, *IEEE Proc.* **88**, 629–829 (May 1994).

[35] Cover story. Radio astronomy: new windows on the Universe. *IEEE Spectrum* (Feb.), pp. 18–25 (1995).

[36] Brunier S. Temples in the sky. Parts I and II. *Sky and Telescope* **85**(2) 19–24 (Feb. 1993) and **85**(2) 26–32 (June 1993).

[37] Napier P. J., Thomson A. R. and Ekers R. D. The very large array: design and performance of a modern synthesis radio telescope. *Proc. IEEE* **71**(11), 1295–1319 (Nov. 1983).

[38] NRAO VLA Home Page: What is the VLA. http//zia.aoc.nrao.edu/via/html/VLAintro.shtml

[39] NMA Status Report 1998–1999.
http://www.nro.nao.ac.jp/~nma/status/status98to99e.html#resolution

[40] Broten N. W. *et al.* Long base line interferometry: A new technique. *Science* **156**, 1592 (1967).

[41] Klemperer W. K. Long baseline radio interferometry with independent frequency standards. *Proc. IEEE* **60**, 602 (1972).

[42] Thompson A. R, Moran J. M. and Swenson G. W. *Interferometry and Synthesis in Radio Astronomy*. Wiley, New York (1986); reprinted by Krieger Press, Melbourne, FL (1991).

[43] Kellerman K. I. and Thomson A. R. The very long baseline array. *Science* **229**, 123 (1985).

[44] Kellerman K. I. and Thomson A. R. The very long baseline array. *Sci. Am.* **258**, 44 (1988).

[45] Napier P. J. *et al.* The very long baseline array. *Proc. IEEE* **82**(5), 658 (1994).

[46] Sade R. S. and Deerkoski L. Tracking and data relay satellite operations in the 1980s. *Proc. AL4A INASA Symp. Space Tracking Data Syst.*, p. 77 (1981).

[47] Levy G. S. *et al.*, Very long baseline interferometric observations made with an orbiting radiotelescope. *Science* **234**, 117 (1986).

[48] Levy G. S. *et al.* VLBI using a telescope in Earth orbit. 1. The observations. *Astrophys. J.* **336** (2, Pt 1), 1098 (1989).

[49] Linfield R. P. *et al.*, 15 GHz space VLBI observations using an antenna on a TDRSS satellite. *Astrophys. J.* 358 (1, Pt 1), 350 (1990).

[50] Edelson C. R. Applications of synthetic aperture radiometry. *Proc. AGARSS '94*, vol. 3, pp. 1326–1328 (1994).

[51] Gaslewski A. J. Technology for spaceborne passive microwave earth remote sensing. *NTC '91*, vol. 1, pp. 271–275.

[52] Le Vine D. M. *et al.* Initial results in the development of a synthetic aperture microwave radiometer. *IEEE Trans. Geoscience Remote Sensing* **28**, 614 (1990).

[53] Le Vine D. M., Griffis A. J., Swift C. T. and Jackson T. J. ESTAR: A synthetic aperture microwave radiometer for remote sensing applications. *Proc. IEEE* **82**(12), 1787–1801 (1994).

[54] Vowinkel B. *et al.*, Airborne imaging system using a cryogenic 90 GHz receiver. *IEEE Trans. Microwave Theory Tech.* **MTT-20**, 535 (1981).

[55] Vowinkel B. Gruener K. and Reinert W. Cryogenic all solid-state millimeter-wave receivers and airborne radiometry. *IEEE Trans. Microwave Theory Tech.* **MTT-31**, 996 (1983).

[56] Ostro S. J. Planetary radar astronomy. *Rev. Mod. Phys.* **65**(4), pp. 1235–1279 (Oct. 1993).

[57] Muhleman D. O., Grossman A. W., Butler B. J. and Slade M. A. Radar reflectivity of Titan. *Science* **248** (25 May), 976–980 (1990).

[58] Muhleman D. O., Butler B. J., Grossman A. W. and Slade M. A. Radar images of Mars. *Science* **253** (27 Sept.), 1508–1513 (1991).

[59] Slade M. A., Butler B. J. and Muhleman D. O. Mercury radar imaging: evidence for polar ice. *Science*, **256** (23 Oct.) 635–640 (1992).

[60] A series of reports on Clementine: Mission to the Moon. *Science* **266** (16 Dec.) (1994).

[61] Nozette S., Lichtenberg C. L., Spudis P., Bonner R., Ort W., Malaret E., Robinson M. and Shoemaker E. M. The Clementine bistatic radar experiment. *Science* **224** (29 Nov.) 1495–1498 (1996).

[62] Billam E. R. Phased array radar and the detection of low 'observables'. *Rec. IEEE 1990 Int. Radar Conf.*, pp. 491–495 (1990).

[63] Baars J. W. M. The measurement of large antennas with cosmic radio sources. *IEEE Trans. Antennas Propag.* **AP-21**(4), 461 (1973).

[64] Knott E. F., Shaeffer J. F. and Tuley M. T. *Radar Cross Section*. Artech House, Norwood, MA (1993).

[65] Skolnik M. I. *Introduction to Radar Systems*. McGraw-Hill, New York (1980).

[66] Richardson D. *Stealth Warplanes*. Salamander Books, London (1989).

[67] Preissner J. The influence of the atmosphere on passive radiometric measurements. *AGARD Conf. Repr. No. 245* (1978).

[68] Currie N. C. and Brown C. E. *Principles and Applications of Millimeter-wave Radar*. Artech House, Dedham, MA (1987).

[69] Miya K., (ed.). *Satellite Communications Technology*. KDD Engineering and Consulting, Tokyo, p. 106 (1981).

[70] Marcum J. I. A statistical theory of target detection by pulsed radars. *Rand Corp Res. Memo RM-754* (Dec. 1947); reprinted: *IRE Trans. Inf. Theory* **IT-6**(2), pp. 59–145 (1960).

[71] Swerling P. Probability of detection for fluctuating targets. *Rand Corp. Res. Memo RM-1217* (Mar. 1954); *IRE Trans. Inf. Theory* **IT-6**(2) (April) pp. 269–308 (1960).

[72] Barton D. K . *Modern Radar System Analysis*. Artech House, Norwood, MA (1988).

[73] Rzemien R. (guest ed.) Special issue on Coherent Radar. *Johns Hopkins APL 1997, Tech. Dig.* **18**(3).

[74] Driscott M. M. *et al*. Cooled, ultrahigh Q, sapphire dielectric resonators for low-noise, microwave signal generation. *IEEE Trans. Ultrasonics Ferroelectrics Frequency Control* **39**(3) 405–410 (May 1992).

[75] Hoisington D. B. Low-probability of detection radar designs challenge tactical ESM capabilities. *The International Countermeasures Handbook*, 11th edn, EW Communications, Inc., Palo Alto, CA, pp. 350–354 (1986).

[76] D. P. Meyer and H. A. Mayer. *Radar Target Detection*. Academic Press, New York (1973).

[77] Cohen M. N. An overview of high range resolution radar techniques. NTC '91, *National Telesystems Conf. Proc.* vol. 1, pp. 107–115 (1991).

[78] Dixon R. C. *Spread Spectrum Systems*. Wiley, New York (1977).

[79] Munday J. and Pinches M. C. Jaguar-V frequency hopping radios system. *IEE Proc., Part F* **129**(3), 213 (1982).

[80] Fourikis N. Novel receiver architectures for ESM facilities. *J. Electr. Electron. Eng., Aust.* **6**(3), 341 (1986).

[81] Ruffe L. I. and Scott G. F. LPI considerations for surveillance radar. *Int. Conf. Radar 1992*, p. 200 (1992).

[82] Reits B. J. and Groenenboom A. FMCW signal processing for a pulse radar. *Int. Conf. Radar 1992*, p. 332 (1992).

[83] Stove A. G. Linear FMCW radar techniques. *IEE Proc., Part F (Radar and Signal Processing)* **139**(5), 343 (1992).

[84] Beasley P. D. L. *et al*. Solving the problems of a single antenna frequency modulated CW radar. *Proc. IEEE 1990 Int. Radar Conf.*, p. 391 (1992).

[85] McGregor J. A., Poulter E. M. and Smith M. J. Switching system for single antenna operation of an S-band FMCW radar. *IEE Proc. Radar, Sonar Navigation* **141**(4), 241–248 (Aug. 1994).

[86] Wirth W.-D. Omnidirectional low probability of intercept radar. *Proc. Int. Radar Conf.*, Paris (1989).

[87]   Koch V. and Westphal R. New approach to a multistatic passive radar sensor for air/space defense. *IEEE AES Systems Magazine.* **10**(1) 24–32 (1995).

[88]   Dunsmore M. R. B. Bistatic radars. In *Advanced Radar Techniques and Systems* (G. Galati, ed.), Ch. 11. IEE Peregrinus (1993).

[89]   Steinberg B. D. *Microwave Imaging with Large Antenna Arrays.* Wiley, New York (1983).

[90]   Heimiller R. C., Belyea J. E. and Tomlinson P. Q. Distributed array radar. *IEEE Trans. Aerospace Electron Syst.* **AES-19**(6) 831–838 (1983).

[91]   Attia E. H. and Abend K. An experimental demonstration of a distributed array radar. *IEEE APS Symp. Dig.* **3**, 1720–1723 (1991).

[92]   McCready L. L., Pawsey J. L. and Payne-Scott R. Solar radiation and its relation to sunspots. *Proc. Roy. Soc. (London)* **190A**, 357–375 (Aug. 1947).

[93]   Herper J. C., Kaiteris C. and Valentino P. A. Combined radar and illumination for sea skimmers (CRISS). *IEEE 1996 Natl. Radar Conf.* Ann Arbor, MI, pp. 243–248 (1996).

[94]   Headrick L. M. Looking over the horizon. *IEEE Spectrum* (July), 36–39 (1990).

[95]   Westinghouse Airships Brochure (1994).

[96]   Dickey F. R. Jr, Labitt M. and Staudaher F. M. Development of airborne moving target radar for long range surveillance. *IEEE Trans. Aerospace Electronic Syst.* **27**(6), 959–972 (Nov. 1991).

[97]   Zingler W. H. and Frill J. A. Mountain Top: beyond-the-horizon cruise missile defense. *Johns Hopkins APL Tech. Dig.* **18**(3) 501–520 (1997).

[98]   Marcotte F. J. and Hanson J. M. An airborne captive seeker with real-time analysis capability. *Johns Hopkins APL Tech. Dig.* **18**(3), 422–431 (1997).

[99]   Fisher A. Secret of perpetual flight? Beam-powered plane. *Popular Science* **248**(1), 62–65 (1998).

[100] Brown S. F. The eternal plane. *Popular Science* **244**(4), 70 (Apr. 1994).

[101] Wardrop B. A Russian experimental high-power, short-pulse radar. *GEC J. Technol.* **14**(3), 1–12 (1997).

[102] Skolnik M., Andrews G. and Hansen J. P. An ultrawideband microwave-radar conceptual design. *Rec. IEEE 1995 Int. Radar Conf.*, pp. 16–21 (1995).

[103] The executive summary of the OSD/ARPA sponsored study of UWB radar. *IEEE AES Magazine* (Nov.), 45–49 (1990).

[104] Threston J. T. and Meinig G. R. AEGIS and the AN-/SPY-1A radar—A description and a quick look report on the first deployment. *Proc. Military Microwaves Conf.*, London, p. 216 (1984).

[105] Thompson G. R. G. A modular approach to multi-function radar design for naval application. *Proc. RADAR-87, IEE Conf. Pub. 281*, pp. 32–36 (1987).

[106] Billam E. R. and Harvey D. H. MESAR an advanced experimental phased array radar. *Proc. RADAR-87, IEE Conf. Pub. 281*, pp. 37–40 (1987).

[107] Fourikis N. Antenna elements and architectures for wide-band multifunction active phased arrays. Invited Keynote Paper, *Int. Conf. Millimeterwave Infrared Technol.*, Beijing, p. 24 (1992).

[108] Fourikis N. and Lioutas N. Novel wideband multifunction phased arrays. *Proc. 16th Int. Conf. Infrared Millimeter*, Lausanne, Switzerland, pp. 499–500 (1991).

[109] Burke M. A. Multifunction/shared aperture systems or smart skins now. *J. Electron. Defence* **14**(1), 29 (1991).

[110] Fourikis N. *Phased Array-based Systems and Applications.* Wiley, New York (1997).

[111] Moore S. A. W. and Moore A. R. Dual frequency multi-function radar antenna research. *IEE 16th Int. Conf on Antennas and Propag.*, (1997).

[112] Hemmi C., Dover T., Vespa A. and Fenton M.-W. Advanced shared aperture program (ASAP) array design. Revolutionary developments in phased arrays. *IEEE Int. Symp. Phased Array Systems and Technology*, pp. 178–282 (1996).

[113] Pigeon E. and Grenier D. Identification of high speed targets facing the radar according to their length and signature. *Canadian Conf. Electrical and Computer Engineering, Conf. Proc.* vol. 2, pp. 417–420 (1994).

[114] Grenier D., Pigeon E. and Turner R. M. High-range resolution mono-frequency pulsed radar for the identification of approaching targets using subsampling and the MUSIC algorithm. *IEEE Signal Proc. Lett.* **3**(6), 179–181 (June 1996).

[115]   Jacobs S. and O'Sullivan J. A. High resolution radar models for joint tracking and recognition. *Proc. 1997 IEEE Natl. Radar Conf.*, pp. 99–104 (1997).

[116]   Garber F. D. Recent advances in automatic radar target identification. *Proc. 1991 IEEE, NAECON*, vol. 1, pp. 353–359 (1991).

[117]   Schindall J. Aircraft detection and identification uses passive electronic-support measures. *Microwave Syst. News* (Oct.) 78 (1986).

[118]   Briemle E. Passive detection and location of noise jammers. *AGARD Conf. Proc. AGARD-CP-488*, p. 33-1 (1990).

[119]   Bell M. R. and Grubbs R. A. Modelling of jet engine modulated radar signal returns for target identification. *IEEE Trans. Aerospace Electronic Syst.*, 73–87 (Jan. 1993).

[120]   Kennaugh E. M. and Moffatt D. L. Transient and impulse response approximations. *Proc. IEEE.* **53**, 893–901 (Aug. 1965).

[121]   Moffatt D. L. Interpretation and application of transient and impulse response approximations in electromagnetic scattering problems. *Report 2415-1*. Depart. of EE, The Ohio State Univ. Electroscience Laboratory, Columbus, OH (Mar. 1968).

[122]   Lin H. and Krienski A. A. Optimum frequencies for aircraft classification. *IEEE Trans. Aerospace Electronic Syst.*, **AES-17**(5) 656–665 (Sept. 1981).

[123]   Rendall D. *Jane's Aircraft Recognition Guide*. Harper Collins (1996).

[124]   Brousseau C., Guiot B., Launay G. and Bourdillon A. A. VHF multifrequency and multi-polarization radar: preliminary results. *IEEE 1996 Natl Radar Conf.*, Ann Arbor, MI, pp. 226–231 (1996).

[125]   Hayes R. D. and Eaves, J. L. Study of polarization techniques for target enhancement, *AF33(615)-2523 AD. 31670 Report A-871*. Georgia Institute of Technology, Atlanta (1966).

[126]   Battles J. (ed.). *Polarimetric Technology Handbook*. GACIAC IIT Research Institute, 10 West 25th Street, Chicago, IL 60616-3799 (1992).

[127]   Ioannides G. A. and Hammers D. E. Optimum antenna polarization for target discrimination in clutter. *IEEE Trans. Antennas Propag.* **AP-27**(3), 357 (1979).

[128]   Swartz A. A. *et al.* Optimal polarizations for achieving maximum contrast in radar images. *J. Geophys. Res.* **91**(B12), 15332 (1988).

[129]   Poelman A. J. Virtual polarisation adaptation: a method of increasing the detection capability of a radar system though polarisation-vector processing. *IEE Proc., Part F* **128**(5), 261 (1981).

[130]   Poelman A. J. Polarisation-vector translator in radar systems. *IEE Proc., Part F* **130**(2), 161 (1983).

[131]   Poelman A. J. and Guy J. R. F. Multinotch logic-product polarisation suppression filters: a typical design example and its performance in a rain-clutter environment. *IEE Proc., Part F* **131**(4),383 (July 1984).

[132]   Hammers E., Fugita M. and Klein A. Applications of adaptive polarisation. *Mil. Microwaves Conf. Rec.*, London p. 383 (1980).

[133]   Giuli D., Fossi M. and Cherardelli M. A technique for adaptive polarization filtering in radars. *Rec. IEEE 1985 Int. Radar Conf.*, p. 213 (1985).

[134]   Brown R. D. and Wang H. Adaptive multiband polarization processing for surveillance radar. *Proc. IEEE 3rd Int. Syst. Eng. Conf.* p. 17 (1991).

[135]   Brown R. D. and Wang H. Improved radar detection using adaptive multiband polarization processing. *Natl. Telesyst. Conf.* George Washington University, Virginia Campus, Washington, DC (1992).

[136]   Chu T. S. Rain induced cross-polarization at cm and mm-wavelengths. *Bell Syst. Tech. J.* **53**(8), 1557 (1977).

[137]   Yamada M. *et al.* Compensation techniques for rain depolarisation in satellite communications. *Radio Sci.* **17**(5), 1220 (1982).

[138]   Fourikis N. Future directions for millimeter-wave systems. Invited Keynote Paper, *6th Int. Conf. Infrared Millimeter Waves*, Lausanne, Switzerland, p. 424 (1991).

[139]   Blanchard A. J. and Jeans B. J. Antenna effects in depolarization measurements. *IEEE Trans. Geoscience Remote Sensing* **GE-21**(1), 113 (1983).

[140]   Ulaby F. T. *et al. Michigan Microwave Canopy Scattering Model*, 22486-T-1 7/88. University of Michigan, Ann Arbor (1988).

[141] Vogel W. J. Terrestrial rain depolarization compensation experiment at 11.7 GHz. *IEEE Trans. Commun.* **COM-31**(11), 1241 (1983).

[142] White W. D. Circular radar cuts rain clutter. *Electronics* **27**, 158 (1954).

[143] Novak L. M. *et al.* Optimal processing of polarimetric synthetic-aperture radar imagery. *Lincoln Lab. J.* **3**(2), 133 (1990).

[144] Holm W. A. Applications of polarimetry to target/clutter discrimination in millimeter wave radar systems. *Proc. SPIE Polarimetry Radar, IR Visible, UV X-Ray* **1317**, 148 (1990).

[145] Curtis R. D. On overview of surface navy ESM/ECM development. *J. Electron. Defense* (Mar.) 31 (1982).

[146] Simpson M. High ERP phased array ECM systems. *J. Electron. Defense* (Mar.) 41 (1982).

[147] DuBose R. L. Chaff systems for ships defense. *The International Countermeasures Handbook*, 11th edn. pp. 343–349. EW Communications, Inc. (1986).

[148] Johnson R. N. Radar-absorbing material: a passive role in an active scenario. *The International Countermeasures Handbook*, 11th edn, pp. 375–381. EW Communications, Inc. (1986).

[149] Harmuth H. F. Use of ferrites for absorption of electromagnetic waves. *IEEE Trans. Electromagnetic Compatibility* **EMC-27**(2), 100 (1985).

[150] Suttle W. R. The application of radar absorbing materials to armoured fighting vehicles (radar absorbing materials). *IEE Colloq. on Low Profile Absorbers and Scatterers*, London, pp. 6/1–6/4 (1992).

[151] Pitkethly M. J. Radar absorbing materials and their potential use in aircraft structures. *IEE Colloq. Low Profile Absorbers and Scatterers,* London, pp. 7/1–7/3 (1992).

[152] Perini J. and Cohen L.S. Design of broad-band radar-absorbing materials for large angles of incidence. *IEEE Trans. Electromagnetic Compatibility* **EMC-35**(2), 223–230 (May 1993).

[153] Zhang Z. Tactical advantages of quasi-wideband phased array radar. *Acta Sinica* **21**(3), 86 (1993).

[154] Jansky K. G. Directional studies of atmospherics at high frequencies. *Proc. IRE* 1920–1932 (Dec. 1932).

[155] Reber G. Cosmic static. *Astrophys. J.* **91**, 621–624 (June 1940).

[156] *Proc. IRE, Radio Astronomy Issue*, **46**(1) (Jan. 1958).

[157] Dicke R. H. and Beringer R. Microwave radiation from the sun and moon. *Astrophys. J.* **103**, 373 (1946).

[158] Bolton J. G. and Stanley G. J. Variable source of radio frequency radiation in the constellation of Cygnus. *Nature (London)* **161**, 312 (1948).

[159] Bolton J. G., Stanley G. J. and Slee O. B. Positions of three discrete sources of galactic radio-frequency radiation. *Nature (London)* **164**, 101 (1949).

[160] Christiansen W. N. and Warburton J. A. The distribution of radio brightness over the solar disk at a wavelength of 21 cm. *Aust. J. Phys.* **6**, 190 (1953).

[161] Covington A. E. and Broten N. W. Brightness of the solar disk at a wavelength of 10.3 cm. *Astrophys. J.* **119**, 569 (1952).

[162] Mills B. Y. and Little A. G. A high resolution system of a new type. *Aust. J. Phys.* **6**, 272 (1953).

[163] Christiansen W. N. and Mathewson D. S. Scanning the sun with a highly directional antenna. *Proc. IRE* **46**(1), 127 (1958).

[164] Ewen H. I. and Purcell M. Radiation from galactic hydrogen at 1420 mc/sec. *Nature* **168**, (1 Sept.), 356–358 (1952).

[165] Muller C. A. and Oort J. H. The interstellar hydrogen line at 1420 mc/sec and an estimate of galactic rotation. *Nature* **168**, 358 (1951).

[166] Pawsey J. L. *Nature*, **168**, 358 (1951).

[167] Burbidge G. Quasars, redshifts and controversies. *Sky & Telescope* (Jan.), 38–43 (1988).

[168] Weinreb S., Barrett A. H., Meeks M. L. and Henry J. C. Radio observations of OH in the interstellar medium. *Nature* **200**, 823–831 (Nov. 1963).

[169] Penzias A. A. and Wilson R. W. A measurement of excess temperature at 4080 Mc/s. *Astrophys. J.* **142** 419–421 (1965).

[170] Hewish A., Bell S. J, Pilkington J. D. H., Scott P. F. and Collins R. A. Observation of a rapidly pulsating radio source. *Nature* **217**, 709 (Feb. 1968).

[171] Cheung A. C., Rank D. M., Townes C. H., Thorton D. D. and Welch W. J. Detection of NH$_3$ molecules in the interstellar medium by their microwave emission. *Phys. Rev. Lett.*, **21**, 1701–1705 (16 Dec. 1968).

[172] Cheung A. C., Rank D. M., Townes C. H., Thorton D. D. and Welch W. J. Detection of water in interstellar regions by its microwave radiation. *Nature* **221**, 626–628 (15 Feb. 1969).

[173a] Verschuur G. L. Interstellar molecules. *Sky & Telescope* (Mar.), 379 (1992).

[173b] http://www.cv.nrag.adu/~awootten/allmols.html

[174] McClintock J. Do black holes exist? *Sky & Telescope* (Jan.), 28–33 (1988).

[175] *Encyclopedia Britannica*, http://www.eb.com:180/cgi-bin/g?DecF+boy/94/H00085.html

[176] Thorsett S. E. The times they are a-changing. *Nature* **367**, 684–685 (24 Feb. 1994).

[177] Fienberg R. T. COBE confronts the big bang. *Sky & Telescope* (July), 34–35 (1992).

[178] Pawsey J. L. and Bracewell R. N. *Radio Astronomy*. Oxford University Press, Oxford (1955).

[179] Perlman M. L, Rowe E. M. and Watson R. E. Synchrotron radiation—light fantastic. *Phys. Today* **27**, 30 (1974).

[180] Shkiovsky L. S. *Cosmic Radio Waves*. Harvard University Press, Cambridge, MA (1960).

[181] Oort J. H. The Crab Nebula. *Sci. Am.* **196**(3), 59 (1957).

[182] Bekefi G. and Barrett A. H. *Electromagnetic Vibrations, Waves and Radiation*. MIT Press, Cambridge, MA (1977).

[183] Gagliano A. and Platt R. H. An upward looking airborne millimeter wave radiometer for atmospheric water vapor sounding and rain detection. *Proc. SPIE, Int. Soc. Opt. Eng.* **544**, 112 (1985).

[184] Dicke R. H. The measurement of thermal radiation at microwave frequencies. *Rev. Sci. Instrum.* **17**, 268 (1946).

[185] Ryle M. and Hewish A. The synthesis of large radio telescopes. *Mon. Not. Roy. Astron. Soc.* **120**, 220 (1960).

[186] Hewish A. The realisation of giant radio telescopes by synthesis techniques. *Proc. IREE, Aust.* **24**(2), 225 (1963).

[187] Pearson T. J. and Readhead A. C. S. Image formation by self-calibration in radio astronomy. *Annu. Rev. Astron. Astrophys.* **22**, 97–130 (1984).

[188] Narayan R. and Nityananda R. Maximum entropy image restoration in astronomy. *Annu. Rev. Astron. Astrophys.* **24**, 127–170 (1986).

[189] Wild J. P. Special Edition: The Culgoora radioheliograph. *Proc. IREE, Aust.* **28**(9) (1967).

[190] Fourikis N. The Culgoora radioheliograph—7. The branching network. *Proc. IREE, Aust.* **28**(9), 315–323 (1967).

[191] Malliot H. A. A cross beam interferometer radiometer for high resolution microwave sensing. *IEEE Aerospace Applications Conf. Dig.*, p. 77 (1993).

[192] Fourikis N. Advanced phased arrays for airborne radiometric imagers. *1996 IEEE Int. Symp. Phased Array Systems and Technology*, Boston MA, pp.435–437 (1996).

[193] Arsac J. Nouveau reseau pour l'observation radiostronomique de la brillance sur le soleil a 9350 Mc/s. *C. R. Hebd. Seances Acad. Sci.* **240**, 942 (1955).

[194] Ruf C. S. *et al.* Interferometric synthetic aperture microwave radiometry for remote sensing of the Earth. *IEEE Trans. Geosciences Remote Sensing* **26**(5), 596 (1988).

[195] Milman A. S. Sparse-aperture microwave radiometers for Earth remote sensing. *Radio Sci.* **23**(2), 193 (1988).

[196] Swift C. T., Le Vine D. M. and Ruf C. S. Aperture synthesis concepts in microwave remote sensing of the Earth. *IEEE Trans. Microwave Theory Tech.* **39**(12), 1931 (1991).

[197] S. Chandrasekhar. *Radiative Transfer*. Oxford University Press, London, 1955.

[198] Cohen, M. H. Radio astronomy polarization measurements. *Proc. IPE* **46**(1), 172 (1958).

[199] Rivera D. J. An S-band solid-state transmitter for airport surveillance radars. *IEEE 1993 Natl. Radar Conf. Rec.*, p. 197 (1993).

[200] Fourikis N. Novel shared aperture phased arrays. *Microwave Optical Tech. Lett.* (20 Feb.) 189–192 (1998).

[201] Brown W. C. The early history of wireless power transmission. http://engineer.tamu.edu/tees/CSP/wireless/newhist2.htm

[202] Jones D. "Daedalus". Star wars 2. *Nature* **376**, 392 (3 Aug. 1995).

[203] Ford M. I. and Onda M. Powering of high-altitude LTA's by surface-to-air microwave transmissions. *AUVSI '96 Proc.*, 363–372 (1996).

[204] Fujino Y., Fujita M., Kaya N., Kunimi S., Ishi M., Ogihara N., Kusakas N. and Ida S. A dual polarization microwave power transmission system for microwave propelled airship experiment. *Proc. 96 ISAP*, vol. 2, pp. 563–566 (1996).

[205] Mullins J. Electronic airship: Masahiko Onda's airship will never run out of fuel because it picks up the power it needs as it goes along. *New Scientist*, **148**(2005; 25 Nov.), 40 (1995).

[206] White T. W. R. A self-steering array for the SHARP microwave-powered aircraft. *IEEE Trans. Antennas Propag.* **40**(12), 1565–1567 (Dec. 1992).

[207] Allnutt J. E. Evolutionary architectures for future multimedia services. *1996 IEE Colloq. What's New in Future Multimedia Communications*, pp. 4/1–4/8 (1996).

[208] Special Issue on Global Positioning System. *Proc IEEE* **87**(1) (Jan. 1999).

[209a] Greiling P. and Ho N. Commercial satellite applications for heterojunction microelectronics technology. *IEEE Trans. Microwave Theory Tech.* **46**(6), 743–747 (June 1998).

[209b] Sturza M. A. LEOs—The communications satellites of the 21st century. *IEEE Technical Applications Conference, Northcon/96*, pp. 114–118 (1996).

[210] J. J. Schuss *et al.* Design of the Iridium phased array antennas. *IEEE Antennas Propag. Soc. Int. Symp.* vol. 1, p. 218 (1993).

[211] Core T. A., Tsang W. K. and Sherman S. J. Fabrication technology for an integrated surface-micromachined sensor. *Solid State Technology* (Oct.) 39–47 (1993).

[212] Howe R. T., Muller R. S., Gabriel K. J. and Trimmer S. N. Silicon micromechanics: Sensors and actuators on a chip. *IEEE Specrum* (July), 29–35 (1990).

[213] Stix G. Micron machinations. *Sci. Am.* **267** 72–80 (1992).

[214] Fujita H. A decade of MEMS and its future. *Proc 10th Annual Int. Workshop on MEMS, An Investigation of Micro Structures, Sensors, Actuators, Machines and Robots*, Nagoya, Japan, pp. 1–7 (1997).

[215] Gabriel K. J. MEMS. *1997 IEEE Aerospace Conf.*, pp. 9–43 (1997).

[216] Nguyen C. T.-C. Micromechanical resonators for oscillators and filters. *1995 IEEE Ultrasonics Symp.*, pp. 489–499 (1995).

[217] Beardsley T. Chilling chips. *Sci. Am.* **276**, 40 (Jan. 1997).

[218] LightSAR Point Design-Radar. http://southport.jpl.nasa.gov/lightsar/facts/97/radar.html.

[219] DRAFT-LightSAR Technology Validation Mission. http://southport.jpl.nasa.gov/lightsar/archive/white.html.

[220] Kirtley J. R. and Tsuei C. C. Probing high-temperature superconductivity. *Sci. Am.*, (Aug.), **275**, 50–55 (1996).

[221] Chu P. C. W. High-temperature superconductors. *Sci. Am.* **273**, 128–131 (1995).

[222] Nisenoff M. and Liang G.-C. MTT-18 Microwave superconductivity: "Cold is better". *IEEE Society MTT* (142), pp. 52–53 (Spring 1997).

[223] Ryan P. A. High-temperature superconductivity for avionic electronic warfare and radar systems. *High T_c Microwave Superconductors and Applications. SPIE Proc* **2156**, 2–12 (1994).

[224] Robertson M. A. Two applications of high-temperature superconductor technology on an airborne platform. *High T_c Microwave Superconductors and Applications. SPIE Proc* **2156**, 13–20 (1994).

[225] Jackson C. M. and Bhasin K. B. High-temperature superconductivity for satellite communications applications. *High T_c Microwave Superconductors and Applications. SPIE Proc* **2156**, 21–26 (1994).

[226] Chaloupka H., Klein N., Peiniger M., Piel H., Pischke A. and Splitt G. Miniaturized high-temperature superconducting microstrip patch-antenna. *IEEE Trans. Microwave Theory Tech.* **39**, 1513 (1991).

[227] Chaloupka H. High temperature superconductor antennas: utilization of lowRF losses and of non-linear effects. *J. Superconduct.* **5**, 403 (1992).

[228] Chaloupka H. *et al.* High temperature superconductor meander antenna. *1992 IEEE–MTT-S Int. Symp. Dig.*, pp. 189–192 (1992).

[229] Zhang D. *et al.* Application of high T_c superconductors as frequency selective surfaces: experiment and theory. *IEEE Trans.* **MTT-41**(6/7), 1032–1036 (June/July 1993).

[230] Mansour R. R. *et al.* Design considerations of superconductive input multiplexers for satellite applications. *IEEE Trans.* **MTT-44**(7), 1213–1228 (July 1996).

[231]   Talisa S. H. High-temperature superconducting space-qualified multiplexers and delay lines. *IEEE Trans.* **MTT-44**(7), 1229–1239 (July 1996).

[232]   Talisa S. H. *et al.* High-temperature superconducting wide band delay lines. *IEEE Trans. Appl. Superconduct.* **5**(2), 2291–2294 (June 1995).

[233]   Wallage S., Thuritz J. L., Hadley P., Lander L. and Mooij J. E. High T_c superconducting CPW bandstop filters. *IEEE Microwave and Guided Wave Lett.* **6**(8), 292–294 (Aug. 1996).

[234]   Talisa S. H., Janocko A., Talvacchio J. and Moskiowitz. Present and projected performance of high-temperature superconducting filters. *1991 IEEE–MTT-S Dig.*, pp. 1325–1328 (1991).

[235]   Talisa S. H. *et al.* High-temperature superconducting four-channel filterbanks. *IEEE Trans. Appl. Superconduct.* **5**(2), 2079–2082 (June 1995).

[236]   Driscoll M. M. *et al.* Cooled, ultrahigh Q, sapphire dielectric resonators for low-noise, microwave signal generation. *IEEE Trans. Ultrasonics, Ferroelectrics and Frequency Control* **39**(3), 405–411 (May 1992).

[237]   Zhou S. A current view of electromagnetic missiles. *1992 IEEE AP-S Int. Symp.*, pp. 85–88 (1992).

[238]   Fulghum D. A. and Anselmo J. C. DAPRA pitches small sats for tactical reconnaissance. *Aviation Week & Space Technology* (9 June) 29–31 (1997).

[239]   Prokop J. S. Pre-competitive industry cooperation in the United States. *ICEMM Proc.* pp. 557–562 (1993).

2

From Array Theory to Shared Aperture Arrays

Good judgment comes from experience; experience comes from bad judgment.

Walter Wriston, CEO Citybank

In this chapter we shall erect the theoretical infrastructure that will allow us to characterize a number of arrays in terms of their topology and the essential characteristics of their RF subsystems. Array theory is shared by all array systems and the analysis of several arrays constitutes a first step toward the derivation of synthesis procedures for arrays having prescribed characteristics. Real world considerations that enable the designer to come one step closer to the realization of arrays are also delineated.

The marginalization of the array sidelobe and grating lobe levels, a central issue of this chapter, is taken up by considering theoretical approaches and real-world constraints imposed by current T/R modules. Marginalization here implies a minimization process consistent with affordable costs attributed to the tight control of significant array electrical and physical parameters. While the requirement for the tight control of the amplitude/phase weights of individual array elements needs no further elaboration, the control of other array parameters such as the physical temperature of its many subsystems, though mandatory and often de-emphasized by electrical engineers, is considered in this chapter.

Other importamt issues are also considered such as the derivations of the conditions under which the designer (i) can safely ignore the effects of mutual coupling between adjacent antenna elements, and (ii) can use programmable phase-shifters or switchable time delays to scan the resulting beam.

If the effects of mutual coupling cannot be ignored, procedures that marginalize their impact on the array radiation pattern are described. Approaches that account for the interaction between antenna elements and their matching structures as well as the phenomenon of array scan blindness are treated in Chapter 3.

Various linear, minimum redundancy, planar, multiplicative and circular/cylindrical arrays are considered before the challenges imposed by broadband arrays are explored. We have already made a case for broadband phased arrays or shared aperture arrays and defined their attractive system characteristics. Work toward other important array systems under development is also delineated in this chapter.

2.1. GENERAL CONSIDERATIONS

Radar arrays meet the dual requirement of having adequate spatial resolutions and power–aperture products. On the other hand, the availability of funds as a function of time has an impact on how the array is populated by antenna elements and T/R modules.

If all funds are available prior to the commissioning stage, the designer has an easy task. Alternatively, enough funds may be available prior to the commissioning stage that the required aperture is only partially populated at first and fully populated when the remaining funds become available.

More realistically, the designer categorizes the requirements into hard/soft (high/low priority) and costs arrays that meet all requirements or the hard requirements only. If funds are constrained, a decision is taken to realize an affordable array that meets all the hard requirements. Given the current financial climate, it is difficult to secure funds to realize systems that meet all requirements.

For radiometric arrays the essential requirements are the array's spatial resolution and sensitivity and the maximization of the baselines between interferometers. The latter requirement is satisfied by operating the array in the aperture synthesis mode. If system requirements dictate that the array will operate in the image synthesis mode, the resulting beam/s are corrected before use.

The following are some additional key topics we shall consider in this chapter:

- The dependence of array grating lobes and sidelobe level on the interelement spacing and when amplitude/space tapers are used.
- The influence of phase/amplitude errors occurring anywhere in the array on the array sidelobe level.
- Conventional and modern approaches to minimize the array sidelobe level; some of the modern approaches are based on real-world considerations related to current T/R modules.
- The important characterizations of working arrays.
- The management of the heat generated by the array's transmitters.

Throughout this chapter topics that are treated extensively in other books are considered synoptically here.

2.2. LINEAR ARRAYS

Linear antenna arrays constitute an excellent starting point to array theory because of the insights they lend into beamforming and the relationship between the array excitation functions and the resulting radiation patterns. We begin by considering arrays where the excitation currents for all antenna elements are the same and the interelement spacings are constant, aperiodic or random before we proceed to minimum redundancy arrays. Next we shall consider arrays where the excitation currents and phases of the antenna elements are chosen to yield radiation patterns having a set of prescribed parameters. Lastly, we shall consider arbitrary arrays that yield low sidelobe levels.

2.2.1. Uniformly Spaced Line Sources of Equal Amplitude

Let us consider the linear array, shown in Figure 2.1a, of N line sources operating in the transmit mode. Each line source is illuminated by a uniform illumination across the aperture of extent, d and the distance between consecutive line sources is s. The resulting far-field pattern of the array, $V(u_1)$, is the product of the far-field pattern corresponding to one line source, $G(u_1)$, and the array geometric factor, AGF, which constitutes the sum of the contributions attributable to the N line sources, or

$$V(u_1) = G(u_1) \times [\text{AGF}] \tag{2.1}$$

$G(u_1)$ can be derived from

$$G(u_1) = \int_{-d/2}^{d/2} F(x) \exp(-j2\pi u_1 x)\, dx$$

where $F(x)$ is the illumination function across the line source, $G(u_1)$ and $F(x)$ constitute a Fourier transform pair, $u_1 = -(\sin \theta)/\lambda$ and λ is the wavelength of operation. If $F(x)$ is a symmetrical even illumination, then $G(u_1)$ is given by

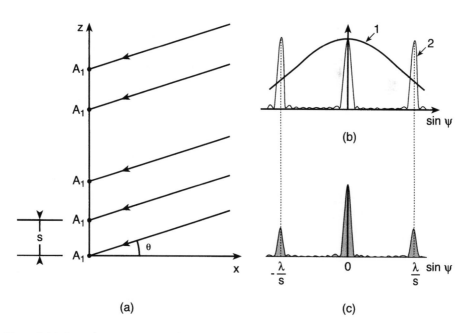

(a)

(b)

(c)

Figure 2.1 A linear phased array of equidistant line sources located at points (•). (a) The array geometry of N line sources. (b) The radiation pattern of the line source when it is uniformly illuminated is marked by 1, while the array geometric function (AGF) is marked by 2. The array specifications are $N = 10$ and $s = 2d$, where d is the linear extent of the line source. (c) The resultant array radiation pattern.

$$G(u_1) = 2 \int_0^{d/2} F(x) \cos(2\pi u_1 x)\, dx \tag{2.2}$$

which is the Fourier cosine transform of the symmetrical illumination. If $F(x) = 1$ the normalized far-field pattern is given by

$$G(u_1)\big|_{\text{uniform}} = \frac{\sin(\pi d u_1)}{\pi d u_1} \tag{2.3}$$

The total phase difference of the fields from adjacent line sources, ψ, is given by
$$\psi = ks \sin\theta + \beta \tag{2.4}$$

where β is the difference in phase excitation between elements and $k = 2\pi/\lambda$.

If we assume the same illumination for all line sources, the sum of all voltages attributable to the line sources, or the AGF, is given by

$$AGF = 1 + \exp j\psi + \exp j2\psi + \exp j3\psi + \cdots + \exp j(N-1)\psi$$
$$= \sum_1^{n=N} \exp j(n-1)\psi \tag{2.5}$$

If the reference point is the physical center of the array and the array beam is pointed toward the x-axis ($\beta = 0$), equation (2.5) is reduced to

$$AGF = \frac{\sin(N\psi/2)}{\sin(\psi/2)} = \frac{\sin(N\pi s u_1)}{\sin(\pi s u_1)} \tag{2.6}$$

and the normalized power of the resulting array radiation pattern, $P(u_1)$, is given by

$$P(u_1)\big|_{\text{uniform}} = \left[\frac{\sin(d\pi u_1)}{d\pi u_1}\right]^2 \left[\frac{\sin(N s\pi u_1)}{N \sin(s\pi u_1)}\right]^2 \tag{2.7}$$

In Figure 2.1b we illustrate the radiation pattern of the line source, marked as 1, and the AGF of an array that has 10 elements ($N = 10$) and when $s = 2d$, marked as 2. The resultant array pattern is shown in Figure 2.1c.

The following insights can be drawn from Figures 2.1b and c and equation (2.7):

- The first bracketed term, the radiation pattern of the line source, varies slowly by comparison to the second bracketed term, the AGF.
- The denominator of the second bracketed term varies slowly with respect to the numerator of the same bracketed term.

The grating lobes of the AGF, illustrated in Figure 2.1b, occur whenever the denominator of the second bracketed term in equation (2.7) goes to zero or whenever

$$\sin(s\pi u_1) = 0$$

$$s\pi u_1 = \pm m\pi \qquad \text{where } m = 1, 2, \ldots$$

which yields $u_1 = \pm 1/s, \pm 2/s, \ldots$ Thus the wider the spacing between line sources the closer the grating lobes are to the main beam.

Between the principal maxima there are the sidelobes, weak maxima followed by minima that occur whenever the numerator of the AGF goes to zero or whenever

$$\sin(N\pi s u_1) = 0$$

$$N\pi s u_1 = \pm n\pi \qquad \text{where } n = 1, 2, \ldots$$

which yields $u_1 = \pm 1/Ns, \pm 2/Ns, \ldots.$

To find the secondary maxima we derive

$$\frac{d}{du_1}[AGF] = 0$$

and obtain the relationship

$$N\tan(\pi s u_1) = \tan(N\pi s u_1)$$

which yields $u_1 Ns = u_1 L = 1.4306$, where L is the array length. The maximum of the first sidelobe of the AGF at this value is -13.26 dB and the envelope of the sidelobe levels follows the $1/u_1$ law. An array of N antennas will have $(N–1)$ nulls and $(N–2)$ secondary maxima.

The normalized AGF or

$$AGF\big|_{\text{norm}} = \frac{\sin[N\psi/2]}{N\sin[\psi/2]} \tag{2.8}$$

can be further simplified. For large arrays the main beam occupies a small solid angle at broadside, thus the sine of the denominator of equation (2.8), can be replaced by its argument. Therefore,

$$AGF\big|_{\text{norm}} \approx \frac{\sin[N\psi/2]}{N\psi/2} = \frac{\sin[L\pi u_1]}{L\pi u_1} = \sin cx \tag{2.9}$$

where L is the total array length and is approximately equal to Ns. This is a significant simplification for there is a large body of work derived from studies related to continuous line sources and filled apertures that is directly applicable to phased arrays.

Let us now derive the conditions for steering the array to a direction θ_0 by inserting a phase shift, β, between adjacent antenna elements. At an angle θ_0, the main beam is formed and θ_0 is in the range $0° \leq \theta_0 \leq 180°$. To accomplish this, β should be adjusted so that

$$\psi = ks \sin \theta_0 + \beta|_{\theta=\theta_0} = ks \sin \theta_0 + \beta = 0$$

or

$$\beta = -ks \sin \theta_0 \tag{2.10}$$

Thus, by changing β, the maximum of the radiation pattern can be directed toward any direction to form a scanning array.

If a linear phased array of omnidirectional antenna elements has an East–West orientation, the resulting main beam and grating lobes will extend from North to South.

A comparison of linear and planar arrays is convenient here:

- A linear array having omnidirectional elements is inexpensive to realize and yields radiation patterns of considerable angular extent. In reality the user of a linear array can define the angular position of a target or a source only in one dimension. For monostatic radar arrays, the target's range defines a sphere.
- Linear arrays 'see' a very large volume of space where many sources/targets/jammers are located at different ranges.
- Pencil beams derived from planar arrays are capable of locating a target in two coordinates. However, the cost and complexity associated with planar arrays are considerable.

The Ericsson's ERIEYE AEW system utilizes two colinear arrays mounted on top of a Fairchild Metro III airplane [1]. The system operates in the 2–4 GHz band and utilizes some 200 T/R modules. The resulting beam can be electronically scanned over either side of the aircraft over ±60° in azimuth. The AEW performance of the system is favorably compared to that of conventional track while scan radars [2].

2.2.1.1. Array Grating Lobes

We have already considered how the beam of a phased array can be scanned within the array surveillance volume and in this section we shall examine the issues related to the resulting grating lobes.

If a linear array is scanned to θ_0, the equation for the resulting array geometric factor is

$$\mathrm{AGF} = \frac{\sin[Ns\pi(\sin\theta - \sin\theta_0)/\lambda]}{N\sin[s\pi(\sin\theta - \sin\theta_0)/\lambda]} \tag{2.11}$$

Grating lobes occur whenever the denominator of equation (2.11) vanishes or whenever

$$\frac{s\pi(\sin\theta - \sin\theta_0)}{\lambda} = \pm\pi, \ \pm 2\pi, \ \pm 3\pi, \dots \tag{2.12}$$

If the angle of the first grating lobe is θ_g, we can deduce

$$\sin\theta_g = \sin\theta_0 \pm \frac{\lambda}{s}$$

or the stringent condition for no grating lobes to be

$$\left|\sin \theta_0\right| - \frac{\lambda}{s} < -1 \tag{2.13}$$

If the array is to be scanned to $\pm 90°$, $\sin\theta_0 = 1$ and $s < \lambda/2$. Thus the condition that the spacing between antenna elements is shorter than $\lambda/2$ is enough to place the grating lobes of the phased array at the horizon.

The impact of this constraint on the many parameters of an active array utilizing solid-state T/R modules is worth considering. At low frequencies, e.g. 1–3 GHz, the $\lambda/2$ criterion is easy to meet when current T/R modules are used and the array has maximum power–aperture product. At these frequencies the power added efficiency, PAE (see section 2.5.8 for a definition of PAE), of solid-state devices is high and the array generates a small amount of heat. Given that the space between adjacent antenna elements is more than adequate, the T/R module can be either MMIC-based or hybrid MIC-based.

As the frequency of operation increases to say 10 or 100 GHz, the PAE of solid-state devices decreases, more heat is generated by the array and the spacing between adjacent antenna elements becomes tighter. This being the case, only MMIC-based T/R modules are used. The heat generated by arrays operating at millimeter wavelengths is considerable and the seriousness of this problem is delineated in section 2.5.8.1.

The $\lambda/2$ spacing is sufficient to position the AGF grating lobes of arrays having regular interelement spacing at the horizon, and this is the criterion adopted by narrowband radar arrays. For bistatic wideband radar arrays, e.g. the OTHRs, the control of the resulting grating lobes is attained by recalling that the resulting array beam is the product of the radiation patterns of the receive and transmit arrays and that the interelement spacing for the receive and transmit arrays need not be the same—see section 2.7.1.1.

As radioastronomy arrays operate in the receive mode only, the criteria for grating lobes are different. For instance, the first-order grating lobes of the multielement crossed radiotelescope realized by Christiansen and his collaborators [3] were present and the instrument was used to derive high-resolution images of the quiet Sun at a wavelength of 21 cm. However, the effects of the resulting ambiguities (due to the presence of grating lobes) introduced negligible errors to the observations because the radio Sun is much brighter than other strong radio sources.

As the array pattern is the product of the AGF and $G(u_1)$, the far-field of the antenna element used, grating lobes can be tolerated if $G(u_1)$ is negligible at the angular position of the grating lobe. This condition is often met if one or more relatively large paraboloids are used in conjunction with an array—see section 2.3.1.

From the foregoing we can deduce that the management of the grating lobes has to be considered in conjunction with the intended array application.

2.2.1.2. The Beamwidth and Bandwidth of Phased Arrays

Let us consider a phased array that uses phase-shifters to scan the resulting beam anywhere within the array surveillance volume. Simple derivations of the beamwidth and bandwidth of an array have been outlined when the array illumination function is of a standard form [4].

If we substitute λ for c/f and introduce f_0 to represent the frequency used to determine the phase gradient into equation (2.11), we obtain

$$\text{AGF } (\theta, f) = \frac{\sin[N\pi s(f \sin \theta - f_0 \sin \theta_0)/c]}{N \sin[\pi s(f \sin \theta - f_0 \sin \theta_0)/c]} \qquad (2.14)$$

At f_0, the AGF(θ,f) reaches maximum at θ_0, and at any other frequency the maximum is reached at an angle given by

$$\sin \theta = \frac{f_0 \sin \theta_0}{f} \qquad (2.15)$$

The –3 dB beamwidth (or bandwidth) is defined as AGF(θ,f) = 0.707, if its maximum is equal to unity. For large arrays we can assume that $\sin(Nx)/[N \sin(x)] \approx [\sin(Nx)]/Nx$ and the half-power points of equation (2.14) are given by

$$\frac{N\pi s}{c} (f \sin \theta - f_0 \sin \theta_0) = \pm \frac{b}{2} \pi \qquad (2.16)$$

where b is the beam broadening factor which depends on the illumination function used. For uniform illumination, $b \approx 0.886$—see section 2.2.2.

The array beamwidth is determined by letting $f = f_0$ and using the approximation

$$\sin \theta - \sin \theta_0 = \frac{\theta - \theta_0}{\cos \theta_0} \qquad (2.17)$$

which is applicable for large arrays near the main beam. The HPBW is therefore given by

$$\begin{aligned} \text{HPBW} = 2(\theta_{3\text{dB}} - \theta_0) &= \frac{0.886c}{Nsf_0 \cos \theta_0} \\[2mm] &= \frac{0.886c}{Lf_0 \cos \theta_0} \qquad (2.18) \end{aligned}$$

where the array length L is taken to be equal to Ns. For an array of length L and frequency f, the HPBW increases as θ_0 increases. For this reason the maximum scanning angle of phased arrays seldom exceeds ±60°. Corroborative support to this design guideline is provided in section 2.3.

Similarly the 3 dB bandwidth can be deduced by letting $\sin \theta = \sin \theta_0$ and solving for $f = f_0$. Hence,

$$\text{Bandwidth (Hz)} = 2(f_{3\text{dB}} - f_0) = \frac{0.886c}{Ns \sin \theta_0}$$

$$= \frac{0.886c}{L \sin \theta_0} \tag{2.19}$$

For other illuminations the beam broadening factors, b, listed in Table 2.1 for different illuminations, can be used to derive the HPBW and bandwidth of the array.

For long phased arrays, and/or substantial scanning angles, the array bandwidth is necessarily narrow. If wideband operation is required, switchable time-delays are used in lieu of switchable phase-shifters.

2.2.1.3. Array Directivity

Directivity is a measure of how well the array directs energy toward a particular direction. If the antenna elements are isotropic, the directivity, D, is solely defined by the AGF. For a linear array operating in the transmit mode, the AGF is defined as the radiated power density in the direction of the main beam maximum divided by the average power density from the array. For a symmetrical array, the array directivity, D, simplifies to

where the spacing between the antenna elements is $\lambda/2$ and A_n are the amplitudes of the

$$D = \frac{\left[\sum_{-n_z}^{n_z} A_n\right]^2}{\sum_{-n_z}^{n_z} A_n^2} \tag{2.20}$$

contributions from the n_z to $-n_z$ antenna elements positioned along the z-axis [5].

The numerator is proportional to the total coherent field squared, whereas the denominator is the sum of the squares of the individual fields from each element.

The directivity is independent of the scan angle and for a uniformly illuminated linear array $D = 2L_x/\lambda$, where L_x is the length of the linear array; this is the maximum directivity obtained by a linear array.

For uniform illumination, the product of the HPBW, expressed in degrees, and directivity at broadside is constant and is given by

$$D \times [\text{HPBW}] \approx 101.53$$

2.2.1.4. Array SNR Gain

We wish to derive the array SNR gain when the array has N antenna elements. To this end let us assume the following:

- The input signal level, proportional to the envelope of the RF signal voltage in the antenna elements, is equal to S_{IN}.
- The input noise power is N_{IN} and the noise from element to element is independent.
- All signals are co-phased by a set of phase-shifters.

Under these assumptions, the output signal voltage, S_{OUT}, is equal to NS_{IN} and the output signal power is equal to $N^2 S_{IN}$. Similarly, the output noise power, N_{OUT}, is equal to NN_{IN}. The array SNR gain, $G|_{SNR}$ is therefore given by the equation

$$G|_{SNR} = \frac{SNR|_{OUT}}{SNR|_{IN}} = \frac{1}{SNR|_{IN}} \frac{S_{OUT}}{N_{OUT}} = N \qquad (2.21)$$

The product of the array G_{SNR} and array beamwidth (at boresight) for a long linear array having an interelement spacing of $\lambda/2$ is equal to 2. The assumptions made here are that the array illumination is uniform and that the length of the long array is $N(\lambda/2)$. Similarly, for a planar array the array SNR gain is equal to 4 when the same assumptions are accepted.

2.2.1.5. Mutual Coupling between Antenna Elements

So far we have assumed that the radiation pattern of each antenna element of an array is identical, an assumption that served the purposes of initiating discussion. In a real array environment, mutual coupling exists between elements, which alters each element's radiation pattern. The electromagnetic environment of an antenna element in isolation is different from that prevailing when the same element is placed near the array center or at the array's perimeter. The presence of surrounding elements alters the current distribution on each element and the field radiated by an excited element is dependent on the induced currents on other elements as well as its own.

The interaction between elements quickly falls to zero as elements become widely spaced. At a separation greater than one wavelength, the interaction is usually considered negligible [6]. In an array where the interelement spacing is half a wavelength, significant couplings exist between an element and its nearest and nearest-but-one neighbors.

For an array having over 1000 elements, the same considerations apply but the majority of the elements are considered to have the same 'embedded' radiation pattern [7]. For the typical arrays we have considered in Chapter 1, comprising 5000 to 10 000 elements, the effects of mutual coupling can to a first approximation be ignored and the essential array parameters are derived by conventional methods.

For small arrays, several methods have been reported that offset the effects of mutual coupling between the antenna elements on the resulting array radiation pattern [6–11]. The approach outlined in references [11] and [9] deserves some further consideration and is based on the following observation. Assuming an array operating in the receive mode, the individual antenna element signal has several components: a dominant component due to the direct incident plane wave and several less dominant components due to scattering of the incident wave at neighboring elements. These components are resolved and scattering is compensated for by linear transformation, which is accomplished by a matrix multiplication performed on the element output signals. As this compensation is scan independent, the matrix is fixed and applies for all required patterns and scan directions.

The above approach has been successfully applied to an eight-element linear array operating at X-band. The measured and theoretical array radiation patterns before and after the application of the compensation for mutual coupling are shown in Figures 2.2a and b respectively [9]. The array was designed to have a theoretical sidelobe level below –30 dB. As can be seen, the compensation used marginalizes the effects of mutual coupling between antenna elements.

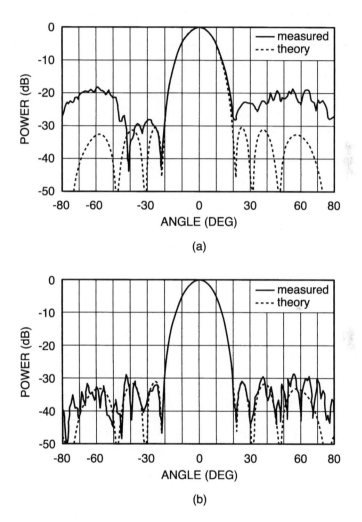

Figure 2.2 Radiation patterns of an eight-element phased array. (a) Measured and theoretical radiation patterns without any compensation for the effects of mutual coupling between antenna elements. (b) Measured and theoretical radiation patterns when compensation for the effects of mutual coupling between antenna elements is implemented. (From [9]; © 1990, IEEE.)

2.2.2. Aperiodic and Random Arrays

The sidelobe level of linear arrays can be reduced by implementing either an amplitude or spatial taper along the length of the array. When an array operating in a transmit mode has an amplitude taper, the T/R modules located in the vicinity of the array center transmit maximum power while the T/R modules located away from the array center transmit powers inversely proportional to their distance from the array center. If a spatial taper is adopted, all T/R modules emanate the same power but the density of antenna elements decreases progressively from the array center to the ends of the array. Both methods entail some power losses as a trade-off for the attainment of low sidelobes.

Table 2.1 lists the magnitude of the first and second sidelobe levels when a line source, of length L, has different illuminations or amplitude tapers. The resulting HPBW factors, b, of λ/L are also listed. While the uniform illumination yields the narrowest beamwidth, the resulting sidelobes are unacceptably high. For other illuminations/tapers the sidelobe level decreases but the beamwidth increases. As can be seen, the cosine squared on a 10 dB pedestal illumination/taper yields reasonable sidelobe levels and a narrow HPBW.

Table 2.1 The essential characteristics of a line source of length L having different illuminations

Illumination	First sidelobe level (dB)	Second sidelobe level (dB)	HPBW factor b of λ/L
Uniform	−13.3	−17.8	0.886
Cosine	−23	−30.7	1.19
(Cosine)^2	−31.7	−41.5	1.44
(Cosine)^2 on a 10 dB pedestal	−20	−24	1
Parabolic	−21.3	−29	1.15
(Parabolic)^2	−27.7	−37.7	1.38

Aperiodic arrays result when the approaches taken to reduce the array sidelobes are deterministic. In more general terms, some aperiodic arrays, such as the minimum redundancy arrays, have the attractors considered in section 2.2.2.2.

In what follows we shall explore several approaches for the minimization of array sidelobes that lend useful insights into phased arrays. The defining characteristic of these arrays is that the interelement spacing is aperiodic and an amplitude taper may or may not be used.

If N uniformly spaced line sources of equal intensity are arranged symmetrically with respect to the origin of a coordinate system, the AGF_u can be written as

$$\mathrm{AGF}_u = \frac{2}{N} \sum_{n,\,\mathrm{odd}}^{N-1} \cos\left(\frac{n}{2}u\right) \qquad n = 1, 3, 5 \ldots, (N-1) \qquad (2.22)$$

where

$$u = \frac{2\pi}{\lambda} \, s \sin \theta \sin \phi$$

and θ, ϕ are the spherical coordinates.

As can be seen, the AGF_u is the sum of cosines that have different frequencies but the same phase. Sidelobes can therefore be reduced if the array elements are shifted slightly and the cosines have different phases.

If each line source is allowed to shift by a small distance, δ_n, the corresponding AGF is given by

$$AGF_n = \frac{2}{N} \sum_{n, \, odd}^{N-1} \cos\left[\left(\frac{n}{2} + \delta_n\right)u\right] \tag{2.23}$$

Using equations (2.22) and (2.23) it has been shown [12,13] that

$$AGF_u - AGF_n = -\frac{2u}{N} \sum_{n, \, odd}^{N-1} \delta_n \sin\left(\frac{n}{2}u\right) \tag{2.24}$$

The evolution of an array having minimal sidelobes therefore involves the following steps:

- The AGF_u corresponding to an array of equally spaced line sources of equal intensity is taken as a starting point of the iteration process.
- The AGF_1 of an array having slightly lower sidelobes is drawn and the fraction of reduction at points near the maxima and minima are noted.
- Equation (2.24) is used to write a set of $N/2$ equations in $\delta_1, \delta_3, \delta_5, \delta_7, \ldots$
- The solution of this set of equations yields the values of $\delta_1, \delta_3, \delta_5, \delta_7, \ldots$, which are used to calculate AGF_2.
- AGF_2 is used in conjunction with AGF_3 to initiate another iteration. Again the sidelobe levels corresponding to AGF_3 are lower than those attributed to AGF_2.
- Fewer than 5 iterations are usually required.

A numerical example will illustrate the reductions on sidelobe levels achieved when an array of 24 nonuniformly spaced line sources of equal intensity is designed and when the same array has a cosine amplitude taper.

Table 2.2 lists the sidelobe levels and half-power beam broadening ratios reported in reference [13] for uniformly/nonuniformly spaced arrays when the arrays had no taper or a cosine taper. As can be seen, the designer can decrease the array sidelobe level by some 10 dB by adopting nonuniform spacing for the array elements. Beam broadening increases as the sidelobe level decreases.

The same procedure can be applied to planar arrays. Given that the minimum spacing between antenna elements that ensures that mutual coupling between elements is negligible cannot be violated, the resulting arrays often have fewer elements than arrays in which the antenna elements are uniformly spaced. This design approach therefore offers the designer another option of achieving low sidelobe levels with fewer antenna

Table 2.2 The sidelobe levels and half-power beam broadening ratios[a] for different arrays

Case	Sidelobe level (dB)	Half-power beam broadening ratio
Uniform spacing, no amplitude taper	−13.26	1
Nonuniform spacing, no amplitude taper	−24.2	1.16
Uniform spacing, cosine taper	−23	1.34
Nonuniform spacing and cosine taper	−32	1.58

[a] The beam broadening factor when compared to that attained by a uniformly illuminated array.

elements. A different approach to low-sidelobe arrays is outlined in reference [14]. The proposed arrays have equispaced elements located near the array center and nonuniform spacing between elements located at some distance away from the array center. With this arrangement the array spatial taper approximated a Gaussian curve. The sidelobe for a 20-element array was about −20 dB and the approach allows the designer some flexibility in the population of the array aperture to meet not only the sidelobe requirement but also the transmitted power requirement.

Wideband aperiodic arrays having a small number of elements can be derived by using multivariable optimal search methods [15,16] to reach a minimal sidelobe level at the highest operating frequency of an array. For a 15-element array, a sidelobe level below −25 dB is reached when amplitude taper is used only for the two extreme antenna elements of the array. Furthermore, the sidelobe level at all lower frequencies remained unaltered.

In Figure 2.3a we illustrate the radiation pattern of a 20-element wideband array when the interelement spacing was varied between half and one wavelength at the

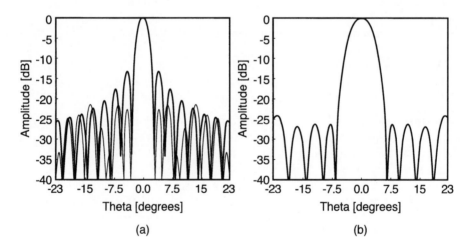

(a)

(b)

Figure 2.3 A 20-element phased array having elements spaced between 0.5 and 1 wavelength. The resulting sidelobe level is minimized at the shorter wavelength of operation and no amplitude taper is used. (a) The initial array geometric factor (solid lines) and the resulting pattern after 13 iterations (lighter lines). (b) The resulting array geometric factor at half the highest frequency of operation.

highest frequency of operation to minimize the array sidelobe level and no amplitude taper is used. The initial and resulting radiation patterns obtained at the highest frequency after 13 iterations are shown in the figure. As can be seen, the array sidelobe level was initially below −13 dB and dropped to below −20 dB after 13 iterations when the angle θ with respect to the array boresight varies between −23° and +23°.

After 26 iterations the sidelobe level of the same array dropped below −27 dB. In Figure 2.3b we show the radiation pattern obtained at half the highest operating frequency after 26 iterations. The array sidelobe level near the main beam is below the −25 dB level. As the frequency decreases, only a portion of the radiation pattern obtained at the highest frequency is seen in the same range of angles. This approach is particularly suitable to wideband phased arrays having a small number of elements. Other techniques that we shall explore in section 2.2.4 are applicable to large arrays.

2.2.2.1. Random Distribution Arrays

An aperiodic distribution of antenna elements can be based on deterministic algorithms (see previous section and section 2.2.2.2) or on the random placement of antenna elements first proposed by Lo [17]. Steinberg further considered random arrays [18] and compared the resulting sidelobes of aperiodic and random arrays. On the average the worst sidelobes of the latter arrays were not inferior to those corresponding to the former [19].

The ratio of the average sidelobe power, ASL, to the mainlobe power, ML, for a random array of N antenna elements having uniform excitation is given by

$$\frac{\text{ASL}}{\text{ML}} = \frac{1}{N} \tag{2.25}$$

The peak sidelobe level of a linear random array, however, is higher than the ASL level by an amount that is proportional to the array length. While it is relatively easy to locate the peak sidelobe level of linear arrays, it is difficult to locate the peak sidelobe levels of planar arrays.

Table 2.3 lists the peak sidelobe level, in dB above the ASL level for linear arrays having lengths of 25 and 5000 wavelengths [20]. Efforts to minimize the resulting peak sidelobes of the random array resulted in the difference sets arrays we shall explore in section 2.2.4.3.

Table 2.3 Peak sidelobe level above the ASL level for linear random and difference sets arrays of different lengths

Array length (λ)	Random array (dB)	Difference sets array (dB)
25	7.7	1.3–4.4
5000	10.2	4.2–7.2

2.2.2.2. *Minimum Redundancy Arrays*

For a given number of antenna elements, minimum redundancy arrays (MRAs) have the highest spatial resolution possible and yield the maximum number of spacings. The first MRAs were used for radioastronomical observations [21–23]. Additionally, the Earth's rotation can be used to obtain more spacings [24] when the array operates in the aperture synthesis mode. MRAs are used for many applications where the highest spatial resolution is required from arrays of finite lengths [23–32] and theoretical work on minimum or null redundancy arrays (M/NRAs) continues unabated [33–36].

In applied science radiometric applications the array length is determined by the dimensions of an airborne platform, while for radioastronomy applications the array length is constrained by costs. Additionally, M/NRAs are used to perform the following radar-related functions:

- Direction of arrival, DOA applications [29]
- Interference cancellation applications [29,30]
- Adaptive beamforming [32]

The canonical array shown in Figure 2.4a forms a beam by coherently adding the powers of its antenna elements. As can be seen, the array has too many redundant spacings depicted in Figure 2.4b. The shortest spacing, for example, occurs six times, the

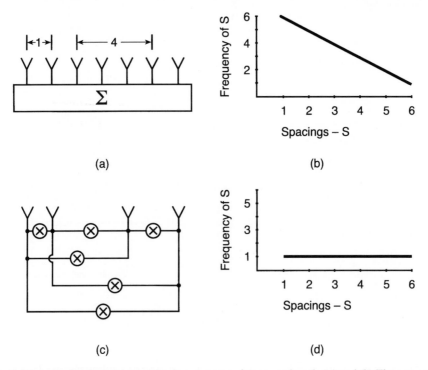

(a) (b)

(c) (d)

Figure 2.4 Canonical and minimum-redundancy arrays of the same length. (a) and (b) The geometry and spatial sensitivity of a canonical array. (c) and (d) The geometry and spatial sensitivity of a minimum redundancy array. The symbol ⊗ is used to denote the correlation of signals received by two antenna elements.

second shortest spacing five times, and the longest spacing only once. For the shortest spacing, $S=1$, while the spacing S between four antennas is 4.

By comparison, the MRA depicted in Figure 2.4c forms a beam by correlating the outputs of the four antennas and all spacings S occur once only, as shown in Figure 2.4d. In the same figure the symbol $\otimes$ is used to denote the correlation of signals received by two antenna elements. In a more general case, each interferometer pair yields a complex number, and the phased array yields many complex numbers that are values of the Fourier transform of the observed brightness distribution function.

As we have seen, canonical arrays are relatively easy to realize, their spectral sensitivity contains a predominance of short spacings, and their resulting radiation pattern has the form of $[\sin(Nx)/N\sin(x)]$. By contrast, the powers received by M/NRAs are correlated, their spectral sensitivity is uniform, and the resulting radiation pattern has the form $[\{\sin(2Nx)/2N\sin(x)\}+$ constant] [22].

While it is relatively easy to define the spacings between antenna elements of M/NRAs when the number of antenna elements is low, e.g. 4 or 5, considerable research has been channeled toward the definition of the spacings between elements of M/NRAs when the number of elements, N, is considerably higher [22,33,34]. Element spacings when N is equal up to 30 have been computed [34]. Table 2.4 lists some MRAs having 3–7 elements defined by their lengths, L, number of elements, N, and interelement spacings, S. In the same table, the redundancy factor defined by N^2/L is listed for the MRAs.

Table 2.4 Some M/NRAs defined by their length, L, number of elements, N, and the interelement spacings

No. of elements N	Length L, units	N^2/L	Interelement spacing
3	3	3	1, 2
4	6	2.67	1, 3, 2
5	9	2.78	1, 3, 3, 2
5	9	2.78	3, 4, 1, 1
6	13	2.77	1, 3, 1, 6, 2
6	13	2.77	1, 1, 4, 4, 3
6	13	2.77	1, 5, 3, 2, 2
7	17	2.88	1, 3, 6, 2, 3, 2
7	17	2.88	1, 1, 4, 4, 4, 3
7	17	2.88	1, 1, 1, 5, 5, 4
7	17	2.88	1, 1, 6, 4, 2, 3
7	17	2.88	1, 7, 3, 2, 2, 2

Most linear radioastronomy arrays have minimum redundancy spacing between the various antenna elements and the compact array of the Australia (radio)Telescope [37], commissioned in1988, is an example of a one dimensional array. Located near Narrabri, about 600 km north-west of Sydney, Australia, the array consists of six 22-m diameter fully steerable antennas with a maximum East–West baseline of 6 km. Five of them are transportable on a 3-km track with 35 stations on which antennas can be positioned. The sixth antenna is located a further 3 km to the west and is transportable to either of two positions 75 m apart. Earth rotation is used to form images of southern sources at several frequency bands and the compact array can be 'connected' to other antennas located in Australia or in other continents to form long or VLBI systems.

The M/NRAs are natural solutions to the problems of improving angular resolution for applications in microwave radiometer Earth remote sensing [25–28]. This is because the maximum number of Fourier spacings are obtained instantaneously for the formation of the snapshot image when the arrays operate in the image synthesis mode.

Recently an aircraft prototype system, referred to in Chapter 1 as ESTAR, operating at 1.4 GHz has been built and flown [28]. Its array consists of eight 'stick' antenna elements and each stick consists of a linear array of eight crossed dipoles. All dipoles on one stick are combined and the output of pairs of stick antennas are cross-correlated to yield a narrow beam of ±6° along the direction of flight, while the beam along the cross-track dimension is ±45°. All cross-correlations are measured simultaneously and the resulting fan beam is swept along the Earth's surface in push-broom fashion [28]—see Figure 1.13b. The measured minimum detectable signal Δt for the above system is 0.53 K and compares favorably with the theoretical value of 0.32 K [27].

The sidelobe levels of MRAs typically vary between –5 and –10 dB when the number of elements varies from 9 to 21 [38]. The sidelobe suppression of MRAs over a range of specified angular distances has been reported in reference [38] by the use of a method outlined in reference [39]—see section 2.2.4. As the number of antenna elements of MRAs is low, the suppression of the sidelobes at all angular distances, on either side of the main beam, is difficult to implement. More specifically for an N-element array, the number of sidelobes one can suppress is N–1. The sidelobe minimization is implemented by deriving optimum weights for each antenna element so that the specified sidelobe level is met.

An 11-element MRA had a sidelobe level of –6.4 dB and was suppressed to –30 dB over the angular ranges of ±13° on either side of the main lobe [38].

Searches for minimum redundant sets with the required overall array length rather than one with the maximum length yield minimum redundant sets having a variety of thinning factors [40].

MRAs and their variants, e.g. minimum missing spacings arrays [41] or minimum hole arrays [42], are members of a family of arrays based on restricted difference bases [42]—see section 2.2.4.3.

2.2.3. Uniformly Spaced Line Sources of Unequal Amplitude

In this section we shall consider arrays that have uniformly spaced line sources symmetrically located about the array center and a symmetrical amplitude taper with respect to the array center. The taper has a variety of shapes and the resulting arrays have different radiation patterns that are defined primarily by the array sidelobe level and by the resulting beamwidth. Other requirements of interest are: (i) the illumination functions should preferably be defined by one parameter, so that optimal arrays can be readily derived; and (ii) the array sidelobe should have a $1/u_2$ envelope, a condition that provides a robust low-Q distribution [43]; here we have defined $u_2 = L(\sin\theta)/\lambda$ and L as the array length.

We have already noted a variety of amplitude tapers defined by analytic functions; we have also listed the resulting first and second sidelobe levels as well as the array HPBWs in Table 2.1. Given that the resulting sidelobes and HPBWs for a variety of illuminations are widely listed, the designer selects the illumination taper that meets requirements. Although these functions are not particularly suitable for phased arrays because they are defined by two or more functions, they do lend a qualitative insight to the illumination functions required for phased arrays.

If the required sidelobe level is known, synthesis procedures exist that yield the required amplitude weights on the array's line sources. The most popular of these procedures, based on the Taylor and Bayliss illumination functions, are widely used and tables exist to aid the practicing engineer to derive the required arrays.

It is seldom that the array requirements are defined in terms of the parameters we have already considered. The additional requirements can be the specification of the sidelobe topology on either side of the main lobe or in the far-out region, and null specifications toward directions where unintentional jammers exist. An interrelated requirement is often the maximization of the main beam efficiency.

As the designer has to meet an ever-increasing set of requirements, more degrees of freedom are needed. The additional degrees of freedom at the designer's disposal can be the positions of the line sources and their excitation defined by amplitude and phase weights. We have already seen that the array sidelobe can be reduced by varying the positions of the array elements and by the application of an amplitude taper. We shall consider in section 2.2.4 some of the mathematical approaches and techniques developed that allow the designer to arrive at an arbitrary array that meets a number of the above array specifications.

Even at this early stage we stress that the derived arrays are of a theoretical nature and that the designer has to consider important real-world considerations that we shall explore in section 2.5 before the arrays are realized.

2.2.3.1. General Considerations

We have considered arrays having line sources of equal amplitude spaced uniformly along the array. The positions of the array elements were symmetrical about the array center and equation (2.22) is a generalized equation of the resulting AGF. If the amplitudes of the symmetrical array elements are $A_1, A_3, A_5, \ldots, A_{N-1}$, equation (2.22) can be rewritten as

$$\text{AGF}\big|_{\text{unequal amp}} = \frac{2}{N} \sum_{n,\,\text{odd}}^{N-1} A_n \cos\left(\frac{n}{2} u\right) \tag{2.26}$$

where $n = 1, 3, 5, 7. \ldots, (N-1)$. By assigning different values for A_n, different AGFs are attained. If the amplitudes of the array elements are proportional to the coefficients of the binomial series [44], binomial arrays result. The coefficients, A_n, of binomial arrays having 3, 4, 5, 6, 7, and 8 elements are:

$$1, 2, 1$$
$$1, 3, 3, 1$$
$$1, 4, 6, 4, 1$$
$$1, 5, 10, 10, 5, 1$$
$$1, 6, 15, 20, 15, 6, 1$$
$$1, 7, 21, 35, 35, 21, 7, 1$$

respectively. When the spacing between antenna elements is $\lambda/4$ or $\lambda/2$, the resulting arrays have no sidelobes [45]. Although it is desirable for an array to have no sidelobes, the ratio of maximum to minimum values of the A_n coefficients is much too high and it increases as the number of elements increases. For an eight-element array, for instance, the ratio is 35:1. Large binomial arrays have the following drawbacks:

- Small errors in the A_n coefficients result in non-binomial array distributions.
- Problems related to mutual coupling between antenna elements arise owing to the large ratios of the A_n coefficients.
- The total power transmitted by the array is low.

As can be seen, the designer pays a high penalty to completely eliminate the array side-lobes. To the extent that these illuminations lend insights into the high price a designer has to pay to eliminate array sidelobes, binomial illuminations are theoretically important.

2.2.3.2. Dolph–Chebyshev Synthesis

The Dolph–Chebyshev synthesis procedure for arrays having prescribed characteristics evolved by Dolph [46] and others [47–50]. According to this synthesis procedure, the designer stipulates the required array sidelobe level and the A_n coefficients are calculated to meet the specification following procedures that are now described in many antenna textbooks. All the sidelobes of the resulting theoretical array are at the same prescribed level and the synthesis procedure has two important attractors, one of theoretical significance and the other of practical value.

(i)   For any array sidelobe level, the resulting arrays have a minimal beamwidth, specified by the beamwidth between the first two nulls; conversely, if the beamwidth is specified, the resulting sidelobe level is minimized.

(ii) The ratio of the maximum to minimum values of the A_n coefficients is not excessive.

Irrespective of the selected sidelobe level, the designer is assured that the array has a minimal beamwidth. Here we recall that minimum redundancy arrays yield the narrowest beamwidth. The second attractor can be appreciated by the following comparisons.

Table 2.5 lists the A_n coefficients of a binomial and a Dolph–Chebyshev 8-element array designed to have a –26 dB sidelobe level [51]. The ratio of the maximum to minimum A_n coefficients is 35:1 and 3.1:1, respectively, a significant difference.

The three essential steps in the design procedure for arrays having the Dolph–Chebyshev illumination are:

Table 2.5 The A_n coefficients for a binomial and a Dolph–Chebyshev array

Binomial array	1	7	21	35	35	21	7	1
Dolph–Chebyshev	1	1.7	2.6	3.1	3.1	2.6	1.7	1

- The rewriting of the AGF equation satisfying the condition that the array has the required sidelobe level.
- The derivation of several equations relating the A_n coefficients of the derived AGF with the coefficients of the well-known Chebyshev polynomials in the region where the functions oscillate.
- The derivation of the array coefficients, A_n, from the derived equations.

Detailed design procedures for Dolph–Chebyshev arrays are outlined in many references including [51] and [45]. As the array sidelobe level does not follow the $1/u_2$ envelope criterion, these arrays are only occasionally used [43].

2.2.3.3. Taylor Syntheses

Let us revisit the array of Figure 2.1a when each line source has an amplitude A_n and ψ is given by the equation (2.4). Under these conditions we can rewrite equation (2.5) as

$$AGF = A_1 + A_2 z + A_3 z^2 + \cdots + A_N z^{N-1} \qquad (2.27)$$

where $z = \exp j\psi = \exp j(ks \sin \theta + \beta)$. The rhs of equation (2.27) is an $(N-1)$ degree polynomial and if its roots are $z_0, z_1, z_2, \cdots, z_{N-1}$, we can rewrite it as

$$AGF = A_n(z - z_0)(z - z_1) \cdots (z - z_{(N-1)})$$
$$= A_n \prod_{n=1}^{N} (z - z_n) \qquad (2.28)$$

Given that $z = |z| \exp j\psi = |z| \angle \psi = 1 \angle \psi$, z lies on a unit circle, usually referred to as the *Schelkunoff circle* [52] and the values of s, θ, and β determine its phase. As s is determined by the requirement to eliminate the array grating lobes, and θ is an independent variable, β is the variable at the designer's disposal.

The Schelkunoff circle is illustrated in Figure 2.5a together with six roots of a generalized distribution. The product of the distances $zz_1, zz_2, zz_3, zz_4, zz_5, zz_6$ for any value of ψ is the measure of the absolute value of the field pattern. As ψ increases from $A(1,0)$, the array radiation pattern is defined. If all the roots are on the circle, the resulting pattern consists of lobes interspersed by deep nulls. If all the roots are not on the circle, the resulting array pattern is devoid of nulls. Lastly, if a set of roots is off the circle followed by a set that is on the circle, we will get a shaped beam in one region followed with sidelobes in the other region. The topology of these zeros on the Schelkunoff circle, therefore, determines the array's geometric factor. Many roots closely spaced on the Schelkunoff circle in the vicinity of ψ_1 ensure low sidelobe levels in the vicinity of that angular distance.

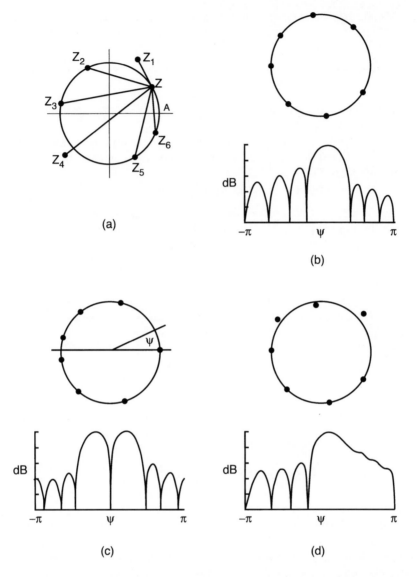

Figure 2.5 (a) The six roots of the array geometric factor on the Schelkunoff circle. (b) The roots and
the resulting sum radiation pattern. (c) The roots and the resulting difference radiation pattern. (d)
The roots and the resulting shaped pattern. ((b), (c) and (d) (From [58]; © 1988, IEEE.)

In Figures 2.5b, c and d, we illustrate the Schelkunoff circle corresponding to a sum,
a difference and shaped radiation pattern, respectively. As can be seen from Figures 2.5c
and d, the array sidelobe levels need not be symmetrical about the main beam and
depend solely on the positions of the array zeros.

Taylor's one-parameter synthesis procedure is based on the following observations:

- The sidelobes of a uniformly illuminated array have the required $1/u_2$ envelope that
  ensures robustness.

- The level of the first few sidelobes corresponding to a uniformly illuminated array is too high but can be decreased by an appropriate rearrangement of the zeros of the array's geometric function around the angular distance where the high sidelobes occur.

Hansen [43] succinctly outlined the design procedure to realize Taylor arrays and here we shall note the essential elements of this procedure. The positions of the close-in zeros are given by

$$U_2 = \sqrt{n^2 + T^2}$$ (2.29)

where T is a single parameter that defines all the array parameters. The resulting pattern is a modified sinc(x) function and is expressed in two forms:

$$F(u_2) = \frac{\sin \pi \sqrt{u_2^2 - T^2}}{\pi \sqrt{u^2 - T^2}} \qquad u_2 \geqslant T$$ (2.30)

$$F(u_2) = \frac{\sinh \pi \sqrt{T^2 - u_2^2}}{\pi \sqrt{T^2 - u_2^2}} \qquad u_2 \leqslant T$$ (2.31)

The transition from the sinc pattern to the hyperbolic form occurs at $u_2 = T$ on either side of the beam. The sidelobe level (SL) in dB and the aperture distribution are given by

$$|SL|_{dB} = 13.26 + 20 \log \frac{\sinh \pi T}{\pi T}$$ (2.32)

and

$$g(p) = I_0 \left(\pi T \sqrt{1 - p^2} \right)$$ (2.33)

where p is the distance from the center of the aperture to either array end as a fraction of half the array length and I_0 is the modified Bessel function of the first kind and order zero.

As the required sidelobe level is known, T can be derived from equation (232); once T is derived, all other parameters of the array are derived from the above equations. Table 2.6 lists the range of values T can take and the corresponding SL level; the aperture excitation efficiency, η, and main beam efficiency, η_b, are also tabulated. From Table 2.6 we observe the following:

- As the beam efficiency increases the excitation efficiency decreases.
- As the array sidelobe level decreases the beamwidth widens.
- The value for T is seldom above 2.2.

Taylor's one-parameter distribution resembles the cosine on a pedestal distribution and both distributions have comparable beamwidths. The additional advantage of Taylor's distribution, however, is that the designer can stipulate any sidelobe level and derive the parameters of the array readily.

Equation (2.32) cannot be readily solved, but the approximate relationship between T and the required sidelobe level is given by [53]

$$\text{SL}\big|_{dB} = c + \sqrt{a^2 + \left(1 + \frac{T^2}{b^2}\right)} \qquad (2.34)$$

where $c = -9.7$, $b = 0.9067$, and $a = 13.26-c$. This approximation results in worst-case errors of only ± 0.1 dB in the calculation of $g(p)$ when the SL is as low as -82 dB. The approximation therefore provides a closed-form calculation of all other array parameters to a considerable accuracy.

The application of Taylor's one-parameter distribution to phased arrays requires the sampling of the derived illumination functions at the appropriate intervals. A more accurate approach involving null-matching has been outlined by Elliott [54].

Table 2.6 The essential parameters of Taylor's one-parameter distribution [43]

$\text{SL}\big|_{dB}$	T	HPBW/2 (rad)	η	η_b
−13.26	0	0.4429	1	0.9028
−20	0.7386	0.5119	0.933	0.982
−30	1.2762	0.6002	0.8014	0.9986
−35	1.5136	0.6391	0.7509	0.9996
−40	1.7415	0.6752	0.7090	0.9999
−45	1.9628	0.7091	0.674	1
−50	2.1793	0.7411	0.6451	1

Taylor's $\bar{n}$ distribution is a two-parameter distribution that offers a modest improvement in efficiency over the one-parameter distribution and narrower beamwidths [55]. Reference [43] and the references therein provide analytic expression for the many array parameters when Taylor's $\bar{n}$ distribution is adopted. Taylor's ideal line source sum pattern is given by

$$F(u_1) = \sin c(\pi u_2) \sum_{n=1}^{\bar{n}-1} \frac{1 - u_2^2/z_n^2}{1 - u_2^2/n^2} \qquad (2.35)$$

The first $\bar{n}$ roots of the sinc function are again substituted by new roots while the remaining roots of the sinc function are retained. Tables relating the parameters of the $\bar{n}$ Taylor distribution are given in reference [43] and in the references therein. As $\bar{n}$ increases, the Dolph–Chebyshev distribution is approached.

The two Taylor distributions are widely used and together with the Bayliss distributions constitute the industry's standards. Other synthesis procedures are outlined in reference [43].

2.2.3.4. Bayliss Synthesis

We have already considered difference pattern distributions often used in conjunction with sum patterns for target tracking. For the one-dimensional case, the requirements are: (i) a null in the direction of a target/source; (ii) two lobes on either side of the target/source; and (iii) low sidelobes in all other directions. So long as the target is in the null of the difference pattern, no error signals are generated. If a target's position is on either lobe, error signals are generated that move the array null to coincide with the target's position.

Approximations to the required difference patterns are many and here we shall consider only an essential set that includes theoretically attractive and the most popular members of the set.

In general, difference patterns are generated when the array is illuminated by skew-symmetrical functions and the sine Fourier transform is used to derive the radiation when the illumination function is known. For a line source of length, a, the resulting radiation pattern $G(u_1)$ is given by

$$G(u_1) = -2j \int_{-a/2}^{a/2} F(x) \sin(2\pi u_1) \, \mathrm{d}x$$

where $u_1 = -\sin(\theta)/\lambda$ and θ is an angle measured with respect to the array boresight axis. If $F(x) = 1$,

$$G(u_1)\big|_{\text{unif}} = a \, \frac{1 - \cos \pi a u_1}{\pi a u_1} \qquad (2.36)$$

While $G(u_1)$ has the general characteristics of the required difference patterns, its first sidelobe is unacceptably high, at the -10.57 dB level. The ideal difference pattern is the Rayleigh function $G(u_1)$ that results when the illumination function is another Rayleigh function, $F(x)$. And $F(x)$ and $G(u)$ constitute a Fourier pair as shown by the expressions

$$F(x)\big|_{\text{Rayleigh}} = x \exp(-\pi x^2) \xrightarrow{\text{Fourier Pair}} G(u_1)\big|_{\text{Rayleigh}} = u_1 \exp(-\pi u_1^2) \qquad (2.37)$$

As Rayleigh and Gaussian distributions tend to zero when $x/u_1 \to \infty$, the resulting radiation patterns have no sidelobes. The Rayleigh/Gaussian functions, however, are not easily approximated and other distributions are required for difference patterns.

Bayliss [56] proposed a difference pattern analogous to the Taylor pattern; the first step of the Bayliss synthesis is to produce a difference pattern by differentiating the Taylor ideal sum source distribution, given in equation (2.35). As the resulting pattern has unacceptably high sidelobes, an iterative process is used to reduce the first sidelobe levels by modifying the first four zeros of the array.

The positions of the first four zeros are usually obtained from tables included in references [43 and 57] while the positions of zeros when $n > 4$ are obtained from

$$z_n = \pm\sqrt{A^2 + n^2} \qquad (2.38)$$

where A is a constant dependent on the required sidelobe level. The range of values of A for different sidelobe levels is listed in Table 2.7 and design procedures to derive Bayliss distributions are outlined in references [57] and [58].

Table 2.7 The Bayliss A values for different sidelobe levels

SL (dB)	15	20	25	30	35	40
A	1.0079	1.2247	1.4355	1.6413	1.8431	2.0415

2.2.3.5. Elliott Syntheses

R. S. Elliott systematized synthesis procedures and formulated approaches that yield distributions meeting several requirements [59]. He also made his extensive knowledge available to a large cross section of engineers/scientists through university courses and courses sponsored by the IEEE. His contributions have therefore shaped developments in this field.

References [58] and [59] are excellent introductions to the subject, while in reference [60] Elliott outlined in detail a generalized pattern synthesis procedure applicable to linear arrays that allows the designer to synthesize array distribution patterns having specified sidelobe topologies on either side of the main beam. In reference [57] he extended the Bayliss synthesis approach for difference patterns, so that difference patterns that have an arbitrary sidelobe topology are realized.

The authors of reference [61] outlined a design procedure that yields shaped patterns of arbitrary sidelobe levels; the design procedure is a development of the generalized procedure outlined by Orchard *et al.* [62].

2.2.4. Arbitrary Arrays

It should come as no surprise that iterative methods used for adaptive arrays can also be used to derive the weights of arrays that meet a set of requirements. Reference [39] describes a simple iterative method that uses sequential updating to ensure that the peak sidelobe levels in the array meet specifications. Computation of each successive pattern is based on the solution of a linearly constrained least-squares problem.

The method proposed in reference [39] can be used for the following range of problems:

1. *Linear equidistanced arrays.* The required radiation pattern is specified in terms of the peak equiripple sidelobe level in well-defined angular distances and the position of the main lobe. An illustrative example is given where the derived radiation pattern is similar to one obtained using the Dolph–Chebyshev method.
2. *Linear equidistanced arrays.* The required radiation pattern is specified in terms of two peak equiripple sidelobe levels in well-defined angular distances on either side of the main lobe.
3. *Linear random arrays.* The required radiation pattern is specified in terms of the peak equiripple sidelobe level in all angular distances but for the range of angular distances where the main lobe is directed.

4. *Linear arrays of nonuniformly spaced nonisotropic elements.* The required radiation pattern is specified in terms of the peak equiripple sidelobe level in all angular distances but for the range of angular distances where the main lobe is directed.

Apart from the diversity and simplicity of the method, convergence occurs after only 3–4 iterations.

Reference [63] describes a simple method that either minimizes the peak sidelobe level or yields prescribed narrow or broad nulls from a linear array by optimizing the distance between its antenna elements. Lastly, reference [64] proposes a method to synthesize linear and circular arc arrays having a prescribed sidelobe topology.

While the various authors claim that their methods are simple, the user has no access to their computer programs, so the immediate use of the proposed methods is restricted to designers who have considerable expertise needed to replicate the required computer programs. User-friendly computer programs that incorporate these modern and powerful methods will greatly facilitate their widespread use among array designers.

2.2.4.1. Genetic Algorithms

Our understanding of naturally occurring evolutionary processes is fairly advanced. These processes involve natural selection and sexual reproduction, aided by random mutations. Individuals that are not fit to survive are eliminated and the fitter individuals produce offspring that are even fitter than their parents. Random mutations, on the other hand, ensure against the development of a uniform population incapable of further evolution.

A diverse range of people have used a combination of cross-breeding and selection to derive better crops, racehorses or ornamental roses for millennia. It is therefore not surprising that scientists and engineers have sought to derive genetic algorithms (GAs) that emulate the evolutionary processes. The attractors of GAs are:

- Once a problem is formulated in mathematical terms and the criteria for selection and reproduction are formulated, the programs reach the global maximum rather than a local maximum, provided certain safeguards are implemented.
- Programmers do not have to 'nurse' their programs by specifying in advance all the features of the problem.
- GAs solve complex problems that their creators do not fully understand [65].

The major shortcoming of GAs is that the algorithms are slow, but supplementary processes are often used to speed up the solutions; alternatively, additional information often accelerates the execution of the programs.

Reference [66] is an indexed bibliography of GAs from 1957 to 1993 and the paper [65] and book [67] of Holland are invaluable references to the field. In reference [68] the concepts and applications of GAs are considered and a delineation of some of the technical shortcomings of GAs is given. Practical issues arising from the application of GAs to optimal design problems in electromagnetics are canvassed in reference [69] and reference [70] is a valuable tutorial on the general theme of designing antenna arrays based on GAs.

We have already seen that aperiodic arrays represent attractive options for wideband arrays and low-sidelobe/grating lobes arrays. For large arrays, nonuniform spacing is

not an attractive proposition to contemplate because the designer has to consider an infinite number of possibilities for the placement of the array elements. Thinning a
uniformly spaced array by turning off, or discarding, some of its elements renders the
specification of an aperiodic array tractable. For an N-element array, thinning has 2^N
combinations and array symmetry reduces this number substantially.

In what follows we shall consider an illustrative example drawn from reference
[71]. One half of a 20-element linear, symmetrical, and equispaced array is shown in the
first row of Table 2.8a and the interelement spacing is $\lambda/2$. The array is mathematically
represented as a string of numbers 1111111111 where '1' stands for an antenna element
occupying the designed position along the array. The maximum relative sidelobe level
(RSLL) corresponding to this array is −13.3 dB. The second and subsequent rows represent aperiodic arrays termed 'genes' numbered from 1 to 7 and the resulting
maximum RSLL, in dB, for each array is listed. A missing element is denoted by a '0'
in the appropriate array position. The aperiodic arrays, resulting from the process
described, are generated by the random arrangement of its elements and the AGF is
derived from the equation

$$f(\vartheta) = 2 \sum_{n=1}^{N} A_n \cos(kd_s \cos \vartheta + \alpha_n)$$

where A_n and α_n are the amplitudes and phase excitations of the antenna elements,
located at distances d_s from the array center. The maximum RSLL for each array is calculated from [71]

$$F(u_0)\big|_{\max} = \left| 2 \sum_{n=1}^{N} a_n \frac{\cos(2\pi n s u_0 + \delta_s)}{f(u_0)} g(u_0) \right|;$$

$$\frac{c_0}{2Ns} \leqslant u_0 \leqslant 1$$

where $2N$ = the number of array elements;
   a_n = weight on each element ($n = 0$, off or $n = 1$, on);
   s = spacing between elements;
   $u_0 = \cos \vartheta$ (ϑ measured from the line passing though the array elements);
   $\delta_s = -2\pi s u_s$ is the steering phase;
   c_0 = constant;
   $g(u_0)$ = the element pattern;
   $f(u_0)$ = the peak of the main beam.

Naturally the calculation of $F(u_0)\big|_{\max}$ excludes the main beam.

The values for the maximum RSLL corresponding to each gene are also shown in the
same table. In Table 2.8b the genes are ordered according to their fitness and genes 0,
1, 2 and 6 are discarded. In Table 2.8c the reproduction of two sets of genes is illustrated. Genes 3 and 5 and 4 and 7 are mated at the random crossover points shown by
an asterisk. Mating involves the exchange of all binary bits to the right of the crossover
line. The resulting genes 3–5, 5–3, 4–7 and 7–4 together with their corresponding
maximum RSLLs are shown in the same table. Lastly, a second selection of the fittest

Table 2.8

(a) Random genes numbers 0 to 7

Gene no.	Array configuration	RSLL$_{max}$ (dB)
0	1 1 1 1 1 1 1 1 1 1	−13.3
1	1 1 1 1 0 1 1 0 0 1	−9.06
2	1 1 0 1 1 1 1 0 0 1	−9.71
3	1 1 1 1 1 1 1 0 0 1	−14.08
4	1 1 1 1 0 1 1 1 1 0	−12.13
5	1 1 1 1 1 1 0 0 1 0	−12.76
6	1 0 1 1 1 1 0 1 0 1	−8.70
7	1 1 1 1 0 1 1 1 0 1	−11.09

(b) First gene selection according to fitness

Gene no.	Half array	RSLL$_{max}$ (dB)
3	1 1 1 1 1 1 1 0 0 1	−14.08
5	1 1 1 1 1 1 0 0 1 0	−12.76
4	1 1 1 1 0 1 1 1 1 0	−12.13
7	1 1 1 1 0 1 1 1 0 1	−11.09

(c) Reproduction of genes at random crossover points (*)

Gene no.	Half array	RSLL$_{max}$ (dB)
3–5	1 1 1 1 1 1 1 *0 1 0	−14.66
5–3	1 1 1 1 1 1 0 *0 0 1	−11.72
4–7	1 1 1 1 0 1 1 1 *0 1	−12.01
7–4	1 1 1 1 0 1 1 1 *1 0	−11.89

(d) Second gene selection according to fitness

Gene no.	Half array	RSLL$_{max}$ (dB)
3–5	1 1 1 1 1 1 1 0 1 0	−14.66
3	1 1 1 1 1 1 1 0 0 1	−14.08
5	1 1 1 1 1 1 0 0 1 0	−12.76
4	1 1 1 1 0 1 1 1 1 0	−12.13

genes from the genes is shown in Table 2.8d together with their corresponding maximum RSLL and mutations 4–7, 7–4, 5–3 and 7 are discarded.

Mutations involve the flipping of one bit selected at random; a '1' is flipped to '0' and a '0' is flipped to '1' in a gene selected at random. How many genes are mutated is negotiable and depends on trade-offs between the number of variables one has to consider and the time taken for the algorithm to converge.

While the array considered was linear and had only 20 elements, genetic algorithms are especially suited to large linear or planar arrays. Similarly, gene reproduction in this example took place between pairs of genes that have comparable fitness to illustrate the process. We can stipulate, for instance, that any of the fittest genes mates with the remaining genes; similarly, we can optimize the array at broadside or over a range of scan angles. Bandwidth of operation is also a variable that can be taken into account. Reference [71] has several examples to illustrate the versatility of genetic algorithms. As can be expected, one has to consider trade-offs between the optimization of many parameters and ascertaining that the algorithm will explore a very large set of solutions on the one hand, and the time the algorithm takes to converge.

The following additional applications of genetic algorithms further demonstrate the versatility and power of these algorithms in solving a diverse set of problems related to phased array synthesis. A genetic algorithm has been used to reduce the sidelobe level of arrays by an optimization process in the quantized phase space [72], a realistic condition. Optimum array pattern nulling [73] and phase-only adaptive nulling [74] with a genetic algorithm have been reported.

There is no doubt that genetic algorithms will gain importance with the passage of time, especially for large array applications.

2.2.4.2. Fractal Designs

Fractal geometry is an extension/generalization of Euclidean geometry and is particularly suitable for describing a variety of self-similar objects that possess structure at all scales. Mandelbrot [75] observed that clouds and the perimeters of islands, for example, possess an inherent self-similarity in their geometrical structure. He coined the word *fractal*, derived from the Latin *fractus* meaning broken or irregular fragments, and established a new geometry based on concepts developed by nineteenth-century mathematicians. The underlying order inherent in fractals provides a simple procedure for creating a variety of complicated and useful geometries that mimic naturally occurring geometric shapes.

The first applications of fractals to problems in condensed-matter physics and computer graphics were followed by applications in biology, economics, music, linguistics, geology, and the engineering disciplines. Books such as *Fractals Everywhere* [76], reflect the wide range of fractal applications in nature.

Non-linear phenomena result when order gives way to chaos. By exploring chaotic conditions and the interface regions between order and chaos one is able to understand a very large set of naturally occurring phenomena. Technical books such as *Universality in Chaos* [77] reflect the theoretical and practical importance of these modern topics, while popular books like *Chaos* [78] and *Complexity* [79] illustrate the widespread interest in the understanding of phenomena that cannot be understood by conventional methodologies.

The prolog to the Special Section on Fractals in Electrical Engineering [80] is a useful introductory paper to the topic and provides a wealth of key references to

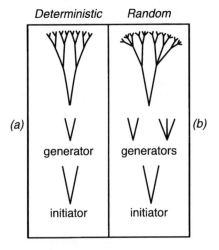

Figure 2.6 Two methods of generating fractal arrays with the aid of initiators and generators: (a) one two-branch generator is used; (b) two generators are used at random. The generators have two or three branches. (From [81]; © 1986, IEEE.)

previous work. Reference [81] is the earliest reference that applies fractals to the design of arrays and Figure 2.6 illustrates how fractal arrays can be realized. In Figures 2.6a and 2.6b we illustrate the initiators and generators for a couple of simple arrays. In the latter figure the 2- and 3-element generators are selected at random. At the tips of each generator antenna elements are placed and the self-similarity between generators and the resulting array is evident. Both configurations of trees occur in nature.

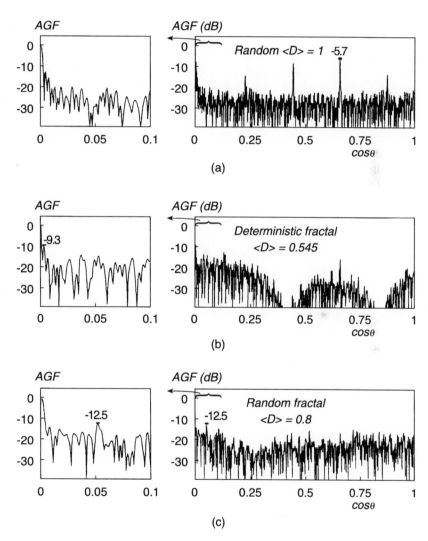

Figure 2.7 The array geometric factor of three linear arrays; all arrays have 180 elements and the total array length is 360 wavelengths. (a) Random array. (b) A deterministic array having 60, three-element subarrays located at positions chosen at random along the length of the array and when the interelement spacing is 0.8 of a wavelength. (c) A random fractal array composed of six different subarrays with element spacings of $0.8, 0.8\sqrt{2}, 0.8\sqrt{3}, 0.8\sqrt{5}, 0.8\sqrt{7},$ and $0.8\sqrt{11}$ wavelengths. D is the average fractal dimension and θ denotes the angle between the observation point and the array axis. (From [81]; © 1986, IEEE.)

What follow are important comparisons between three arrays explored in reference [81]. In Figure 2.7a the radiation pattern of a linear random array having 180 elements and overall length of 360 wavelengths is shown. Random arrays have the following attractive characteristics:

- Robustness with respect to element position errors and failures;
- Low average sidelobes

They nevertheless have high peak sidelobe levels and the peaks occur at angular positions not known *a priori*. Two high sidelobes are seen in Figure 2.7a, the highest being at a –5.7 dB level. While it is relatively easy to detect the peak sidelobes when the array is linear, peak sidelobes are hard to detect when the array is planar. More explicitly, one has to calculate the array radiation pattern at many azimuth angles, in order not to miss the peak sidelobes that can occur at angles not known *a priori*.

In Figure 2.7b we illustrate the radiation pattern of a deterministic fractal array that consists of 60 three-element subarrays placed at 60 randomly selected positions and where the interelement spacing in each subarray is 0.8 of a wavelength. As can be seen, the highest sidelobe level occurs near the main lobe and it is at the –9.3 dB level. The radiation pattern of the subarray becomes the envelope of the overall radiation pattern. As can be expected, the array sidelobe level can be more evenly distributed if the spacing between the elements of the subarrays varies.

In Figure 2.7c the radiation pattern of a random fractal array is shown. It consists of subarrays where the interelement spacings are

$$0.8, \quad 0.8\sqrt{2}, \quad 0.8\sqrt{3}, \quad 0.8\sqrt{5}, \quad 0.8\sqrt{7}, \quad \text{or} \quad 0.8\sqrt{11}$$

and are selected so that the nulls and secondary maxima corresponding to each subarray do not coincide. The average sidelobe level is comparable to that corresponding to the random array, but the peak sidelobe level is at a –12.5 dB level.

The fractal dimension D for each array is also shown in Figure 2.7 and is defined as [81]

$$D = \frac{\log N}{\log \varrho} \tag{2.39}$$

where N is the number of elements in one subarray and ϱ is defined as [81]

$$\varrho = \frac{\text{average element spacing of random initiator}}{\text{average element spacing of random subarray generator}}$$

Large values of D indicate that the sidelobe energy is less concentrated near the main lobe and that the sidelobe level is less controlled, while lower values of D indicate the converse.

In reference [82] arrays resulting from Cantor sets are explored and the resulting maximum array sidelobe level is calculated. In references [81 and 83] frequency-independent features of self-similar fractal arrays are proposed, while Weierstrass arrays, having fractal radiation patterns, are proposed in reference [84]. Given that the Weierstrass arrays have high sidelobes, the arrays are not attractive for radar applications but are for remote sensing, plasma physics, and ionospheric modification applications [84].

2.2.4.3. Combinational Approaches

We have already considered the attractive characteristics of random arrays and approaches to reduce their peak sidelobe level in the previous section. Here we shall consider other approaches to achieve the same goal by using arrays based on combinational sets. Leeper [20] first pointed out that arrays based on difference sets can have peak sidelobe levels significantly lower than those corresponding to random arrays. Comparisons between linear random arrays and arrays based on difference sets have been illustrated in Table 2.3. As can be seen, one stands to gain some 6 dB by designing an array based on difference sets; since then Leeper's work has been verified and expanded [85,86] to include two-dimensional arrays and perfect binary arrays (PBAs) [87–89] have been shown to be related to a class of difference sets [90]. PBAs have the property that all out-of-phase values of their periodic autocorrelation function are equal to zero. Wider applications of difference sets include radar and communications [91] as well as frequency-hopping radars [92].

Minimum redundancy arrays and nonredundant arrays with minimum missing lags or holes are special cases of difference base arrays. Given N elements there are $N(N–1)/2$, pairwise separations. If no redundancies and holes are allowed, the aperture of a linear array of length L is equal to $N(N–1)/2$. If redundancies and/or holes are allowed, then the array length is given by

$$L = \frac{N(N–1)}{2} – R + H \qquad (2.40)$$

where R is the number of redundancies and H the number of holes. If $H = 0$, and R is minimized, the array length is maximized; this is the case for minimum redundancy arrays. Minimum hole arrays are equivalent to Golomb rulers. A Golomb ruler of length L has only K marks on the ruler, where $K < L$. If the marks are arranged such that all distances 1, 2, . . ., L can be measured with the ruler, then we have a K-element restricted difference basis for L. The popular articles [93–95] lend several insights into the problem of finding restricted difference bases.

2.3. PLANAR ARRAYS

We require to extend the methodology we have developed to analyze linear arrays to planar arrays. Without any loss of generality, we shall consider the rectangular array shown in Figure 2.8 with spacings between antenna elements of s_x and s_y along the x- and y-directions. Square arrays are just special cases of rectangular arrays. We use the usual spherical coordinates and assume that the excitations along the rows and columns are separable.

The AGFs of the planar array, having M by N antenna elements along the x- and y-directions are

$$\text{AGF}\big|_X = \sum_{m=1}^{M} I_m \exp j(m-1)(ks_x \sin\theta \cos\phi + \beta_x) \qquad (2.41)$$

and

$$\text{AGF}\big|_Y = \sum_{n=1}^{N} I_n \exp j(n-1)(ks_y \sin \theta \sin \phi + \beta_y) \qquad (2.42)$$

where β_x and β_y are the progressive phase shift between antenna elements along the x- and y-directions, respectively. The required array AGF is therefore

$$\text{AGF}\big|_{\text{ARRAY}} = \text{AGF}\big|_X \; \text{AGF}\big|_Y \qquad (2.43)$$

If all the amplitude excitations I_n and I_m are equal, then

$$\text{AGF}\big|_{\text{ARRAY}} = \frac{1}{M} \frac{\sin(M\psi_x/2)}{\sin(\psi_x/2)} \frac{1}{N} \frac{\sin(N\psi_y/2)}{\sin(\psi_y/2)} \qquad (2.44)$$

where

$$\psi_x = ks_x \sin \theta \cos \phi + \beta_x$$
$$\psi_y = ks_y \sin \theta \cos \phi + \beta_y$$

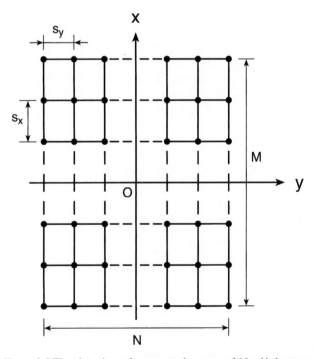

Figure 2.8 The plan view of a rectangular array of $M \times N$ elements.

The conditions to form a beam at $\theta = \theta_0$ and $\phi = \phi_0$ can be derived from equations

$$ks_x \sin \theta_0 \cos \phi_0 + \beta_x = 0 \quad \text{and}$$
$$ks_y \sin \theta_0 \sin \phi_0 + \beta_y = 0$$

from which we can derive

$$\tan \phi_0 = \frac{\beta_y s_x}{\beta_x s_y} \tag{2.45}$$

and

$$\sin \theta_0 = \frac{\beta_x^2}{(ks_x)^2} + \frac{\beta_y^2}{(ks_y)^2} \leq 1 \tag{2.46}$$

We reiterate here that the above conditions are valid for beam formation when the excitations of the array columns and rows are separable.

While the grating lobes along the x- and y-axes are governed by the equations derived when linear phased arrays were considered, the grating lobes on any other planes are not easily derived. Von Aulock [96] proposed a useful way of visualizing the movement of the grating lobes as the array is scanned in two dimensions and his approach has been accepted widely [97]. The methodology developed is therefore used to ensure that the grating lobes do not enter the visible space for the required scanning angles.

If the elements are placed at positions forming equilateral triangles, the number of antenna elements that fully populate the array is decreased by approximately 16% compared to the case where the elements are placed on a square grid [96,97]. For both cases the condition applies that the grating lobe will not enter the visible space at the maximum scan angle.

When compared to the square lattice, the triangular lattice of the antenna elements has the following attractive characteristics:

- The array is more economical to realize because it utilizes fewer T/R modules.
- The effects of mutual coupling between antenna elements are less severe.
- Heat dissipation problems are more manageable.

Arrays having a square lattice, however, offer maximum power–aperture products at increased cost and more serious heat dissipation problems.

Ultimately, working arrays are designed with detailed knowledge of the array antenna elements and their dimensions; thus the appropriate packing scheme is chosen to meet a diverse range of requirements.

Arrays designed on the assumption that the excitations along the principal planes, xz and yz, are separable have radiation patterns along these directions in accordance with the designs adopted. The resulting beams and sidelobe levels along planes $\phi \neq 0°$ and $\phi \neq 90°$ will, however, be substantially different. If the sidelobe levels along the principal planes are, for instance, −32 and −33 dB, the resulting sidelobe level at other planes will be −65 dB; similarly, the resulting beam is broadened and the directivity is lowered in other planes. If there is a requirement for equal sidelobes in all planes, the design procedure outlined in references [98,59] is followed.

We seek to determine how the 3 dB beam area, B, of a planar array varies as a function of θ and ϕ. It has been shown that B, at a scan angle θ_s, is [99]

$$B = \Theta_1 \Theta_2 \sec \theta_s \qquad (2.47)$$

where Θ_1 and Θ_2 are the 3 dB beamwidths in the broadside direction of the x-directed and y-directed linear arrays, respectively. Equation (2.47) holds so long as the array is not scanned to directions close to the horizon. Similarly the peak directivity of the array, D, is

$$D = D_1 D_2 \cos \theta_s \qquad (2.48)$$

where D_1 and D_2 are the peak directivities of the x-directed and y-directed linear arrays, respectively. When the array is uniformly illuminated, the product of its HPBW area, B_0, and its directivity is given by [99]

$$DB_0 = 32\,400 \qquad (2.49)$$

where B_0 is expressed in degrees.

2.3.1. A Systems View of Grating and Sidelobes

As we have seen, the grating lobes of a canonical array introduce positional ambiguities if the array operates in the receive mode. If the array operates in the transmit mode, power is wasted toward directions other than the direction the array is pointed to.

In this section we shall consider grating lobes from a systems view, for in the final analysis even grating lobes have to be considered in trade-offs between various array parameters. More specifically we shall explore the conditions under which grating lobes can be tolerated in the receive mode; and alternative ways of eliminating grating lobes.

We have already considered crossed arrays used for radioastronomical observations in Chapter 1. The multielement crossed radiotelescope realized by Christiansen and his collaborators [3] was designed initially to map the quiet Sun. We have already noted that the first-order grating lobes of this radiotelescope, on either side of the main beam, were present. However, as there are no sources of comparable flux density, the positional ambiguities created by the grating lobes introduced negligible errors to the observations.

At a later date, the same radiotelescope had to be modified to take non-solar observations. While the original radiotelescope was left intact, reflectors were added to the East–West (EW) and North–South (NS) arrays. The EW and NS arms of the array had the following pertinent parameters:

Frequency of operation	1420 MHz
Wavelength, λ,	21 cm
Number of antennas	32
Diameter of the antennas	18 ft (5.5 m)

Interelement spacing, s,	40 ft (12.2 m)
Maximum spatial resolution	3′ arc

Figure 2.9a illustrates the main lobe and the two grating lobes of the resulting array radiation pattern. Suppose we add one reflector, of diameter D, at a distance $s_1 = ms$ from the center of the last antenna of the array and along the EW direction. If the powers received by the E-W array and the reflector are multiplied, the resultant power, $P(\theta)$, is given by

$$P(\theta) = A_1(\theta) A_2(\theta) \frac{\sin Nx \cos (2mx + \theta_1)}{\sin x}$$

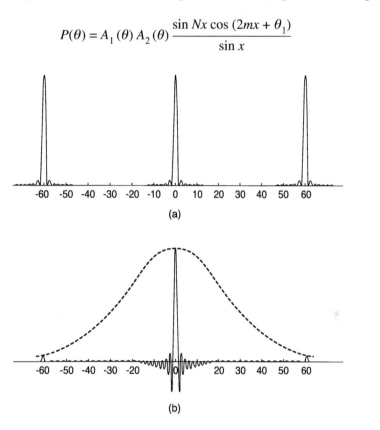

(a)

(b)

Azimuth Angle (minutes of arc)

Figure 2.9 (a) The radiation pattern of an array consisting of 32 elements when the interelement spacing is 58 wavelengths. (b) The array radiation pattern of the compound interferometer when the output of the same array is multiplied with a 60 ft antenna placed 116 wavelengths from the array end. (From [100]; © 1963, IREE (Aust.).)

if we ignore constants of proportionality [100]. Here θ is measured with respect to the boresight axis of the array, $A_1(\theta)$ and $A_2(\theta)$ are the voltage responses of the array antennas and that of the additional antenna, respectively, and $x = \pi s (\sin\theta)\lambda$. The angle θ_1 can be made equal to zero by special adjustments and for the special case of $m = N/2$ (the center of the additional antenna is located half a unit spacing beyond the end array antenna),

$$P(\theta) = A_1(\theta)\, A_2(\theta)\, \frac{\sin 2Nx}{\sin x} \tag{2.50}$$

Equation (2.50) shows that the resultant power of the compound interferometer is similar to that derived from an array of twice the length of the original array. Theoretically the diameter of the additional antenna has to be 40 ft in diameter and 20 ft away from the end antenna. In the interest of gaining sensitivity, the designers opted for a 60 ft antenna at a distance 80 ft from the end antenna. The resulting power diagram is shown in Figure 2.9b, where the main beam has a HPBW of 1.5′ and the grating lobe is attenuated by 12 dB.

Compound interferometers were first proposed by Covington [101] and similar considerations hold for the NS arm of the crossed array. If the additional antenna elements have a substantial diameter, grating lobes can be safely neglected. Radar engineers are familiar with multiplicative arrays used in bistatic radars where the resulting patterns of the receive and transit arrays are multiplied—see section 2.7.1.1.

From the foregoing considerations it is clear that grating lobes of phased arrays have to be considered in the context of specific applications. For some applications, e.g. radar applications, extraordinary measures have to be taken to decrease their amplitude. For other applications the effects caused by array grating lobes can be marginalized.

We have already explored several ways of minimizing the sidelobe level of arrays and the unacceptably high premium the designer has to pay to render the sidelobe level close to zero. The question that arises is: What is an acceptable low sidelobe level? The AWACS has a one way sidelobe level ranging from –40 to –50 dB and these systems are considered as having ultra-low sidelobe levels. We shall adopt these sidelobe levels as the benchmark for our considerations. As the required sidelobe level is decreased, the designer has to control not only the electrical parameters of the T/R modules and beamformers but also the mechanical tolerances related to the position of the array antennas. The effects of mutual coupling between antenna elements also have to be taken into account.

2.3.2. Hansen Synthesis—Circular Distribution

Hansen's synthesis procedure [102] is similar to that proposed by Taylor but is applicable to circular distributions. The starting point is the observation that for a uniform illumination only the first sidelobes of the resulting pattern, having the form of $J_1(u_2)/u_2$, need lowering. The two forms of the resulting patterns are

$$F(u_2) = \frac{2I_1(\pi\sqrt{H^2 - u_2^2})}{\pi\sqrt{H^2 - u_2^2}} \qquad u_2 \leqslant H \tag{2.51}$$

and

$$F(u_2) = \frac{2J_1(\pi\sqrt{u_2^2 - H^2})}{\pi\sqrt{u_2^2 - H^2}} \qquad u_2 \geqslant H \tag{2.52}$$

where $J_1(u)$ is the Bessel function of the first kind and order 1, $I_1(u)$ is the modified Bessel function of the first kind and order 1, and H is the one parameter which defines all the pertinent parameters of the distribution shown in Table 2.9. The SL in dB and the aperture distribution are given by

$$|SL|_{dB} = 17.57 + 20 \log \frac{2I_1(\pi H)}{\pi H} \tag{2.53}$$

and

$$g(\rho) = I_0 \left(\pi H \sqrt{1 - \rho^2} \right) \tag{2.54}$$

where $\rho = r/R$, R is the radius of the aperture, and r is the radial distance from the aperture center. The range of values for H is between zero and about 2.2.

Table 2.9 Pertinent parameters of the Hansen one-parameter distribution [43]

SL (dB)	H	HPBW/2 (rad)	η	η_b
17.57	0	0.5145	1	0.8378
25	0.8899	0.5869	0.8711	0.9745
30	1.1977	0.6304	0.7595	0.993
35	1.4708	0.6701	0.6683	0.9981
40	1.7254	0.7070	0.5964	0.9994
45	1.9681	0.7413	0.539	0.9998
50	2.2026	0.7737	0.4923	1

A hyperbolic approximation to equation (2.53) can again be defined; more explicitly, the parameter H is defined by

$$H = b \sqrt{\frac{(SL - c)^2}{a^2} - 1} \tag{2.51}$$

where $c = -11.5$, $b = 1.1796$, and $a = 17.57 - c$. Knowing the SL level, one can calculate H from equation (2.55) and $g(\rho)$ from equation (2.54).

2.3.3. Taylor $\bar{n}$ synthesis—Circular Distribution

The Taylor $\bar{n}$ circular source distribution [103] offers a modest improvement in efficiency and beamwidth over the Hansen one-parameter distribution just as observed for the linear distributions. Again the starting point is the $2J_1(\pi u_2)/(\pi u_2)$ function which corresponds to a uniform illumination. The positions of $\bar{n}$ zeros on either side of the main beam are modified to produce the desired sidelobe level and the radiation pattern is given by

$$F(u_2) = \frac{2J_1(\pi u_2)}{\pi u_2} \prod_{n=1}^{\bar{n}-1} \frac{1-(u_2^2/z_n^2)}{1-(u_2^2/u_n^2)} \tag{2.56}$$

where μ_n are the zeros of $J_1(\pi u_2)$. The pattern zeros are given by

$$z_n = \pm\sigma\sqrt{A^2 + (n-\tfrac{1}{2})^2} \qquad 1 \leqslant n \leqslant \bar{n} \tag{2.57}$$

and σ, the dilation factor, is given by

$$\sigma = \frac{\bar{n}}{\sqrt{A^2 + (\bar{n}-\tfrac{1}{2})^2}} \tag{2.58}$$

Extensive tables of the Taylor circular $\bar{n}$ distributions are given in reference [104]. Most of the other synthesis approaches considered for linear arrays are also applicable to planar arrays. Design procedures for Bayliss circular arrays and a description of a discretizing technique for rectangular grid arrays are outlined in reference [59].

2.4. CIRCULAR/CYLINDRICAL ARRAYS

The most basic circular array is illustrated in Figure 2.10, where its elements are placed on the perimeter of a circle and the incoming wavefront is received by N antenna elements. As θ changes, another set of N elements along the perimeter of the array receives the incoming radiation. For the array to form a beam, either time delays or phase-shifters are connected between the N elements and the summing point. Thus, as different sets of N antennas are connected to the summing point, the array 'looks' toward a different azimuth angle, θ.

If certain conditions are observed, circular arrays can support several phase modes and wideband operation is theoretically possible [105]. Several techniques of exciting the required modes are outlined in [105 and the references therein].

As couplings can occur between neighboring omnidirectional antenna elements and between antenna elements diametrically located antenna elements having some forward directivity marginalize the latter couplings.

The advantages of circular arrays are:

- 360° azimuth scan coverage is easily attained.
- The resulting beamwidth is invariant as the scan angle varies.
- The connections between antenna elements and a central summing point have equal lengths.
- Wideband operation is possible if sufficient phase modes are appropriately generated [105].

The most serious drawbacks of circular arrays are:

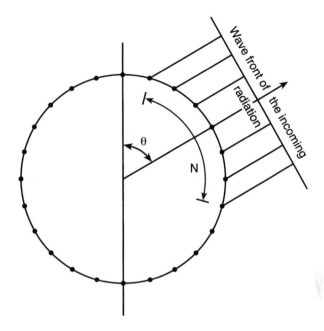

Figure 2.10 An *N*-element circular array when *M* elements receive the incoming radiation. As θ changes, a different set of *M* elements receive the incoming radiation.

- No elevation information is available.
- High sidelobes that can be minimized—see section 2.4.2.
- As only a few antenna elements are used at the time, array sensitivity is low.
- If *N* is high, the designer has to provide a switching–commutation subsystem that is difficult to realize.

A cylindrical array that consists of several rows of circular arrays provides elevation information and increases the array sensitivity.

In what follows we shall consider several recent circular and cylindrical arrays. While each array illustrates an interesting property, all arrays utilize conventional subsystems. The technological risk of realizing circular/cylindrical arrays is therefore minimal.

2.4.1. 2–18 GHz Circular Arrays

In Figure 2.11 a wideband electronic support measures (ESM) system operating over the 2–18 GHz frequency range is shown [106]. The band of operation is split into a low, 2–7.5 GHz, and high, 7.5–18 GHz, band and the two circular arrays and their associated electronic subsystems accommodating the two bands are shown in Figure 2.11a. Both circular arrays consist of 32 antenna elements and one of them is shown in Figure 2.11b. Bicones are used to shape the beam pattern in the elevation plane and enhance the bearing accuracy in a multipath environment. Each antenna element is a microstrip wideband tapered slotline antenna further considered in section 3.6.5.2. The outputs of all the antenna

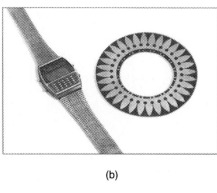

(b)

(a)

Figure 2.11 Two circular arrays used in an ESM system operating in the 2–18 GHz band (Source: Dr Stig Rehnmark, Anaren Microwave Inc.). (a) The low-high-band arrays and associated electronics are shown in the lower/upper part of the photograph, respectively. The low/high bands extend from 2 to 7.5 GHz and from 7.5 to 18 GHz, respectively. (b) The 32-antenna-element circular array. Each antenna element is a microstrip, wideband, tapered slotline antenna, considered in some detail in section 3.6.5.2.

elements of each circular array are fed into a 32-input/7-output Butler matrix that provides phase progressions at its outputs that are proportional to integer multiples of the incident bearing angle, b. Comparisons between the reference channel and the output RF channels produce phase measurements proportional to $2b$, $4b$, and $8b$. While the $1b$ measurement provides the unambiguous phase/bearing relationship over the 360° FOV, the $8b$ measurement produces the required bearing accuracy of better than 2° rms. The $2b$ and $4b$ measurements provide suitable hand-off points between the coarse and fine measurements.

The resulting ESM system is remarkable considering its frequency coverage and bearing accuracy.

2.4.2. Radioastronomy Circular Arrays

In the radioastronomy context, circular phased arrays have a number of attractive characteristics:

- A circular array having an odd number of antenna elements has no redundant baselines, and has therefore the maximum number of baselines. If the number of elements is N, the number of available baselines is approximately N^2 [107].

- The distribution of baselines is completely circularly symmetric [107].
- The connecting transmission lines between the antennas and the summing point are equal.

We have already considered the Culgoora Radioheliograph in Chapter 1 and in this section we shall focus attention on the synthesis procedure used to reduce its sidelobes [108,109].

The AGF, $F(u, \phi)$, of a circular array having N antenna elements is given by [108]

$$F(u, \phi) = N[J_0(p) + 2J_N(p) \cos(N\phi) + 2J_{2N}(p) \cos(2N\phi) + \ldots]^2 \qquad (2.59)$$

where $p = \pi D u$, $u = \sin \theta / \lambda$, D is the diameter of the circular array and the angles θ and ϕ are the zenith and azimuth angles. J_N are the Bessel functions of the first kind of order N. Here we have also assumed that $N/2$ is even. The first term of the rhs of equation (2.59) defines the mainlobe and sidelobes of the array, while the subsequent terms define its grating lobes, which are azimuth dependent.

The designer aims at eliminating the grating lobes of a circular array with the aid of the radiation pattern of the antenna elements, or by positioning the grating lobes to angular distances outside the instantaneous FOV of interest. In what follows we shall explore the technique used to minimize the sidelobe level of the AGF.

The profile of the radiation pattern, $P(r)$, of the main beam is deduced from equation (2.59) to be

$$P(r) = \mathrm{AGF}|^2_{\mathrm{circular}} = J_0^2(2\pi a r) \qquad (2.60)$$

where r is the angular displacement in radians from the boresight axis of the array and, a is the radius of the array measured in wavelengths.

As the second maximum of the $J_0(x)$ function has a value of 0.4026, the first sidelobe of the circular array is only −7.9 dB below the main lobe; a reduction of the sidelobe level is therefore essential for many applications.

While a reduction of the sidelobe level for linear arrays is attained by an amplitude taper in the illumination of the array, for circular arrays an appropriate phase taper is required to decrease their sidelobe level.

The sidelobe reduction technique is based on the following observations:

(a) If all antenna elements are connected to a central summing point in phase, the resulting radiation pattern has the form of $J_0(2\pi a r)$—equation (2.60).

(b) Circularly symmetrical patterns are generated by inserting a phase shift, Φ_n, between the antenna elements and the central summing point. Any antenna element can be a reference where the phase shift is zero and Φ_n increases uniformly around the circle. If the $\Phi_n|_{\max} = 2\pi k$, where k is an integer, the resulting far-field radiation patterns are generated [108,109]:

$$P(r)_k = J_k^2(2\pi a r) \qquad (2.61)$$

(c) By a judicious choice of the amplitudes of the $J_k^2(2\pi a r)$ terms and signs, the designer can synthesize the required radiation pattern.

For radar and radioastronomy applications, the implementation of the J_k^2- synthesis procedure requires the generation of a suitable set of $J_k^2(2\pi a r)$ patterns, the weighted sum of which has sidelobes well within the required specifications. The derived AGF is therefore given by

$$F(r) = \sum_{k=0}^{\infty} t_k J_k^2(2\pi ar) \tag{2.62}$$

where t_k are the weighting factors that are determined to attain the required $F(r)$. To attain the resolution limit, the t_k, values are [109]

$$t_k = 1, \ 0, -\frac{2}{3}, 0, -\frac{2}{15}, 0, -\frac{2}{35}, 0, -\frac{2}{63} \quad \dots \text{ when } \quad k = 0, 1, 2, 3, 4, \dots \tag{2.63}$$

Although the series of the rhs of equation (2.62) is theoretically infinite, it has been shown [108] that the series can be truncated within the range defined by $N/2 > k > 0$. For the Culgoora Radioheliograph, $N=96$, 49 terms are sufficient to approximate the series. In practice the number of terms can be further reduced to between 10 and 20 without appreciably affecting the shape of the resulting polar diagram.

A good compromise between resolution and low sidelobes is afforded by the pattern [109]

$$F(r) = 0.3\Lambda_1(4\pi ar) + 0.7\Lambda_3(4\pi ar) \tag{2.64}$$

where

$$\Lambda_q(z) = \frac{2^q q! J_q(z)}{z^q}$$

in which case

$$t_k = 1, \quad 0.7, \quad -0.48, \quad -0.42, \quad -0.24, \quad -1, \dots \tag{2.65}$$

In Figure 2.12a we illustrate the phases Φ_n required between the antenna elements of an 8-element circular array and the central summing point. The values of Φ_n required to generate the $J_0(x)$, $J_1(x)$, and $J_2(x)$ radiation patterns are given in Table 2.10.

Table 2.10 Φ_n values of an 8-element circular array to generate the $J_0(x)$, $J_1(x)$ and $J_2(x)$ rediation patterns

Angle	$J_0(x)$	$J_1(x)$	$J_2(x)$
$\Phi_{0/360}$	0	0/360	0/720
Φ_{45}	0	45	90
Φ_{90}	0	90	180
Φ_{135}	0	135	270
Φ_{180}	0	180	360
Φ_{225}	0	225	450
Φ_{270}	0	270	540
Φ_{315}	0	315	630
Φ_{360}	0	0/360	0/720

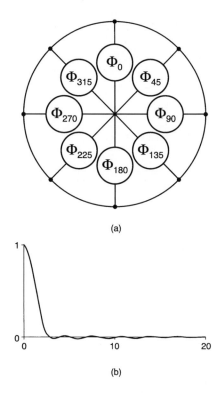

(a)

(b)

Figure 2.12 (a) The required Φ_n phases to generate the $J_0(x)$, $J_1(x)$. . . and $J_k(x)$ radiation patterns from an 8-element circular array. To generate the $J_0(x)$ radiation pattern all phases are equal and the $J_1(x)$ and $J_2(x)$ patterns are generated when a progressive phase amounting to 360° and 720° is inserted between the array elements and the summing point at the center of the circular array respectively. (b) The resulting radiation pattern of the 94-element Culgoora array when the synthesis procedure referred to in the text is adopted. (From [109]; © 1967, IREE (Aust.).)

In Figure 2.12b we illustrate the resultant radiation pattern attained by adopting the latter synthesis procedure for the Culgoora 96-element Radioheliograph. The resultant main beam has low sidelobes and can be used for imaging the radio Sun.

A variant method for the reduction of the sidelobes of circular arrays by adjusting the phases of the array has been described in reference [110] and the references therein. *Inter alia* the millimeter array (MMA) is to be dedicated to studies related to star formation and molecular clouds, planetary science, and astrochemistry. The essential characteristics of the proposed MMA are [111]:

Maximum array diameter	3 km
Range of wavelengths of operation	9–0.8 mm
Number of transportable antennas	40
Antenna diameter	8 m
Spatial resolution	0.07λ arc seconds
Total collecting area	2010 m^2

The array site is to be located at a high altitude so that the atmospheric attenuation at the naturally occurring windows at these wavelengths is negligible over long periods of observing times. As its spatial resolution is comparable to that of the Hubble telescope, radio images derived from this millimeter-wave radiotelescope will be compable to those attained with the Hubble telescope.

2.4.3. A One-Octave Cylindrical Array

The cylindrical array described in references [112,113] operates over an octave band from 3 to 6 GHz and is used in the receive mode to measure the direction of arrival (DOA) of incoming radiations; the same array can be used as a subsystem to an ESM system.

The array's physical dimensions are 40 inches (1016 mm) diameter and 22.5 inches (571.5 mm) high. It consists of 2048 vertically polarized elements arranged in 128 columns and 16 rows. It electronically scans eight simultaneous and contiguous beams 360° in azimuth and has the following additional characteristics:

- ±10% beamwidth variation
- Azimuth sidelobes (near main lobe)   –20 dB
- Cross-over between adjacent beams   2.5–4.5 dB
- Elevation coverage to   25°
- Gain variation over 360° azimuth scan   ±0.5 dB

The array's elements enter the column networks designed to produce the desired elevation pattern. From the column networks the signals pass through T/R modules before they enter a two-dimensional Luneberg lens. The array utilizes eight single-pole 8-throw (SP8T) switches to commutate the 64 antenna elements for azimuth scanning. Signals of different DOA are then accommodated on the remaining 64 positions of the Luneberg lens.

The array is an experimental model and options exist to widen the bandwidth of operation and/or use the array in the transmit mode.

2.4.4. 3D/2D Cylindrical Arrays

A 3D array provides range, azimuth and elevation information and represents a maximum complexity design. As the 2D array provides range and azimuth information only, it represents a medium complexity design. The arrays can be used for radar applications such as air-traffic control and surveillance.

The descriptions of these arrays are necessarily sketchy [113] and we shall focus attention on aspects particularly important to cylindrical arrays. To simplify the switching–commutation subsystem, phase-shifters are used to scan the beam ±15° in azimuth; the commutation is further simplified by dividing the array aperture into three 120° sectors. With this arrangement the commutator need only provide for the commutation of four 30° sectors, enabling it to be realized using simple 4×4 transfer switches.

The 2D array does not use column network phase-shifters and arrangements can be made for the array to have either several elevation beams or fine beam scan in azimuth.

Despite the sketchy descriptions of the above arrays, their realizations constitute a significant step toward the acceptance of cylindrical arrays as serious contenders for low-cost radars or ESM systems that operate in environments where the targets are at short to medium ranges.

In reference [114] a cylindrical active phased array radar is presented as a low-cost, efficient, and versatile solution suitable for operation in the tactical environment.

2.5. CHARACTERIZATION AND REALIZATION OF ARRAYS

So far we have considered phased arrays from a theoretical point of view and made several simplifications. This approach served the purpose of initiating discussion and introduced the reader to several topics; it is now appropriate to consider and characterize phased arrays with as few simplifications as possible.

The issues we shall address in this section are related to:

(a)   Array architectures and interconnect options
(b)   The derivation of the array's (i) gain over temperature (G/T) ratio, (ii) noise figure, (iii) intermodulation distortion; and (iv) dynamic ranges
(c)   The approaches taken to implement array amplitude tapers when current T/R modules are used
(d)   The impact of random and quantization errors on the array performance
(e)   Array power supplies
(f)   The generation and management of heat in arrays

Of all these issues, only issue (c) needs some elaboration here. When current T/R modules operate in the transit mode, maximum efficiency is attained when the final power amplifier (FPA) operates at maximum power. Here efficiency includes the device efficiency and the power dissipated in the subsystem. One way of implementing an amplitude taper is to have amplifiers that have a range of power outputs at maximum efficiency. This issue is problematic not so much because there are no technological solutions to hand but because the present technological solutions involve increased costs—see section 4.2.3. Another way to meet the requirement for an amplitude taper by using current T/R modules is to implement a spatial taper. We shall explore the trade-offs that this and other approaches offer in this section.

The above-mentioned problem in relation to current T/R modules, though significant now, is not likely to be a permanent issue for phased arrays. An inexpensive solution is to be hoped for in the not too distant future because of its importance. Until then, designers have the several options we shall consider in this section.

2.5.1. Active, Passive, and Hybrid Arrays

Requirements, costs, and technological issues often dictate the way an aperture designated for a phased array is populated. The fundamental issue the designer considers is whether the array is designated to yield one beam or many. In this chapter we shall assume that only one beam is required and we consider multibeam arrays in Chapter 5.

The active array shown in Figure 2.13a is often favored by designers because it has minimal losses between the antenna elements and its FPA/LNA, and a variety of signal processing functions can be performed after the LNA without any degradation of the array sensitivity. One of the important signal processing functions is beamforming.

There are, however, a number of issues we have to consider before we wholeheartedly accept the active array as the answer to all requirements. For a start, one assumes that substantial powers can be generated by the transmitters of the active array. While this assumption is true for solid-state FPAs operating at centimeter wavelengths, the power available from solid-state devices operating at millimeter wavelengths is minuscule—see Chapter 4. Vacuum tubes, on the other hand, tend to be bulky and offer

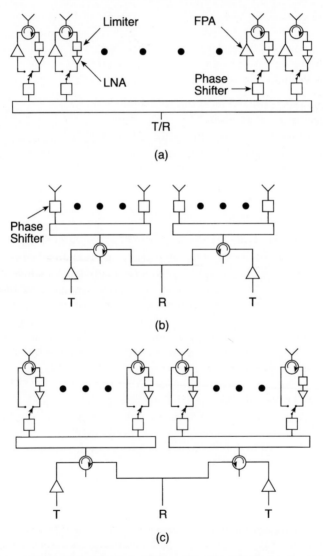

Figure 2.13 Three variants of phased arrays: (a) the active array; (b) the passive array; (c) the hybrid array.

considerable powers at millimeter wavelengths; they are therefore used in conjunction with passive arrays, shown in Figure 2.13b.

Another issue is related to costs. Passive arrays, utilizing a small number of vacuum tubes, were less expensive than active arrays having a high number of solid-state T/R modules for some time. Active arrays utilizing MMIC-based T/R modules, and operating at centimeter wavelengths, are now not only affordable but also cost-competitive. A couple of issues closely associated with costs are:

- The device lifetime; the MMIC-based T/R module is characterized by high mean time between failure (MTBF) when compared with its vacuum-tube counterparts—see Chapter 4; and
- The key MMIC-based T/R module parameters can be controlled to a degree that depends on the manufacturing processes involved; furthermore, the same parameters can be tightly controlled as the manufacturing processes are better understood—see Chapter 4.

Another issue of importance is the vulnerability of the array front-ends to electromagnetic missiles. It is well known that some vacuum tubes can withstand high power levels. The electrostatic amplifier (ESA) for instance has a noise figure of 1.4 dB from L-band through to Ku-band and can withstand up to 500 kW of input power without any additional protection [115]. MMIC-based amplifiers, on the other hand, require protection afforded by a limiter, which in turn introduces some additional loss; the combined losses of the T/R switch and limiter, for instance, range from 0.5 to 1.5 dB [116]. Current research is related to quantifying the effects on and protection of MMIC T/R modules from high-power electromagnetic pulses.

Some additional comparisons between active and passive arrays are worth considering. The passive arrays require high-power phase-shifters, whereas the phase-shifters used by the phased array have a low power rating. Similarly, the passive arrays utilize high-power-rating power dividers and circulators. By contrast, active phased arrays utilize low-power-rating circuits and deliver high power on target. Components having high power ratings are bulky and the maintenance of a passive array where high power is distributed all over the array is costly. These considerations clarify the significant attractive attributes of active phased arrays.

The question that arises naturally is how many antennas can be connected in parallel in a passive array. Every antenna increases not only the array gain but also the array losses.

The array gain of N half-wave dipoles is given by

$$G_A = \frac{4\pi A_{eff}}{\lambda^2} \; N = \frac{4\pi(\lambda/2)(\lambda/2)}{\lambda^2} \; N = \pi N \qquad (2.66)$$

The effective area of a half-wavelength dipole is here conservatively assumed to be $\lambda/2$ by $\lambda/2$.

The line losses for a square array of patch antennas can be approximated by $(N^{1/2}-1)$ d/λ where d is the interelement spacing in both dimensions [117]. In our considerations we have ignored the losses due to the power dividers, because these losses are not as high as the line losses; and the losses due to the phase-shifters, because one can add the latter losses to the former losses.

At broadside the gain, G, for a square aperture of area A having N antenna elements and $\lambda/2$ spacing is given by [117]

$$G_P = N\pi 10^{-\gamma} \qquad (2.67a)$$

where

$$\gamma = (N^{1/2} - 1)\ \frac{\alpha_{dB/\lambda}}{20} \qquad (2.67b)$$

If we set $dG/dN = 0$ we shall define the conditions where the array gain is maximum; this leads us to the expression

$$N_{max} = \left(\frac{40}{\alpha \ln 10}\right)^2 \qquad (2.68)$$

Clearly there is no point in increasing N beyond N_{max}. Even if α is as high as 1 dB/λ, N_{max} is equal to 301.

In a more realistic case, other losses associated with the antenna elements have to be taken into consideration and this topic will be revisited in the next chapter in some detail. The COBRA DANE radar system utilizes a passive phased array that has 96 sub-arrays and each subarray has 160 antenna elements. A traveling-wave tube (TWT) is used to feed the 160 antenna elements via a 1:160 power divider [118]; with this approach the losses are kept at a reasonable level.

Recently the availability of compact mini TWTs offers the designer another option to consider. Smaller-sized passive arrays can be fed by mini-TWTs so that:

- The array transmits sufficient powers at centimeter and millimeter wavelengths
- Losses are further minimized.
- The array has an improved graceful degradation characteristic.

We shall consider mini-TWTs, other vacuum tube microwave transmitters/amplifiers, and the microwave power modules (MPMs) in Chapter 4.

In Figure 2.13c we illustrate the hybrid array, where the array is active on receive and passive on transmit. A hybrid array can also be active on transmit and passive on receive. An array operating at millimeter wavelengths can, for instance, have the former configuration; the array will have considerable output power and the losses on receive are minimized. The above arrangements afford the designer several options to consider before arriving at an array type that will meet requirements.

2.5.2. Front-end Architectures

While the required array power–aperture product and spatial resolution dictate the radar array area, the array volume depends on the following considerations:

- Array options: passive, active or hybrid; we have already noted that conventional passive arrays tend to be bulky.

- Array frequency of operation and transmitted power requirements.
- Array sensitivity and scanning requirements.
- The type of the transmission lines used to distribute signals in an array; we have already noted that high-power-rating waveguides are bulky and heavy.
- Array power supplies—see section 2.5.8.
- Front-end array architectures, which we shall consider in this section in some detail.

The central issues for front-end architectures are:

- The minimization of the insertion losses of components/subsystems at the front-end of arrays.
- The plethora of beamforming options that active phased arrays offer because the effects of losses incurred after the T/R module are marginalized.

Beamformer architectures yielding one beam are considered in section 2.5.5.3 and architectures yielding multiple beams are considered in Chapter 5.

While loss minimization is a significant attractor for active phased arrays, the realizations of the extremely versatile beamformers, considered in Chapter 5, which often significantly increase the array sensitivity, constitute the strongest attractors of active arrays. It is not an exaggeration to state that the availability of affordable active arrays, based on MMIC T/R modules, triggered innovations in beamformers that significantly increase the array sensitivity.

2.5.2.1. Brick Architecture

The brick architecture is illustrated in Figure 2.14a; the antenna elements and the other subsystems have an orientation that is parallel to the longitudinal direction of the array and are stacked along a transverse direction. With this architecture, the power received by the antenna elements is coupled either electromagnetically or by electric connections to the remaining T/R components assigned to each antenna element. The brick architecture consists of several 'sticks' connected to a back plate and electrically connected to the RF, logic, and DC manifold. The ends of the sticks, which have different lengths, are connected to a cooling manifold. The cost of interconnecting the assembly of 2000 T/R modules is high, the array typically weighs hundred of kilograms and is approximately 30 cm long [124,125]. Newberg and Wooldridge, the authors of the same references, state that 'The "stick" assembly of T/R modules is not conformal and requires a myriad of electrical and mechanical connections that help make active arrays unaffordable.'

As late as 1994, most of the phased arrays under development or planned for near-term production used the brick architecture [119]; the ground-based radar (GBR) is an archetype for these architectures. The architecture allows the designer to use any of the following antenna options:

- conventional microstrip narrow band dipoles; or
- wideband printed dipoles that have 40% instantaneous fractional bandwidth [120]; or
- tapered slotline antennas that can operate over 1–3 octaves [121,122]—see Chapter 3.

If space is available, the considerable volume that the antenna elements and T/R modules occupy contributes toward easing the thermal problems.

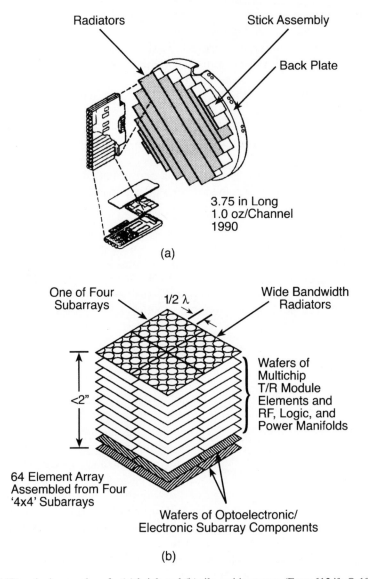

Radiators   Stick Assembly

Back Plate

3.75 in Long
1.0 oz/Channel
1990

(a)

One of Four   1/2 λ   Wide Bandwidth
Subarrays   Radiators

Wafers of
Multichip
T/R Module
<2"   Elements and
                                 RF, Logic, and
                                 Power Manifolds

64 Element Array
Assembled from Four
'4x4' Subarrays

Wafers of Optoelectronic/
Electronic Subarray Components

(b)

Figure 2.14 Practical examples of a (a) brick and (b) tile architectures. (From [124]; © 1993, IEEE.)

With this architecture, full polarization information is attained by an eggcrate structure [123], shown in Figure 3.21a. With this arrangement, half of the antennas receive
one polarization while the other half receive the orthogonal polarization. As can be seen,
the eggcrate structure is viable but difficult to implement. For the brick architecture,
replaceability of components or modules is practical.

2.5.2.2. Tile Architecture

The tile architecture is illustrated in Figure 2.14b; the antenna elements and other sub-systems, e.g. phase-shifters and T/R modules, have an orientation that is parallel to the transverse array aperture and are stacked along a longitudinal direction. With this architecture, the power received by the antenna elements located at the top of the tile structure is usually electromagnetically coupled to the second layer, which can consist of phase-shifters or active elements depending on whether the array is active or passive. Similarly, the third and other layers follow in order to complete the remaining signal processing functions assigned to the array.

A version of a proposed tile architecture, illustrated in Figure 2.14b, utilizes multi-chip module MCM packaging, digital technology, photonic signal manifolding, and true time delay for the realization of affordable, compact, lightweight and wideband phased arrays. The array consists of any number of subarrays, four of which are illustrated in Figure 2.14b, interconnected to populate a given aperture. For comparison, the weight of a 2000-element array using this new technology is estimated to be less than 34 kg and less than 50 mm in depth. The ratio of depths for the brick and tile architectures is about 6:1 [124,125] and this ratio is corroborated by Scalsi *et al.* [126]. Furthermore, the resulting array conforms to the skin of an aircraft and can meet the space and weight restrictions that apply to missiles, spacecraft, aircraft, and mobile ground-based platforms. Additionally, narrowband/wideband phased arrays having a tile architecture can accommodate a dual-polarization capability easily [127]. Dual-polarization narrowband patch antennas are often used and dual-polarization wideband flared slot antennas can be used for wideband arrays—see Figure 3.21b

As the resulting array is compact, the thermal problems have to be better managed for high-power arrays. While the arrays having the tile architecture are affordable, the replaceability of components may be difficult [127].

2.5.5.3. Single-Beam Beamforming Architectures

Beamforming architectures define the ways the RF power of the N antenna elements of passive/active arrays are summed to form one or more beams. Here we recall that the antenna elements of an active array are followed by T/R modules, while the antenna elements of a passive array are followed by programmable phase-shifters and the beamforming takes place after the T/R modules or the phase-shifters.

As can be expected, beamforming architectures depend on the front-end array architectures used and whether the array is to generate one or more independent and simultaneous antenna beams. Figure 2.15 illustrates the five conventional array interconnects that result when one beam is required. The planar corporate feed, illustrated in Figure 2.15a, is suitable for the tile architecture. The electromagnetically coupled patch antennas, shown in the upper tile, couple the received power to the feed patches that are in turn connected to the other layers of the tile where the additional T/R functions are performed. On the last layer of the tile, the array interconnects shown generate one beam. The outputs of the T/R modules, denoted by filled circles, are connected to a summing point, S, via equal lengths of transmission lines.

The volume corporate feed, illustrated in Figure 2.15b, is suitable for the brick

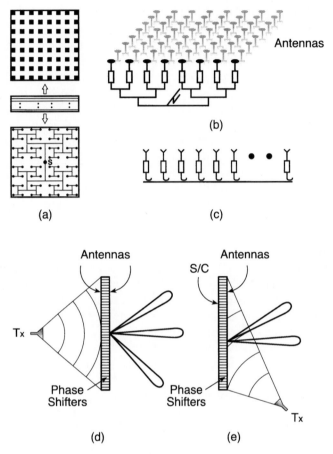

Figure 2.15 Beamformer architectures or interconnects capable of yielding one antenna beam. (a) Planar corporate feed. (b) Volume corporate feed. (c) Series-fed time delay. (d) Space-fed array, lens type. (e) Space-fed, reflect array.

architecture of active/passive architectures. In the figure an active array is shown where the antenna elements are connected to T/R modules before the received powers are summed in a summing point via equal-length lines; the summing points for each column are then connected to one point, again via equal length lines. This is by far the most popular interconnect approach because a planar phased array can consist of several columns each having a different length, so that the resulting aperture has the required geometric shape.

The series or source feed, illustrated in Figure 2.15c, is used in many waveguide and microstrip realizations forming rows or columns of a planar phased array.

In an effort to eliminate the connecting lengths of lines between individual antenna element–T/R combinations (or antenna elements/programmable shifter combinations) and the summing point, a space-fed architecture results in two realizations: the lens-fed and reflect array illustrated in Figures 2.15d and e, respectively. In the lens-fed version, both faces of the array are populated by pairs of antenna elements and each pair, located on either side of the rectangle shown, is connected to a programmable phase-shifter (or a T/R module). The array is steered to the required direction by the appropriate settings of the phase-shifters; thus the rectangle shown acts as lens, and the programmable phase-shifters take into account the differences in path lengths between the feed shown and the pairs of antenna elements along the length of the rectangle.

Although the reflect array shown in Figure 2.15e operates on the same principles as the lens-fed version, its programmable phase-shifters are terminated in shorted circuited (s/c) transmission lines and the offset geometry of the feed horn is chosen to minimize the obstruction of the incoming/outgoing radiation. The receive feed horn can be placed near the transmit horn shown in Figures 2.15d and e.

The Patriot radar is an example of a space-fed radar array [128], it was effective during the Gulf War and enjoyed wide publicity in the press. It is a multifunction mobile radar designed for tactical air-defense missions and operates at C-band.

The resulting arrays are simpler than their more conventional versions but are not volumetrically attractive for airborne applications where space is at a premium. A multimode feed of a space-fed array readily yields the sum and difference beams and the T/R switch or duplexer normally used in conventional interconnect architectures is eliminated; multimode feeds are considered in section 3.3.5. Losses and costs are therefore minimized and the complexities of deriving the difference beam are overcome. The derivation of several independently steerable antenna beams is, however, problematic with space-fed arrays.

We have already considered a beamforming architecture capable of yielding several staring/scanning beams from an array operating in the receive mode in Figure 1.1. The beamformer architecture can be seen either as an extension of the architectures considered in this section or as an architecture radically different. More beamformer architectures capable of yielding several antenna beams are considered in Chapter 5.

2.5.3. Array G/T Ratio and Noise Figure

The derivation of the *G/T* ratio and the noise figure of a realistic array is required so that design trade-offs can be studied systematically. Recent developments in wideband arrays utilizing photonics-based beamformers have increased the urgency of the need for the definition of these two important array parameters. Here realistic arrays are arrays in which all of the losses, including losses due to amplitude tapers, are taken into account.

While beamformers will be considered in some detail in Chapter 5, it is important to note the pertinent characteristics of photonics-based beamformers here. After suitable low-noise amplification, the RF is translated to optical wavelengths and programmable delays (instead of phase-shifters) in the form of fibers are used before a wavelength translation back to RF is implemented. The great attractions of these beamformers are wide instantaneous bandwidths over several octaves and light weight. However, the losses incurred in the wavelength translations range from 30 to 40 dB and efforts to reduce these losses are in train. This being the case, the need to define the *G/T* ratio and noise figure of a realistic array is of practical and theoretical importance. Here we shall follow the approach taken in reference [116] and consider the related issue of defining the dynamic ranges of realistic arrays in the next section.

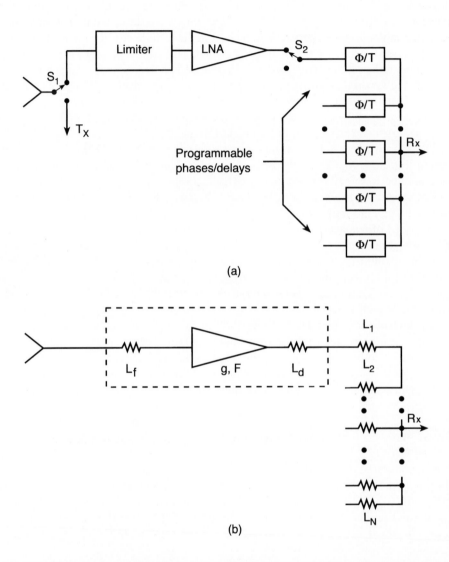

Figure 2.16 Block diagrams of the receivers of a radar phased array. (a) The limiter protects the LNA from high EM powers and switches S_1 and S_2 are the T/R switches. The programmable phases or delays are inserted before the summing point where the beam is formed. (b) The LNA has a gain g and noise figure F. The losses associated with S_1 and the limiter are represented by the resistor L_f while the losses associated with S_2, programmable phases/delays, and subarray combiner are represented by resistor L_d. The resistors $L_1, L_2, \ldots, L_N$ represent losses due to the array amplitude taper. (From [116]; © 1993, IEEE.)

In Figure 2.16a we illustrate the receive portion of a radar array that consists of the antenna element, the limiter, two T/R switches S_1 and S_2, and a LNA. The limiter protects the LNA from high microwave powers and the LNA has gain g and noise figure F. In Figure 2.16b we represent the losses in the system by resistors. These losses are:

L_f The front-end losses attributed to the antenna element, the limiter, and the switch S_1.

L_d The downstream losses attributed to S_2, the phase-shifters or time-delays, and the subarray combiner.

L_N The losses due to the array amplitude taper.

While the antenna is pointed to temperature T_i, all other subsystems are at room temperature, or T_0. For simplicity, the random amplitude and phase errors are not taken into account here and all N channels are assumed identical except for the individual weights denoted by losses L_N, in Figure 2.16b. Additionally, edge effects and mutual couplings between antenna elements are neglected.

In what follows we shall adopt the analysis outlined in reference [116]. If s_i is the input signal to each array element, the total signal output is given by

$$S_o = \frac{s_i g}{L_f L_d} \left[\sum_{k=1}^{N} \sqrt{1/L_k} \right]^2 \tag{2.69}$$

Similarly, the output noise n_o is

$$n_o = \sum_{k=1}^{N} n_k \tag{2.70}$$

where

$$n_k = \frac{kT_i Bg}{L_f L_d L_k} + \frac{kT_0 B(L_f-1)g}{L_f L_d L_k} + \frac{kT_0 B(F-1)g}{L_d L_k} + \frac{kT_0 B(L_d-1)}{L_d L_k} + \frac{kT_0 B(L_k-1)}{L_k}$$

The SNR is therefore deduced to be

$$\frac{s_o}{n_o} = \frac{Ns_i}{kTB} \frac{\left[\sum\limits_{k=1}^{N} \sqrt{1/L_k} \right]^2}{N \sum\limits_{k=1}^{N} (1/L_k)} = \frac{Ns_i}{kTB} \eta \tag{2.71}$$

where

$$T = T_i + T_0 \left[L_f F - 1 - \frac{L_f}{g} + \frac{NL_f L_d}{g} \frac{1}{\sum\limits_{k=1}^{N} (1/L_k)} \right] \tag{2.72}$$

and η is the well-known illumination efficiency.

If G_e, is the antenna element gain, the array gain, $G = NG_e\eta$ and the array G/T ratio is given by

$$\frac{G}{T} = \frac{NG_e\eta}{T} \tag{2.73}$$

and the array noise figure (NF) by

$$\text{NF} = \frac{kTB}{kT_0B} = L_f F - \frac{L_f}{g} + \frac{NL_f L_d}{g} \frac{1}{\displaystyle\sum_{k=1}^{N}(1/L_k)} \tag{2.74}$$

As can be seen, if g, the gain of the LNA, is high, the last two terms of equation (2.74) can be neglected and the array NF in dB is given by

$$\text{NF} = L_f + F$$

For an optimum design one therefore has to (i) minimize L_f; and (ii) ensure that g is high enough to render the second and third terms of equation (2.74) negligible.

Two typical examples illustrate how high g has to be to marginalize the contributions of L_f and L_d [116]:

Example 1   Consider that NF =2 dB, L_f =1.5 dB, and L_d = 2 dB. If g = 25 dB, the impact of L_d on the array noise figure is negligible.
Example 2   Consider that NF =2 dB, L_f varies from 0.5 to 1.5 dB, and L_d = 35 dB. g has to be at least 40 dB to render the impact of L_d on the array noise figure negligible.

If non-resistive networks are used to implement the amplitude taper, the term

$$\sum_{k=1}^{N}\left[\frac{1}{L_k}\right] \text{ tends to } N$$

in equations (2.72) and (2.74).

2.5.4. System Dynamic Ranges

In Figure 2.17 the input power, P_{IN}, versus output power, P_{OUT} of a typical fiberoptic link system is shown [129]. In the same figure the 1dB compression point of the system and the third-order intermodulation intercept point resulting from a two-tone inter-modulation measurement are shown. The compression dynamic range (DR) and spurious-free dynamic range (SFDR) are defined in terms of the following parameters:

• The input versus output power plot.
• The total noise power when the system bandwidth is 1 MHz.

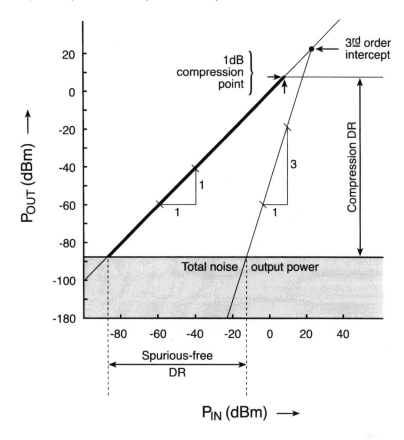

Figure 2.17 The input power, P_{IN}, versus output power, P_{OUT} of a fiber optic link system. The spurious-free dynamic range (SFDR) and compression (DR) are defined in terms of the total noise output
power, the 1 dB compression point and the third-order intercept point. The total noise output power
is calculated when the bandwidth is 1 MHz and the derived SFDR and compression DR are 73.8 dB
MHz$^{2/3}$ (113.8 dB Hz$^{2/3}$) and 95.9 dB MHz$^{2/3}$ (155.9 dB Hz), respectively [129].

- The measurements of the 1 dB compression point and the two-tone intermodulation
  distortion measurement defining the third-order intercept point.

The derived SFDR and compression DR for a typical system operating at 900 MHz are
73.8 dB MHz$^{2/3}$ (113.8 dB Hz$^{2/3}$) and 95.9 dB MHz (155.9 dB Hz), respectively.

2.5.5. Modern Array Synthesis Procedures

We have already broached the subject of implementing an amplitude taper for active
arrays utilizing current T/R modules in section 2.5. As it is expensive to use transmitters that emanate arbitrary powers at maximum efficiency, we shall restrict the types of
transmitters we are to employ in an array to say five or six and explore how best we can

approximate the required amplitude taper in an array. The design objective here is to achieve a minimal sidelobe level by using a staircase approximation by using five or six types of T/R modules. The constraints in the number of different T/R modules are imposed by cost only. Other approaches to achieve low sidelobe levels by utilizing current T/R modules are also explored.

2.5.5.1. The Lee Approach

Several authors have appreciated the difficulties and costs of producing T/R modules that (i) are identical in all respects including amplitude and phase tracking; and (ii) yield gains or power outputs that are arbitrary, to meet the amplitude taper requirements, at maximum efficiency. Lee [130] accepted a five-step approximation to the required taper and reported a −36 dB sidelobe level, a significant result. Without any loss of generality, Lee assumed an elliptic array area that was partitioned into five zones, shown in Figure 2.18, and each zone was populated by T/R modules having the same power output. An approximation to a Taylor −35 dB illumination is attained and convergence is easily reached with the aid of

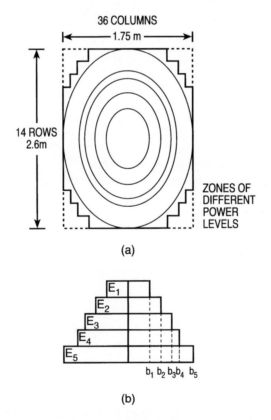

(a)

(b)

Figure 2.18 The optimum approximation of an amplitude taper across an array when current T/R modules are used, according to Lee. (a) Plan view of the five zones within which the same T/R modules are used. (b) Side view of the extent of the zones and the power levels of the T/R modules used. (From [130]; © 1988, IEEE.)

a gradient method. The parameters $E_1, E_2 \ldots E_n$ and $b_1, b_2 \ldots b_n$ are the variables. If the sidelobe level is set unrealistically low, e.g. –45 dB, convergence is not possible. In Figure 2.18 the extent of the five zones and the normalized power of each zone are shown. The attained minimum sidelobe level is not exceeded in any azimuth angle.

In the absence of theoretical work to establish a relationship between the attainable global minimum sidelobe level and the number of zones used, Lee's work can be used as a baseline case from which other approaches can evolve.

Table 2.11 The relative powers and extent of the five zones used by Lee [130]

E_1/b_1	E_2/b_2	E_3/b_3	E_4/b_4	E_5/b_5
1/ 0.34	0.74/0.52	0.52/0.68	0.36/0.78	0.2/1

In Table 2.11 the relative powers and extent of the five zones used by Lee [130] are listed; as can be seen, the power changes from one level to the next are approximately 3 dB. An additional stage of amplification can therefore accommodate the above requirement.

Mailloux and Cohen [131] proposed that the transition from one quantized level to the next be smoothed by using statistical thinning that effectively lowers the resulting sidelobe level. The authors used a three-level staircase approximation to a Taylor amplitude distribution designed to approximate a –50 dB sidelobe level and explored three methods of minimizing the resulting array sidelobe level after the establishment of the optimum radii at which the transitions from one level to the next occur. The resulting sidelobe levels ranged from –42.5 to –47 dB. Although these sidelobe levels are impressive, one can theoretically attain lower sidelobe levels by increasing the number of steps of the staircase approximation [131].

It is anticipated that Lee's synthesis procedure and its derivatives will be widely used until the cost of T/R modules capable of implementing amplitude tapers across arrays is lowered.

2.5.5.2. The Willey Approach

In a remarkable paper Willey [132] proposed:

- spatial tapers as alternatives to amplitude tapers for active phased arrays; and
- the population of a given aperture by separate transmit and receive arrays in 1962.

For randomly distributed arrays, spatial tapers are implemented if the density of antenna elements per unit area is maximum near the center of the array minimum near the periphery of the array area, and the transition from maximum to minimum density follows a well-defined taper function. Spatial tapers represent an economical and attractive solution to the problem of realizing low-sidelobe active arrays because all T/R modules are the same, and no powers are dissipated in the system (owing to the taper).

In Willey's terminology the conventional array utilizing an amplitude taper, termed the reference array, has U antenna elements and the number of elements of the array utilizing a spatial taper is R. The ratio R/U therefore represents the degree of thinning of the latter array. When R is too low or when the specified sidelobe is too low, the array sidelobe level at angular distances away from the main lobe tends to rise above the side-lobe level of the first sidelobe.

If the specified sidelobe level (SL, in dB) is not to be exceeded, the following equations should hold for the linear and planar arrays, respectively:

$$SL\big|_L = -10 \log \frac{R}{2} + 10 \log Q \tag{2.75}$$

and

$$SL\big|_P = -10 \log \frac{R}{4} + 10 \log Q \tag{2.76}$$

where

$$Q = \left(1 - k \frac{R}{U}\right)$$

respectively, and k is related to the interelement spacing of the reference array by the relationship $kd = \lambda/2$ (k is usually equal to unity).

Having determined R, the designer has to judiciously place the R elements on the linear/planar array. For a linear array the area under the required illumination function $f(x)$ is divided into a number of equal segments and equal current sources, or elements, are assigned to each segment. With this procedure $f(x)$ is approximated by a space taper.

For circular arrays the array is divided into annular rings of width w equal to the element spacing of the reference array. The illumination function $f(\rho)$, is integrated over each annular ring and over the total aperture and the number of elements per ring is defined by

$$\frac{\text{Integral over the ring}}{\text{Integral over the aperture}} = \frac{\text{Number of elements in the ring}}{\text{Total number of elements}} \tag{2.77}$$

where ρ is the radial distance from the array center.

In an example cited in [132], a planar array realized by following the above procedure had a maximum sidelode of –25 dB (Taylor distribution) when $k = 1$ and R_P/U =0.199. As can be seen, a respectable sidelobe level is achieved even when the array is severely thinned.

For a planar thinned array, having an area A, the resulting gain is given by [133]

$$G_{\text{thinned}} = (\text{Element gain}) \frac{R_P^2}{U}$$

$$= \frac{4\pi A}{\lambda^2 U} \frac{R_P^2}{U} = \frac{4\pi A}{\lambda^2} \left(\frac{R_P}{U}\right)^2 \tag{2.78}$$

regardless of the positions of the antenna elements.

Skolnik *et al.* [134] reported a maximum sidelobe level of –35 dB when the aperture had a Taylor distribution.

The COBRA DANE [118], a phased array-based radar, has the following characteristics:

Diameter	95 ft (29 m)
Peak transmitted power	15.4 MW
Frequency band	L-band
Number of antenna elements	19 403
R/U	0.558

Its antenna elements are so distributed on the array aperture as to form a spatial taper. It was deployed in 1977 on one of the Aleutian islands and its main function was to provide early warning of intercontinental ballistic missiles (ICBMs); the essential feature of the radar is to provide high range resolution observations of targets at long ranges, of the order of 1853 km.

While system considerations often define how an aperture is populated, it is useful to explore different ways of populating the aperture. When an aperture is thinned, it is possible to separate the receive from the transmit function in such a way as to fully utilize the aperture [133]. With this arrangement some antenna elements are connected to transmit-only modules while the remaining antenna elements are connected to receive-only modules. In the original proposal the transmit and receive arrays were interleaved and occupied the same aperture area.

This approach offers the following attractive features compared to conventional arrays:

- The receive and transmit arrays can have different spatial tapers.
- The receive and transmit arrays can occupy the same or different areas; the resulting beams can therefore be similar or widely different.
- As the front-end T/R switch is eliminated, valuable space is saved and cost decreases.
- The mutual coupling problem between antenna elements is not as severe; here the comparison is made with respect to arrays employing T/R modules.
- The heat emanated by the transmit modules does not easily migrate to the low-noise receive modules.

The requirement of an extra set of phase-shifters is a drawback of this approach; here we have assumed that one set of phase-shifters is used by the conventional array.

If a spatial taper is implemented by using the same T/R modules over the array aperture, the resulting sidelobe levels will be at the designed level. This being the case, the taper cannot be varied and this is a limitation for applications where one requires different sidelobe levels with the passage of time. It is often the case, for instance, that a sidelobe level of –X dB is acceptable at the commissioning stage of the array but is no longer acceptable at a later date. The comparison is here made with respect to an ideal case where the array taper can be varied without any losses. Other approaches to meet the requirement for different amplitude tapers are considered in section 2.5.5.4.

If the array has low sidelobe levels in the transmit mode, third parties are prevented from intercepting its emanations; similarly, low sidelobe levels in the receive mode increase the radar's resistance against intense jamming.

2.5.5.3. The Frank–Coffman Approach

The central element of the Frank–Coffman approach [135,136] is an 'active hybrid' array that is affordable when the cost of T/R modules is high. With reference to Figure 2.19a, the central portion of the proposed array is an active array that consists of 2500 antenna elements and an equal number of T/R modules. The remaining annulus of the aperture consists of 7500 antenna elements forms a passive array where one T/R module is connected to eight antenna elements via individual phase-shifters for each element.

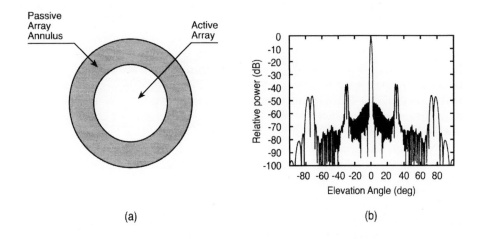

(a) (b)

Figure 2.19 The active-hybrid array proposed by Frank and Coffmann. (a) The central part of the array is occupied by an active array while the annulus (shaded) is occupied by a passive array. (b) The resulting array radiation pattern, illustrating the first and second grading lobes due to amplitude quantization. (From [136]; © 1994, IEEE.)

The total number of T/R modules used by the active hybrid array is approximately 3440 while the corresponding number of T/R modules for a conventional array would be 10 000. Thus the savings offered by the proposed array are significant.

The amplitude taper for the array was chosen to yield sidelobes at the −50 dB level. In Figure 2.19b the radiation pattern of the proposed array is shown: although the sidelobe level is at the designed level, amplitude quantization (AQ) lobes, which we shall consider in section 2.5.6.2, occur at the angular distances that are related to the subarray size and their maxima are at the −39 dB level. As can be seen there is merit in subarraying at the annulus of the array where the taper is severe and the power in the AQ lobes is diminished. If the subarrays have four antenna elements (2×2), the AQ lobes will move further away from the main beam but more T/R modules are needed. Techniques used to marginalise the AQ lobes are considered in section 3.8.1.

2.5.5.4. Other Approaches

An interesting variant to the approach Willey proposed results when the central portion of an aperture is occupied by conventional T/R modules and the remaining annulus is occupied by interleaved transmit-only and receive-only modules. With this arrangement the transmitted power is higher than that of the Willey array. Several tapers require near maximum power in the central portion of the array.

In Chapter 4, approaches to implement amplitude tapers will be explored. For some applications we have to meet the additional requirement to transmit enough power to explore targets located within a certain range only, so that third parties located beyond that range do not intercept the radar transmissions. Here we shall consider a proposal to implement a variety of amplitude tapers, and variation of the maximum array transmitted power with negligible loss of power.

In this proposal [137] the transmitter for each T/R module consists of M MMIC transmitters and a conventional spatial field combiner [138] that combines the outputs of the M transmitters before delivering the resulting power to the antenna element. Maximum power from a particular module results when all M transmitters deliver their powers to the antenna element; the delivered power, however, can be less than the maximum power if only a number of MMIC transmitters, fewer than M, are activated, so that a desired amplitude taper is implemented. As the insertion loss of the spatial field combiner is negligible, the power loss in implementing the amplitude taper is minimal. With this arrangement the array transmitted power is also variable.

This proposal is particularly suitable for arrays operating at low frequencies where the interelement spacing is sufficient to accommodate the spatial field combiner. Space limitations ultimately limit the maximum array power.

The population of the array aperture by receive-only and transmit-only modules offers a set of interesting possibilities to the designer that we shall explore in section 2.7.4.7. It is clear that many interesting options exist for the designer to consider.

2.5.6. Random and Quantization Errors

It is known that the gain of a reflector antenna increases as the frequency of operation increases up to a point beyond which the reflector's gain decreases as the frequency of operation increases. The highest frequency of operation is defined by:

• the random phase errors due to the manufacture of the reflector's surface and its supports; and
• systematic errors on the reflector's antenna mainly due to gravity loading and thermal distortions.

Phased arrays also suffer from random phase and amplitude errors due to mechanical causes, e.g. the tolerances associated with the positioning of the antenna elements and the phase and amplitude tolerances associated with the antenna feeds, the T/R modules, and the beamformers. As it is expensive to realize phased arrays that have no amplitude and phase errors, we have to consider the impact of random errors on the array's performance. Cost considerations are always interwoven with the realization of high-quality working arrays. The long-term objectives of production engineers are therefore to produce T/R modules that are affordable and have amplitude and phase tolerances that decrease as the knowledge base of the processes involved broadens and deepens.

We have already seen that amplitude quantization errors give rise to high array grating lobes. Amplitude, phase, and delay quantization errors can be likened to the systematic errors encountered in reflectors only because of their severe impact on the array performance.

2.5.6.1. Random Errors

Several studies [139–145] have defined the impact of random array errors on the sidelobe level, pointing error, and directivity of phased arrays. As the number of array elements increases, the impact of random errors on the array directivity and pointing errors tends to be not significant. In this section we shall focus attention on the impact of random errors on the array sidelobe level only.

Phase and amplitude errors that occur at different parts of an array direct a fraction of the available energy, σ_T, from the main beam and distribute it to the sidelobes. For small independent random errors, σ_T, is given by

$$\sigma_T^2 = \sigma_\Phi^2 \big|_{net} + \sigma_\alpha^2 \big|_{net} \tag{2.79}$$

where $\sigma_\Phi|_{net}$ and $\sigma_\alpha|_{net}$, the net rms phase and amplitude errors, in radians and in ratios, respectively, are given by

$$\sigma_\Phi|_{net} = \left[\Phi_1^2 + \Phi_2^2 + \Phi_3^2 + \ldots \right]^{1/2} \tag{2.80}$$

and

$$\sigma_\alpha|_{net} = \left[\alpha_1^2 + \alpha_2^2 + \alpha_3^2 + \ldots \right]^{1/2} \tag{2.81}$$

where Φ_1 and α_1, Φ_2 and α_2, Φ_3 and α_3, ... are the rms phase and amplitude errors respectively occurring in different subsystems of the array. The energy due to errors is radiated into the far field with the gain of the antenna element pattern.

To determine the mean-squared sidelobe level (MSLL) of the array of N antenna elements it is necessary to compare the energy due to errors with the peak of the pattern. More explicitly the MSSL in dB is given as [141]

$$MSSL|_{dB} = 10 \log \left[\frac{\sigma_T^2}{\eta N(1 - \sigma_T^2)} \right] \tag{2.82}$$

where η is the aperture efficiency of the array. If $\sigma_T^2 \ll 1$, equation (2.82) can be rewritten as

$$MSSL|_{dB} \approx 10 \log \left[\frac{\sigma_T^2}{\eta N} \right] \tag{2.83}$$

Phase and amplitude errors in different subsystems of a phased array are often expressed as upper and lower values in degrees and decibels, respectively; suitable conversions are therefore needed before one can use equations (2.79)–(2.81). It has been shown that for a uniform error density function spread over an interval I, the rms error is $I/(12)^{1/2}$. So if a subsystem of a phase array has phase errors defined as $\pm 2°$, the range is $4°$ or 0.698132 radian or 0.02 rms radian.

Similarly we require to convert a range between an upper and lower value expressed in dB into σ_α; to this end the following equation is used:

$$\sigma_\alpha = \frac{10^{\text{upper}/20} - 10^{-\text{lower}/20}}{\sqrt{12}}$$

Thus a ± 1 dB tolerance will convert into an rms ratio of 0.0666.

The bulk of amplitude and phase errors in an array are due to the phase-shifters that can be introduced either at RF or at IF. The first option is straightforward but yields relatively larger errors. If the phase-shifters are introduced at IF, the addition of a downconverter and a phased-stabilized local oscillator is required, but the resulting errors are lower. In this section we shall derive useful guidelines for the designer to consider and explore the technological issues related to the latter option in Chapter 4.

Table 2.12 lists the derived MSSL of active arrays when the number of T/R modules and the magnitudes of random phase/amplitude errors vary. Typical errors for conventional T/R when the phase-shifters are introduced at RF are ± 1 dB and $\pm 5°$ or ± 0.25 dB and $3°$[145]. For convenience we have assumed that the beamformers introduce the same set of errors as the T/R modules. The resulting MSSL are listed in Table 2.12 when the array efficiency is assumed to be 70%.

As can be seen, the higher the number of array elements, the lower the resulting MSSL. If a small array having only 100 elements is required to have low MSSLs, the random errors have to be much lower than those tolerated when the array is large. In the examples we have considered the small array should have errors of ± 0.1 dB and $\pm 1°$ before the resulting MSSL is comparable to that corresponding to an array having 4000 antenna elements. If a small array is required to have low MSSLs, the option of introducing the phase-shifters at the IF is therefore attractive.

As the errors are random, the designer should ascertain that the maxima of any

Table 2.12 The array MSSL in the presence of random phase and amplitude errors as a function of the number of array antenna elements

N	$\pm \alpha_1$ (dB)	$\pm \alpha_2$ (dB)	$\pm \Phi_1$ (deg)	$\pm \Phi_2$ (deg)	MSSL (−dB)
100	1	1	5	5	37
	0.25	0.25	3	3	44.7
	0.1	0.1	1	1	53.8
400	1	1	5	5	43
	0.25	0.25	3	3	50.7
1000	1	1	5	5	47
	0.25	0.25	3	3	54.7
4000	1	1	5	5	53
	0.25	0.25	3	3	60.7

sidelobes are not above a desired level. To this end it is useful to assume that the resulting radiation of an array symmetrically illuminated consists of:

(i)   The ideal pattern, which is the sum of cosine functions and is *unfluctuating*. In communication engineering terms this pattern can be likened to the *signal*.
(ii)   A mean floor of random sidelobes having a uniform average power in angular space when considered in correlation intervals of approximately the array half-power beamwidth; this component of the radiation pattern can be likened to Gaussian noise.

The designer's aim is to ensure that no sidelobe can exceed the specified level to a high probability, e.g. 0.85 to 0.95.

The similarity between the problem in hand and a problem solved by Rice [146] allowed Hsiao [147,148] to use the theoretical framework already established by Rice. Kaplan [149], derived a graph that allows the designers and manufacturers to study the trade-offs available to them so that the resulting sidelobe level does not exceed the required specified level to a given probability.

2.5.6.2. Amplitude, Phase, and Delay Quantization Errors

In this section we shall consider the impact of amplitude, phase, and delay quantization errors on the array radiation pattern. Our treatment will be brief, however, because these topics have been explored in some detail in many references, including the book [150].

Amplitude quantization (AQ) lobes occur because the amplitude control of the array illumination is implemented at the subarray level rather than at the antenna element level. By analogy with conventional array theory, grating lobes for an array having rectangular subarray lattices will occur in sine space at locations given by [136]

$$u_Q = u_0 \pm p \; \frac{\lambda}{s_x} \quad v_q = v_0 \pm q \; \frac{\lambda}{s_y} \quad p, q = 1, 2, \ldots \tag{2.84}$$

where s_x and s_y are the inter-subarray spacing in the x and y directions rather than the interelement spacing more typically associated with these equations. As can be seen the positions of the AQ lobes can be altered by varying the number of subarray elements, a procedure that changes s_x and s_y.

If we restrict attention to subarrays, of length L, formed along the x-direction, the nulls of the radiation pattern corresponding to the subarray, are at angular distances $\pm p \, (\lambda/L)$; for contiguous subarrays $L = s_x$ so the positions of the nulls coincide with the positions of the grating lobes. When all subarrays are uniformly illuminated, the grating lobes have negligible power levels. However when the subarray illumination has a suitable taper to lower the array sidelobes, the main beam and grating lobes are broadened so that the resulting array radiation pattern has split grating lobes resembling the monopulse radiation patterns. This is clearly seen in the grating lobes of the array radiation pattern illustrated in Figure 2.19b.

If the array has m subarrays of n elements, then the peak power of the AQ lobes, G_p, is given by [151]

$$B_0^2$$

$$G_p = \frac{1}{m^2 n^2 \sin^2(\pi p/n)} \tag{2.85}$$

where $p = \pm 1, \pm 2, \pm 3, \ldots$ and B_0 is the beam broadening factor or the ratio of HPBWs of the broadened beam corresponding to a uniformly illuminated array. For certain applications the AQ lobes can be objectionable, even at the −40 dB level, if the lobes enter the visible space at some scanning angle.

Modern phased arrays utilize digital phase-shifters to direct the beam anywhere within the array surveillance volume. The inserted phases are therefore only approximations of the required phases.

If an N-bit phase-shifter is used, the least significant phase, ϕ_0, is given by the equation

$$\phi_0 = \frac{2\pi}{2^N} \tag{2.86}$$

These quantization errors give rise to array grating lobes resembling those due to AQ but without the characteristic monopulse feature. Let u_0 and u_s be the direction cosines of the intended and actual directions the array is pointed. The worst case occurs when the error build-up and its correction are entirely periodic, in which case the array is divided into virtual subarrays and each subarray has the same phase error gradient. This occurs when the distance between subarray centers is given by [151]

$$|M \, \Delta\phi - M \, \Delta\phi_s| = \frac{2\pi s}{\lambda} \, M|u_0 - u_s| = \frac{2\pi}{2^N} \tag{2.87}$$

where M is the number of antenna elements in the virtual subarray and s is the interelement spacing. If $u_0 - u_s$ is known, M can be derived from equation (2.87). The virtual subarraying induces grating lobes to form, and the power in each lobe is given by [151]

$$P_{\text{GL}}|_p = \left[\frac{\pi}{2^N} \frac{1}{M \sin(p'\pi/M)} \right]^2 \tag{2.88}$$

where $p' = p + (1/2^N)$ and $p = \pm 1, \pm 2, \pm 3, \ldots$. In Table 2.13 we have tabulated the power in the first grating lobe as a function of M and N; as can be seen, $P_{\text{GL}}|_p$ can reach a high level of approximately −14 dB, when $M = 2$ and $N = 3$. As the number of bits used increases, the first grating lobe power decreases dramatically. The power in the second grating lobe is even lower.

It is often economical to steer the beam of a large array such as the COBRA DANE [118], operating over a small fractional bandwidth, by the use of time delays inserted at the subarray level, and phase-shifters at the array element. The resulting grating lobe power P_D due to delay quantization is given by [151]

$$P_D|_{\text{GP}} = \frac{\pi^2 X^2}{\sin^2 \pi[X + p/M]} \tag{2.89}$$

where

$$X = \frac{u_0 s}{\lambda_0} \frac{\Delta f}{f_0}$$

Table 2.13 The first grating lobe power (in –dB) of an array as function of N and M

	M				
	2	4	6	8	10
N					
3	13.97	17.92	18.57	18.80	18.90
4	20.11	23.57	24.15	24.35	24.45
5	26.17	29.39	29.94	30.13	30.44
6	32.20	35.31	35.84	36.03	36.11

If the array fractional bandwidth $\Delta f/f_0$ is much less than 0.1, the power in the grating lobes is negligible.

The COBRA DANE system has a 95 ft (30 m) aperture that is divided into 96 sub-arrays each having 160 antenna elements. If the array is pointed 22° in elevation off the aperture boresight, the top subarray receives a 36 ns time delay with respect to the bottom subarray. With this arrangement the incoming power is collimated over a bandwidth of 200 MHz [118].

2.5.7. Realization Aspects of Important Proposals

In Chapter 1 we considered the concepts upon which several important systems or proposals for future systems are based and in this chapter the realization aspects of three important proposals are explored. In section 2.7.4 detailed realization aspects of wide-band shared aperture systems are explored in some detail, while in this section we explore the realization aspects of a typical wireless power transmission (WPT) system and two realizations of bistatic radars utilizing illuminators of opportunity. The relatively long development times required to produce workable systems based on a continuous revision of the original proposals is stressed.

The concept of transmitting power at microwave frequencies originated with W.C. Brown in the 1960s. To this day no working system exists, but several prototypes considered in Chapter 1 have emerged and proposals are continually updated.

The original parameters of the SHARP system are detailed in reference [152] and are collated in Table 2.14. In the proposal the lightweight plane flies in a circle of radius several times the diameter of the rectenna. The proposed system is designated to provide communication and surveillance functions over the resulting footprint, but WPT has wider space-to-ground, space-to-space, and ground-to-ground applications [153].

Table 2.14 The pertinent characteristics of the proposed SHARP system [152]

Frequency of operation:	2.45 GHz (alternative frequency 5.8 GHz)
Transmitting phased array	250, 5 m antennas, densely populating a circle 85m in diameter
	Transmitted power by each antenna: 5 kW
Unmanned lightweight airplane	
At a height of	21 km
Rectenna diameter	30 m
Power flux density at 21 km	500 W/m^2
Derived DC power	35 kW
DC power for electric motor propulsion	25 kW
DC power for payload electronics	10 kW
Footprint on Earth	1000 km diameter

The issues of importance are:

- System considerations: What is an optimal frequency of operation for a pseudo-satellite?
- Rectenna realization.
- The selection of suitable and affordable transmitters.
- Self-cohering/self-steering schemes for the transmit array.

If D_{gr} and d_{rec} are the diameters of the antennas used on the ground and that of the rectenna, the equation

$$D_{gr} \approx \frac{4}{\pi} \frac{\lambda}{d_{rec}} h$$

can be derived from diffraction theory, where λ is the wavelength of operation and h is the height of the pseudo-satellite above ground—see also [154].

Taking into account the naturally occurring transmission windows through the atmosphere, systems operating at 35 and 94 GHz are worth considering. Assuming operation at 35 GHz and rectenna diameters of 4, 10, 20, and 30 m, the corresponding D_{gr} is 55, 21.6, 10.8, and 7.2 m, respectively. If a 94 GHz system is considered and the rectenna diameter takes the same range of values, the corresponding D_{gr} are 20, 8.12, 4.06, and 2.7 m, respectively. As can be seen, the D_{gr} and d_{rec} shrink considerably as the frequency increases from 2.45 to 35 and 94 GHz.

The designer therefore has a range of options to consider for an optimal system when the weather conditions of a certain locality of operation are known. Similarly, the d_{rec} and D_{gr} are selected on considerations related to how transportable the system has to be.

The ground antenna can either be a steerable filled aperture or a phased array comprising many antenna elements; for the latter case, a self-steering/self-cohering system is required.

The SKYLINK concept [155] explores a 35 GHz system and the unmanned aerial vehicle (UAV) height can range from 1500 m or 30 000 m depending on the application. The options for transmitters range from magnetrons, traveling-wave tubes (TWTs), gyro-devices, arrays of solid-state oscillators, and electrostatic-accelerator free-electron masers (EA-FEMs) [154]. EA-FEMs are considered as efficient sources of millimeter-waves for power beaming from a ground station and for plasma heating in tokamak fusion reactors [154, and references therein]. Power levels of 1 MW at 150 and 300 GHz are being targeted. The microwave power module (MPM) can also be added to this list of transmitter options for powering pseudo-satellites.

The problems related to rectennas operating at 2.54 GHz have been addressed in reference [152]. Dual-polarization rectennas are used and the rectifying diodes are chosen to dump spurious oscillations caused by the incoming radiation. Hybrid rectennas seem affordable up to frequencies of 94 GHz [155]. Above this frequency, expensive MMIC rectennas meet the dimensional accuracy required for their manufacture.

As the techniques used for radioastronomy are inadequate, self-cohering/self-scanning schemes have to evolve before the powering of pseudo-satellites becomes a reality. This is a crucial area of development and a promising concept [156] will be explored in some detail in section 5.7.4.

It is now appropriate to re-visit the concept of utilizing WPT to disable or harm airborne targets. Power densities of the order of 500 W/m^2 can be attained over an area normally occupied by the rectenna located at a height of 21 km from densely populated phased arrays of transmitters, and higher power densities can be expected from the same arrays transmitting higher powers. Similarly, the dimensions of the area occupied by the rectenna are negotiable. As the power density on targets increases, the threat to electronic equipment on board aircraft is increased proportionally.

If a sparsely populated phased array of transmitters is considered, however, the array gain is substantially decreased (see equation (2.78)) and the resulting power density decreases proportionally. The threat to electronic equipment on board airplanes is therefore confined to densely populated phased arrays transmitting substantial powers.

The detection of objects by bistatic radars using illuminators of opportunity was first proposed in 1934 [157] and since that date numerous reports of experimental systems using TV/FM station transmissions have been reported (e.g. [158] in 1986). In November 1998 a commercial bistatic radar system utilizing FM/TV stations appeared to be available [159]. It simply takes that long before ideas become commercially available systems. As there are some 55 000 FM radio and television stations worldwide, the available system has considerable appeal. The designers of the system put the unintentional jammers, FM and TV transmissions, to good use for the detection of various targets in a covert manner. The system descriptions are sketchy but a phased array is used in conjunction with special software and algorithms backed with parallel processing capabilities.

The attractors of using the GPS and GLONASS constellation of satellites as illuminators of opportunity are [160]:

- Time and phase synchronization, a major drawback of bistatic radar can be easily established.
- Forty-eight satellites transmit coherent circularly polarized signals in L-band and form a worldwide system of cooperative illuminators.
- Three physical phenomena are observed: (i) blockage of one or more space vehicle signals; (ii) diffraction of one or more space vehicle signals; and (iii) reflection of one or more space vehicle signals.

The availability of other constellations of LEO/MEO satellite systems designated to perform communication functions further increases the usefulness of this approach. However, the power levels available from the GPS/GLONASS systems is low at −130 dBm. Several targets such as aircrafts and an antitank missile have been detected and future work is directed toward the extraction of range and Doppler data using new, state-of-the-art receivers, dedicated software and omnidirectional as well as directional antennas [160].

This is a promising and challenging area of multidisciplinary area research encompassing signal processing, dedicated algorithms, RF technology, and knowledge of the GPS/GLONASS systems.

2.5.8. Heat Management and Power Supplies

Power supplies and the management of the heat emanated by the T/R modules and their power supplies are not topics popular with electronics engineers and scientists. Some authors even omit these topics from general considerations on phased arrays. The assumption is often made that these topics have long been successfully dealt with.

In this section we contend that recent research in power supplies and thermal management has resulted in novel solutions that are far superior to the conventional solutions offered in the past; furthermore we hold that:

- continuing research in these fields will result in additional benefits for the designer of phased arrays; and
- the design and realization of power transmitters and their power supplies is ideally a cooperative venture between the array designer and experts in the disciplines of modern power supplies and heat management.

2.5.8.1. Heat Management

The power-added efficiency (PAE) of solid-state devices and tubes is a measure of the efficiency by which a device converts DC or RF input power to higher RF output. Typically the PAE of solid-state devices varies between 10% and 80% when the devices operate at several tens of GHz and a few GHz, respectively.

The power dissipated, P_D, when a nominal output RF power, P_{out}, is generated is given by the equation

$$P_D = P_{out} \left| PAE^{-1} - 1 \right| \tag{2.90}$$

When $P_{out} = 10$ W and the device PAE takes the values of 10%, 50%, and 80%, P_D equals 90, 10, and 2.5 W. As can be seen, a lot of power has to be dissipated in order to generate 10 W of RF when the device PAE is only 10%. Taking into account the $\lambda/2$ spacing criterion between antenna elements, P_D per m^2 increases at a prodigious rate as the frequency of operation increases and device PAE decreases. These trends can be seen in Table 2.15, where the PAE takes the values of 10%, 50% and 80%, $P_{out} = 10$ W, and the frequency of operation is 3 and 100 GHz. The table clearly demonstrates the seriousness of the heating problem that the array designer faces. Air cooling is

Table 2.15 P_D, and P_D/m^2 as a function of PAE and frequency of operation (P_{out}=10 W)

PAE (%)	P_D (W)	P_D/m^2 at 3 GHz (kW/m^2)	P_D/m^2 at 100 GHz (MW/m^2)
10	90	36	40
50	10	4	4.4
80	2.5	1	1.1

appropriate for low-power arrays, while liquid cooling is more appropriate for high-power arrays. Water, for instance, is 1000 times denser than air and its heat capacity is 4 times greater. The major issue for concern is the provision of safeguards so that electric power is switched off when the cooling system fails. For the latter method of cooling, the thermal interfaces between the modules and the cooled channels need to be optimized also.

In reference [161] several efficient methods of cooling are described; liquid-cooled plates for power supplies and transformers in a radar power supply system are described in the same reference. It was experimentally verified that the MTBF of the subsystems of the ultrareliable radar (URR), increase with cooling [162].

2.5.8.2. Power Supplies

Power supplies designated for radar applications should be highly efficient and reliable. High-efficiency power supplies reduce the heat management problem, occupy less space, and are lighter. High reliability is required to match the reliability of the MMIC-based T/R modules.

In a typical conventional linear power supply the AC is at 60 Hz and the available voltage is transformed and rectified to 300 V DC before it is converted to the three voltages required to power the T/R modules [163]. An alternative cryopower system consists of a cryogenic superconducting generator that generates a 50 V AC at 1000 Hz, a rectifier, and a smoothing capacitor. The latter system has the potential of reducing the mass and volume of the conventional system by 50% [163]. By cryocooling the switching devices and capacitors to optimum temperatures one can potentially improve the savings to 80%. The switching devices considered here are high-voltage MOSFETs and the capacitors are of the ceramic multilayer type [163]. While the on-resistance of MOSFETs and the equivalent series resistance (ESR) of the capacitor decrease as their physical temperature decreases from 300 to 77 K, a compromise temperature of 150 K is found to yield an economic solution to the problem under consideration [163].

Switched-mode power supplies (SMPSs) offer the designer similar attractive characteristics and are used in applications where size and heat dissipation are important considerations. Their AC input, usually at 50/60 Hz, is rectified, filtered and converted to an AC having a frequency of a few tens of kHz before it is rectified again.

With reference to Table 2.16, the maximum efficiency, power, and weight densities of SMPSs are significantly higher than those corresponding to the linear power supplies. The synergy of these three parameters constitutes a formidable attractor for SMPSs. The increased efficiency is attained by the use of fast recovery time Schottky diodes instead of conventional rectifier diodes [164]. Similarly, the increased power and weight densities are due to the reduction in size of the various components operating at 20 kHz (and above) compared with the linear power supply operating at 50/60 Hz. Both types of power supplies have comparable reliability, as measured by their MTBF.

Table 2.16 A comparison of linear and switched mode power supplies [164]

Parameter	Linear discrete	SMPS discrete	SMPS high density
Maximum efficiency (%)	50	90	90
Maximum power density (W/in^3)	1	3	30
Maximum weight density (W/lb)	20	40	400
Max MTBF (h)	30 000	20 000	30 000

Power supplies for pulsed solid-state radars present several challenges. Very high peak currents are generated when low-voltage supplies power high-power transmitters. With current technology, these high peak currents are provided by energy storage capacitors located close to the RF power amplifiers. The function of the power supply is then to provide a precision recharging circuit. As the RF amplifiers are driven to saturation, small variations in the collector voltage cause small but significant variations in the amplitude and phase of the transmitter's power. These variations in turn influence the moving-target indicator (MTI) improvement factor, which is defined as

$$\text{MTI improvement factor} = \frac{\text{Uniform average of the system's output SNR}}{\text{Input signal-to-clutter ratio}}$$

over all target radial velocities of interest [165]. While the MTI improvement factor depends on the performance of many subsystems [161], here we shall consider the contributions of the power supply only. Quantitatively the instabilities of amplitude, A, and phase, Φ, are given by

$$A = 20 \log (A/\Delta A) \tag{2.91}$$

and

$$\Phi = 20 \log (1/\Delta\Phi) \tag{2.92}$$

where ΔA and $\Delta\Phi$ are the small but significant variations on the amplitude and phase of the transmitter caused by the power supply. $A/\Delta A$ is expressed as the ratio of the nominal supply voltage, V, to the small increment, ΔV, present on the supply voltage. $\Delta\Phi$, in radians, is the phase increment due to ΔV and is deduced from the pulling factor, P, often expressed in degrees per volt. Equations (2.91) and (2.92) can therefore be transformed to

$$A = 20 \log (V/\Delta V) \tag{2.93}$$

and

$$\Phi = 20 \log (1/0.0175P\Delta V) \tag{2.94}$$

If $V = 40$ V, $\Delta V = 5$ mV, and $P = 4$ degrees/V, $A = 78.06$ dB and $\Phi = 69.1$ dB.

If $\Delta V = 10$ mV, $A = 72.04$ dB and $\Phi = 63.1$ dB. As can be seen, the voltage of the power supply has to be highly regulated to maximize the MTI improvement factor.

Although SMPSs have the attractive features we have already outlined, their output voltage is not as easily controlled to the precision required; the output of linear power supplies, however, is easily regulated to the precision required. Often the bulk of the required power is obtained from SMPSs and a linear regulator is 'piggy-backed' to provide the final millivolts of charge to the energy capacitors with a high degree of precision [166]. In the same reference, techniques for meeting the requirements of variable radar duty cycle and PRF are outlined.

The same requirements apply when the radar utilizes vacuum tubes that require high-voltage power supplies; reference [167] describes power supply topologies that meet various requirements. The series resonant power supply combined with a patented control scheme has demonstrated a high degree of regulation, e.g. ±0.05% for capacitive loads of equal or higher than 20 nF.

If one supply is used to power a significant number of T/R modules, the designer has to ensure that the reliability of all power supplies is high. Some redundancy in power supplies is therefore used to buy insurance against power supply failures. Utilizing a number of redundant power supplies in a variety of configurations, an airborne active element array has a mean time before catastrophic failures (MTBCFs) of 15 209 hours [164].

The microwave power module (MPM) consists of a solid-state medium-power amplifier followed by a traveling-wave tube (TWT), which generates 100 W of power in the 6–18 GHz band. The MPM, considered in Chapter 4, utilizes a switched power supply or electronic power conditioner (EPC). Great advances in the miniaturization of the EPC have been reported, for example a reduction in volume of 10–20 times compared to currently available supplies [168]. The inverter frequencies are 150 or 300 kHz for the two approaches taken for its realization and efficiencies of about 90% have been reported [169]. Interestingly enough, the EPC occupies about 81% of the MPM volume. Further reductions in the volume of the EPC are possible and we shall re-visit this topic in Chapter 4.

The switched-mode power supplies (SMPSs) of the ultrareliable radar had a switching frequency of 50 kHz [162] and most radioastronomy systems utilize SMPSs.

From the foregoing considerations we may conclude that the issues related to power supplies designated for radar phased arrays are far from trivial and that continuing R&D will yield power supplies that are increasingly smaller, more reliable, and extremely efficient.

2.6. AFFORDABLE AND LOW-COST ARRAYS

This section is a roadmap of the different approaches being pursued to reduce the cost of realizing phased arrays. Approaches to reduce the total cost of MMIC-based T/R modules are delineated in Chapter 4, while other approaches aimed at array cost minimization are outlined in section 2.6.1. Although the resulting arrays are affordable, there is still room to realize even lower-cost arrays.

If one could eliminate all the phase-shifters of an array and their control subsystems, a substantial saving would result. Even if half of the phase-shifters are eliminated, considerable savings are attained. In Chapter 5 we shall consider arrays that scan the resulting beam in one dimension using phase-shifters and a RADANT system to scan it in the other dimension. In section 3.4 we shall consider Luneberg and Rotman lenses that employ no phase-shifters.

We have already considered self-cohering systems in Chapter 1. Apart from not having any phase-shifters, these systems do not require knowledge of the target's bearings. Communications between a flying aircraft and satellite can therefore take place by the use of a self-cohering array and similar considerations apply when a pseudo-satellite is powered by a network of closely packed transmitters on the ground. Simple and ingenious self-cohering systems are considered in Chapter 5. The development of techniques for realizing self-cohering arrays or 'intelligent arrays' constitutes an active and promising research area. More importantly, the low cost of realizing these arrays constitutes a significant attractor for phased arrays.

2.6.1. Approaches to Affordable Radars

In an excellent paper, Skolnik [170] delineated several approaches that result in affordable arrays. Here we shall summarize approaches aimed at array cost minimization without any deterioration of performance; indeed in many cases performance is improved by the affordable arrays.

The minimization of losses in critical RF subsystems results in efficient and affordable systems. Losses attributed to power dividers, duplexers (T/R switches), and limiters have to be minimized in affordable systems. The Russian-designed *Grill Pan* array provides an interesting comparison with western-designed radar arrays [170, 171]. The array operates at X-band, has 10 000 antenna elements, and is designed for aircraft and tactical ballistic missile defense. The array is space-fed and utilizes ESAs. Apart from protection from high-power EM radiation, the low-noise ESAs eliminate the limiters used in conjunction with MMIC-based modules and space-fed arrays utilizing multimode monopulse horn feeds have the following attractors:

- The use of one receiver and one transmitter, instead of using one T/R module for each element.
- The elimination of duplexers.
- More straightforward derivation of sum and difference beams.
- The minimization of the transmission losses usually associated with arrays using corporate-feed.

The total two-way RF loss from transmitter to receiver (excluding propagation loss) is reported to be about 3 dB and compares well with the 7–12 dB loss encountered in western operational phased array radar systems [170,171].

We have already noted that systems employing one receiver and one transmitter do not exhibit graceful degradation and the derivation of several independent beams is problematic. Additionally we reiterate here that space-fed arrays are particularly attractive for ground-based radars where space is not at a premium.

Other factors that lower the cost of the Russian array under consideration are:

• The use of separate radars for the surveillance and tracking functions, an approach that also improves its performance in tracking multiple targets in realistic environments—see section 2.7.4.3.

• The use of Faraday-rotator reciprocal phase-shifters with orthogonal circular polarizations for transmit and receive mode. Each phase-shifter is capable of 360° phase shift in azimuth/elevation and separate coils are used for the horizontal and vertical phase shifts. The resulting beam is therefore driven to the required position by separate row and column commands.

What follows are other approaches to realize affordable radar arrays.

Given that radar array systems have more than one face, cost minimization schemes involve several options and trade-offs worth considering. For a start, one has to deduce the number of faces the system ought to have to satisfy requirements and then consider the options of whether all of the array's faces operate simultaneously or in a sequential mode to meet cost and performance constraints. In reference [172] the optimal number of array faces for 360° horizon surveillance is deduced when (i) the total number of T/R modules is constant; (ii) the detection performance is the same in all beam positions; and (iii) the horizon scan time is used as a measure of effectiveness. Based on these criteria, the authors of reference [172] deduced that a three-face system is optimum when active arrays operate in a simultaneous or sequential mode, and a similar conclusion is reached when passive arrays operate in a sequential mode. However, when passive arrays operate in a simultaneous mode, the higher the number of array faces the shorter is the horizon time scan. As these criteria are not unique, the citation of other studies based on different criteria are included in the same reference.

To minimize the system cost of a four-face array, one transmitter is used on a time sharing basis [173]. To minimize leakages between arrays, each array operates at a different frequency and the four frequencies are interleaved in time. Given that the transmitter chain constitutes a significant portion of the module cost, the proposal can result in substantial cost savings.

A trainable array is an electronically steerable phased array that can be mechanically positioned, as needed, to cover a 90° (or wider) sector. A system comprising a simple mechanically rotating radar operating at one frequency to cue two trainable arrays operating at another frequency to the sector of interest is a more affordable and effective solution to a four-face array [170]. For their shipboard air-defense radar systems the Russians use trainable arrays almost exclusively [170].

When an array has to perform the tracking and surveillance functions at two widely separated frequencies, cost and complexity increase when a multifunction array is used [170]. If a compromise frequency is used, the two functions are not performed efficiently. In reference [170] it is claimed that it is less expensive to have two arrays operating at the preferred frequencies than to have one multifunction array operating at two widely separated frequencies. While it is too early to compare costs, the indications are that the cost of the shared aperture array performing the radar functions at widely separated frequencies compares favorably with the cost of a system comprising several apertures—see section 2.7.4.5. Development costs related to shared aperture arrays are, however, bound to be considerable.

The other aspect of interest is the platform on which the systems being compared operate. For ground-based systems, multiple systems are worth considering. However, in the next section we shall explore *inter alia* the many attractive features of wideband multifunction arrays that can perform the functions necessary for an aircraft's or ship's survival when space is at a premium.

Lastly, computer software can constitute a significant fraction of the array cost [170]. The lowering of these costs therefore impacts on the total system cost.

2.7. WIDEBAND ARRAYS

Wideband arrays have been in use over several decades and the total bandwidth of OTHRs, ESM systems, and some phased array-based radiotelescopes extend over several octaves. While the instantaneous fractional bandwidth of OTHRs and that of the radiotelescopes is only a few percent, the instantaneous bandwidth of ESM systems is wider and in some cases equals the total bandwidth.

Arrays currently under consideration afford the user new and exciting possibilities to:

- Perform many radar/radar-related functions at optimum frequency bands either on a time sharing basis or in parallel.
- Realize the radio equivalent of the Schmidt optical telescope capable of operating at several frequency bands simultaneously.

The technological developments that support wideband arrays are in the fields of high-quality wideband/dualband/multiband microstrip antennas; reflector antennas capable of yielding several low-sidelobe staring beams; wideband T/R modules; cryogenically cooled, wideband MMIC-based LNAs; wideband and versatile beamformers; and the promise of massive computer power at reasonable costs.

The problems related to the management of array grating lobes over wide bandwidths and when the arrays have substantial scan angles are challenging. We have already considered random/aperiodic arrays and shall examine other approaches that meet requirements in this section. A valuable knowledge base on wideband arrays derived from EW systems can be used to realize future wideband arrays. In this section we shall consider existing wideband arrays before we explore proposals for wideband shared aperture arrays.

2.7.1. Over-the-horizon Radar Systems

We have already considered OTHRs in Chapter 1 and noted that apart from their military uses, some of them are used for applied science applications, e.g. the monitoring of wind patterns over vast stretches of the ocean. A recent additional application is the interdiction of drug traffickers and illegal immigrants.

Pre-1990 developments and R&D undertaken in the United States, the then Soviet Union and the United Kingdom and Australia are outlined in reference [174]. Cold-war applications of the first OTHR built by the Naval Research Laboratory (NRL) included the detection and tracking of an aircraft out to ranges of 2000 nautical miles, in the early 1960s. Other OTHRs followed and demonstrated the additional capabilities of ballistic missile launch detection and ocean storm tracking [174]. In the early 1990s the relocatable OTHR (ROTHR) was built and tested in the United States [174]. By 1996 two ROTHRs, one located in Virginia and the other in Texas were operational [175]; their primary functions are to detect and track incoming aircraft and ships at long distance from the radar, at locations near a conflict. As the sites of conflict are not known *a priori*, the ROTHR is a useful adjunct to US defense capabilities. Since 1989 the US Air Force and Navy has permitted the National Oceanic and Atmospheric Administration (NOAA) limited ocean monitoring tests with their OTHR systems, some of which have been deactivated with the end of the cold war [176].

The Australian OTHR system located in Alice Springs, Northern Territory, dates back to the 1970s and is a low-budget system known as Jindalee, the Aboriginal word for 'bare bones'. At the time of writing, two OTHRs are being built, one located in Longreach, Queensland, and the other near Laverton, Western Australia, for the surveillance of the sea–air gap extending from the north-east to the north-west of the Australian continent. Upon completion, the two OTHRs will form the Jindalee Operational Radar Network (JORN), with its coordination center in Salisbury, South Australia. The original Jindalee OTHR may also be used in the JORN.

JORN is a unique network of OTHRs that is an essential part of Australia's defense. It was scheduled for completion in 1997 [177,178] at a cost of about A$1 billion, but problems arising from contractual and project management difficulties postponed its completion [177,178]. It is now estimated that the JORN will not be operational before the year 2001 and that its cost will exceed its original contract cost [178].

In Table 2.17 are listed the generic characteristics of a typical receive array of an OTHR system. The bistatic radar operates in an FMCW mode to maximize the transmitted power and the separation between the receive and transmit arrays typically ranges between 80 and 160 km. The transmit array usually utilizes wideband log-periodic antenna elements, while the antenna elements of the receive array consist of monopoles. Both arrays are linear and their length varies between 1 and 3 km. As can be seen, the system can perform the surveillance and detection functions over a very wide area: 60° in azimuth when the radar range is between 1000 and 3500 km, which amounts to about 5.9 million km^2.

Table 2.17 Generic characteristics of a typical receive array of an OTHR system

Linear array length (km)	1–3
Angular resolution (degrees)	0.5
Azimuth angle coverage	60°
Array elements	Twin-monopoles
Mode of operation	FMCW
Tx–Rx separation (km)	80–160
Total system bandwidth (MHz)	3–30
Launch elevation (degrees)	1–30°
Radar range (km)	1000–3500
Signal over atmospheric noise ratio (dB)	60
Typical transmit power (kW)	100

To ensure that the array 'looks' over one half of the hemisphere, suitably phased twin monopoles are used as the array antenna elements; the ROTHR, for instance, uses 372 twin-dipoles [176]. When the transmit power is of the order of 100 kW, the signal is at least 60 dB over the atmospheric noise. While this specification is required for military tasks, low-power, low-cost OTHRs specifically built for ocean monitoring are an attractive possibility [176].

The attractors of OTHRs are:

- Low-cost wide-area surveillance capability.
- Given that ocean backscatter is from water gravity waves with dimensions comparable to those of the radar wavelengths, the waves can be visualized and studied to provide a sea and surface diagnostic.
- The resonant scattering region of ships and large aircraft are in the HF band. As these targets move, their returns can be detected by measuring the Doppler shifted frequencies.
- OTHRs illuminate aircrafts from angles where their RCS is not minimal. The system's probability of detection is therefore increased.

The obvious disadvantages of OTHRs are:

- Operation in the HF band, which is heavily occupied by broadcasts.
- System vulnerability to jamming.
- Coarse spatial resolution: tens of kilometers compared to tens of meters (or better) afforded by microwave radars.
- The nuisances introduced by the ionosphere and quiet or active Sun—see section 2.7.1.1.
- As the OTHR looks down on its targets, a large-amplitude backscatter echo from the Earth and/or sea is produced at the same range as that of the desired target. Elaborate signal processing techniques and algorithms are therefore used to extract the target echo from the backscatter echo.

Detailed information about near-surface ocean currents is required for applications in effective fisheries management, pollution mitigation, search and rescue operations, and climate studies. Using one OTHR, one current component can be mapped. The use of two OTHRs allows the mapping of the surface-current vectors. In near coastal environments, pairs of shore-based HF radars have been used to map surface currents over an area of a few hundred km^2 [179]. With the use of a pair of OTHRs having an overlap region, a map in the Florida Straits, a sea area of 70 000 km^2, has been reported with a resolution of 10 km [176]. As this area represents only 2% of the area covered by these radars, this technique of mapping truly vast ocean areas is extremely useful. More observations are reported in references [175] and [180]. The coverage overlap region of the three OTHRs in the Australian continent is considerable.

2.7.1.1. Phenomenological/Technological Issues

The ionosphere extends from 50 km to at least 400 km above the Earth's surface and its electron density is mainly affected by emanations from the quiet and active Sun; its electron density in turn determines radio wave transmissions. When an oblique-incidence wave travels through an electron density gradient that increases with altitude, its path is bent away from the vertical and the wave will reflect back to Earth if the gradient is large enough. The lower the frequency, the smaller the required gradient.

The altitude of the maximum electron density varies modestly as a function of time when compared to the variations of the electron density. Thus the illumination of a given region on Earth by a fixed transmitter is not difficult if the transmission frequency is changeable. As the electron density of the ionosphere cannot be predicted, the optimum frequency to illuminate a certain region is found experimentally. OTHRs therefore have a real-time subsystem that determines the optimum frequency of transmission, which in turn is used by the radar system. Other factors that determine the transmission frequencies of radars are spectrum occupancy by HF transmissions and emanations from auroras.

The important technological issues associated with OTHRs are the optimization of the monopoles used in the receive array over the required frequency range, and the management of the system's grating lobes and sidelobes.

It is well known that monopoles are low-cost, robust, narrowband antennas. To meet the requirement of operation over wide bandwidths, their length can be adjusted [174]; alternatively, the tuning of monopoles ensures that their matching efficiency is optimum within the required bandwidth. We shall explore the tuning of monopoles with the aid of low-loss circuits and microprocessors in the next chapter.

The management of the sidelobes and grating lobes over a frequency range of 10:1 is of particular importance for OTHR arrays. As we have outlined several methods of managing array sidelobes, we shall focus attention on the management of the array's grating lobes. With a bistatic radar, the designer is free to select the interelement spacing for the receive and transmit array elements to be different, so that the resulting grating lobes are minimized after the two array radiation patterns are multiplied.

Cottony [181] and Cohen [182] proposed design procedures for arrays having acceptable grating lobes when the number of antenna elements per array is low. Extensions of their work to planar arrays are possible. In both approaches the assumption is made that for each array the interelement spacing is constant. Other possible approaches are:

- To adopt transmit and receive arrays where the interelement spacing is constant but different for the two arrays and the receive array has a prescribed radiation pattern. The sidelobe level of the receive array over the regions where the grating lobes of the transmit array occur is minimized.
- To adopt a receive array where the interelement spacing is aperiodic. With this arrangement the system's grating lobes can be made acceptable even when the interelement spacing of the transmit array is constant.
- To adopt transmit and receive arrays having aperiodic spacings and iteratively determine optimal spacings that result in acceptable system sidelobes/grating lobes over the required frequency range and scan angles.

For linear aperiodic arrays it is relatively easy to ascertain theoretically where the maximum grating lobes occur and take corrective measures. As can be seen, the possibilities are many.

Lastly, the spatial resolution of OTHRs can be improved by longer arrays.

2.7.2. Futuristic Approaches to Wide Area Surveillance

Setting the recent Australian experiences aside, OTHRs provide a plethora of useful products at reasonable cost for military and non-military users. For non-military applications less sensitive systems will be economically even more attractive.

In the military context the surveillance of wide areas at reasonable cost is important for many countries. The ROTHR will provide the same products to the users over any region of conflict. Airborne/spaceborne radars and SAR systems are complementary but expensive systems. LightSARs considered in section 1.11.2 are less expensive versions of their conventional counterparts and constellations of MEO/LEO satellites performing the surveillance function on a worldwide basis have been considered in section 1.11.5.1.

Are there any serious competitors to OTHRs? In the short term, bistatic radars using TV/FM transmitters of opportunity seem worth considering in areas where there are such transmitters—section 2.5.7. For the longer term, pseudo-satellites and bistatic radars using illuminators of opportunity such as the emissions from GPS/GLONASS satellites as well as other LEO/MEO constellations of satellites dedicated for communication purposes are worth considering. The long development times required have to be taken into consideration. The returns, however, are considerable and seem within reach.

2.7.3. Other Wideband Systems

In this section we shall briefly outline several experimental wideband systems for the purposes of establishing a baseline of previous art.

In section 1.3.3 we considered the bands of operation for the VLA and the NMA. The arrays are capable of operating over several bands at centimeter and millimeter wavelengths. At a given time, however, the system instantaneous bandwidth, centered at a frequency within one of the bands of operation, is narrow.

An experimental 96-element array operating from 3.5 to 6.5 GHz with a scanning capability of ±60° has been reported [183] and its antenna elements were open-ended waveguides forming an approximate triangular configuration. The methods of impedance matching over the required scan angles and frequency range are described and wideband phase-shifters are used.

A phased array system operating in the L-, S-, and C-bands has been realized and its fractional bandwidth at the three bands were 10%, 20%, and 20%, respectively [184]. The array consisted of three arrays of interlaced radiating elements.

An experimental phased-array that operated over the frequency band of 2.5–10 GHz has been realized by interlacing two sets of Archimedean spiral antennas [185]. The band was split into the bands 2.5–5 and 5–10 GHz and each band was covered by one set of Archimedean spirals.

A shared aperture phased array has been realized by stacking three different-sized Archimedean spirals in layers, thus creating a three-dimensional array of spirals [186]. The array operated over a two-octave bandwidth and the blockage due to the partial overlap of elements from the different layers of the array was minimal.

2.7.3.1. Electronic Warfare Arrays

As radars operate at widely separated frequencies, EW systems and arrays have to operate over wide bandwidths. The maximization of the effective isotropic radiated power (EIRP) for ECM arrays is a prime requirement. Given the choice of either combining the powers of several transmitters conventionally or combining the powers of many transmitters with the aid of phased arrays, one would prefer the latter option because it offers the advantage of aperture gain. If the location of a monostatic radar can be determined accurately, surgical jamming, derived from a planar array, can direct all available power to the radar's receiver. With this arrangement the jamming power is not detected by ESM systems. Often a linear array generating a fan beam is used because: (i) the DF system on board planes is not sophisticated, hence the target position is not known accurately; and (ii) the decreased complexity of a linear array, when compared to the planar array.

For airborne operations typical power levels for the different types of jamming are:

- 1–10 kW for self-protection; and
- 30–300 kW or higher for stand-off [187]

The latter type of jamming is used to protect an aircraft from radar detection. Fail-safe jamming is omnidirectional jamming often used when: (i) an aircraft flies among many threats, or (ii) it is required to jam a bi/multistatic radar and the position(s) of the receiver(s) is not known.

The generation of jamming powers approaching 1 MW in the I/J bands, accomplished by a combination of multiple TWT amplifiers and high-gain phased array antennas, has been reported [187, 188].

An active phased array architecture has been shown to be a valid alternative to the traditional ECM power transmitter based on TWT in the H/J band. The criteria used were power consumption and dissipation. The solid-state solution was by far the most preferable [189].

Ideally, the ESM system has an array with the following characteristics:

- It operates over a wide bandwidth to maximize its POI in the frequency domain.
- It yields several beams to maximize its POI in the spatial domain and several nulls to protect it from jammers.
- It is equipped with several tunable band rejection filters to protect it from jammers in the frequency domain.

Jammer excision in the frequency or spatial domain also enables the ESM to detect weak emissions that would otherwise be masked by high-power jammers.

The efficient de-interleaving of the incoming pulses intercepted by an ESM system depends heavily on specific measurable parameters such as the DOA and emitter frequencies, which can be used as discriminants for the sorting strategies and algorithms. If the DOA is used as a discriminant, the ESM system offers discrimination of radars using frequency agility, pulse jitter and/or spread-spectrum techniques [190]. As the accuracy with which the DOA is measured increases, the sorting algorithms become more efficient. Multibeam arrays (MBAs) have been used in conjunction with ESM facilities [187, 188]. When a MBA is used by the ESM and ECM subsystems of an EW system, fail-safe jamming or jamming toward two or more directions can be implemented.

The ESM and ECM subsystems are often integrated to increase the survivability of high-value platforms. The threat of antiship cruise missiles (ASCMs) that can be launched from many platforms, including submarines, shortened the time the victim ship has to take defensive action to less than 30 seconds. The reaction times of the ESM/ECM defense therefore had to be shortened to cope with this short warning span [188].

The available raw power on a platform is bounded and space is limited. The survivability of both aircraft and ships therefore depends on decisions made in quasi-real time during a tactical engagement.

2.7.3.2. A Precision ESM System

We have already considered a wideband ESM system consisting of two circular arrays of wideband antenna elements in section 2.4.1. Its rms accuracy in the measurement of a target's azimuth bearing accuracy was 2°. *Inter alia*, higher accuracy in the target's bearings in elevation and azimuth are required for improved anti-air warfare capabilities, especially to low altitude targets; multisensor correlation; and improved de-interleaving capabilities.

Anaren's precision ESM prototype shown in Figure 2.20 is particularly suited to pin-

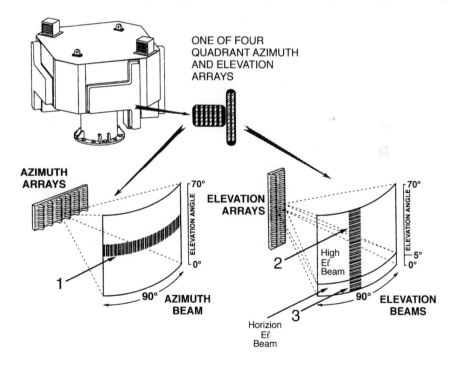

Figure 2.20 A precision wideband ESM system consisting of four quadrant azimuth and elevation arrays. **1**: The azimuth beam has a 90° × 70° FOV and is resolved into 4096 resolution elements of 0.02°. **2**: The high-elevation beam has a 90° × 65° FOV and is resolved into 4096 0.02° resolution elements of 0.02°. **3**: The horizon beam has a 90° × 5° FOV and is resolved into 512 resolution elements of 0.02°. (Courtesy of Dr Stig Rehnmark, Anaren Microwave Inc.)

pointing low-altitude targets in naval and ground-based scenarios with severe elevation multipath [191]. As can be seen, each of the four quadrants of the system covers a 90° × 70° segment in azimuth and elevation, respectively. The azimuth arrays consist of six 32-element arrays, while the low-elevation array consists of three 32-element arrays connected to a vertical 64-element antenna array. All antenna elements are wideband tapered slotlines. The essential system parameters are:

- Frequency range 7.5–18 GHz; and
- Rms bearing accuracy in elevation 0.17° and in azimuth 0.16° at 9 GHz in a multipath environment.

The system has an azimuth and a low-elevation multibaseline interferometer for angle measurement in each active quadrant and the phenomenal accuracy over 90° of bearing is attained because the system uses multiple-baseline interferometric techniques [192] combined with quadrantial azimuth beamwidth antennas. The use of interferometer elements with shaped beams is an additional contributing factor in suppressing the multipath reflections, a requirement that is readily met by the antenna elements used but not by wideband spiral antenna elements [191].

2.7.3.3. The Role of FPA Non-linearities in Wideband Arrays

In a wide-scan, active transmit array operating over three octaves, non-linearities associated with the FPAs can influence the array radiation pattern. In a preliminary study of linear arrays, three cases have been considered: (i) one beam is transmitted at one frequency; (ii) two beams at different frequencies are transmitted through the same FPAs; and (iii) one beam with two transmit carriers is transmitted through the same FPAs [193].

For the first case the resulting beams due to the fundamental frequency and its second and third harmonics are coincident, but the distributions due to the second and third harmonics are squared and cubed versions of the distribution pattern due to the fundamental frequency.

For the second case similar comments apply, but the second- and third-order intermodulation products, either sum or difference, are in general offset in beam angle and have squared and cubed amplitude tapers. Some band limiting is needed to dramatically decrease the number of radiated signals.

For the third case let us assume that the carrier frequencies are f_1 and f_2 and that a single beam is derived pointed to the direction θ_0. The angle of the nth-order intermodulation product $\theta_{n\text{-order}}$ is given by

$$\theta_{n\text{-order}} = \sin^{-1}\left(\frac{nf_0\sin\theta_0}{f_{\text{IM}}}\right)$$

where f_{IM} is the frequency of the particular intermodulation product. There is no angular dispersion at broadside and the angular dispersion increases in proportion to the sine of the scan angle. This method of analysis is to be extended to planar arrays and further verification of the beam pointing angles of the intermodulation beams for an active array would be useful [193].

Considering many system aspects and detailed knowledge of the components of a wideband array operating over, say, three octaves, designers of EW systems divide the band of interest into one-octave bands for further processing. The bands can be separated with the aid of multiplexers located after the antenna elements. In view of the above analysis, this approach seems eminently suitable for wideband radar arrays.

2.7.3.4. Proposals for Second-generation Radiotelescopes

The evolution of radiotelescopes followed the evolution of optical telescopes. Optical telescopes of ever-increasing sensitivity having a spatial resolution of one arc second, defined by the effects of the 'dirty atmosphere' on the quality of the images, evolved with the passage of time. City lights and high water vapor content in the air impair the operation of optical telescopes, hence the move to install optical telescopes at high-altitude sites and away from city lights.

In a parallel development, the realization of the wide FOV telescope, or Schmidt optical telescope, in the 1920s paved the way for several telescopes having wide FOV, e.g. a few degrees by a few degrees. Schmidt telescopes operating from the northern/southern hemisphere observatories provide maps of the cosmos at adequate spatial resolution, inexpensively. Recently the Hubble Space Telescope, orbiting the Earth, lowered the resolution limit to 0.07 arc seconds or 70 milli-arc seconds [194].

In radioastronomy, phased arrays replaced the early radiotelescope that used a single reflector and the spatial resolution of the arrays improved as the dimensions of the arrays increased. A maximum spatial resolution of sub-milli-arc seconds has been reached by radiotelescopes utilizing the OVLBI technique.

Radiotelescopes are affected by industrial terrestrial noise and interference from broadcasting and radar stations, hence the move to install radiotelescopes away from large cities and in valleys shielded from unwanted radiations. If operation at millimeter wavelengths is intended, the millimeter-wave telescopes are located at high-altitude sites, side-by-side with optical telescopes.

Similarly phased arrays having one wide field of view or multiple fields of view have been proposed ([195] and references therein)—see also section 3.3.2.1. The proposed

Table 2.18 The essential parameters of systems designated to observe the four important OH transitions and of the GPS systems

System	Frequency (MHz)	Polarization	Bandwidth
OH 'line' receivers	1612, 1665, 1667, 1720	100% linear or circular	± a few MHz
Differential GPS	1237 1575	Circular	± 20 MHz [196]

radiotelescopes are in effect the radiofrequency equivalents of the Schmidt optical telescopes. The other requirement for observing celestial sources at many frequency bands simultaneously has also been addressed in the same section and reference. Radiotelescopes meeting the last two requirements can be used to accelerate the acquisition of the required observations of the cosmos; to observe variable sources; and to search for evidence of extraterrestrial intelligence.

The recent completion of the GPS constellation of satellites and the coming of other LEO/MEO satellites is posing a difficult problem for current and future radiotelescopes. A comparison of the frequencies assigned for the GPS system and the four important transitions of the hydroxyl (OH) radical, listed in Table 2.18, will qualitatively illustrate one aspect of the problem. As can be seen, the 1612 OH transition is only 37 MHz away from the 1575 MHz frequency of the GPS system; taking into account the bandwidths of the two systems, and the relative strengths of the two signals, the OH line receivers will need a high degree of interference rejection.

As more constellations of LEO/MEO satellite systems are expected to come into operation [197], at many frequency bands listed in Table 1.16, the radiofrequency interference (RFI) problem is expected to become a central design issue for current and future radiotelescopes. Irrespective of where radiotelescopes are located, they are vulnerable to RFI. The mitigation of the effects these transmitters on the very weak signals emanated from distant sources and other galaxies is therefore mandatory. Work in this area and in other areas related to the optimum realizations of future radiotelescopes is outlined in sections 3.3.2.1 and 3.10.

2.7.4. Shared Aperture Arrays

In this section we shall explore approaches for shared aperture systems and the fundamental considerations related to proposals and prototype systems. Several diverse proposals and prototypes drawn from the military and civilian sectors will complement our account of shared aperture systems and illustrate the opportunities afforded by the novel approaches. In pursuing these aims, we shall refer the reader to references related to important work that falls outside the scope of this coverage.

There is broad agreement that the L- and X-bands are considered as optimum frequency bands for the radar surveillance and tracking functions, respectively. Without any loss of generality, we shall assume that the nominal frequencies are 1 and 10 GHz.

As agreement on the shared aperture concept is not universal, we shall outline and consider the dissenting views even-handedly. We shall contrast the arguments for shared aperture systems capable of performing the functions necessary for the survival of a platform with the arguments for separate systems performing the same functions. It is stressed that this topic is in an evolutionary state and that the reported work is of an exploratory nature.

2.7.4.1. The Arguments for Systems Integration

The proliferation of advanced sensor and communication systems on board military platforms has led to a multitude of systems. A US Navy Aegis cruiser, for instance, has over

100 antennas and the number is expected to rise as new systems are added [198]. In the first instance these antennas increase the total RCS of the platform. A reduction of the number of antennas used through systems integration is therefore highly desirable.

If the designer allows all these independent systems to run from the finite power supply of the platform, serious system problems can be experienced in a dynamically changing environment where the system response time is shrinking with the passage of time. For aircraft, the available area for apertures accommodating several independent systems is also finite, so integration is imperative.

Whether the integration of the systems on board a platform leads to one shared aperture accommodating the many functions necessary for the survival of the platform or to several apertures, RFI experienced when receiving and transmitting systems operate simultaneously is a serious concern. The present solution to marginalize the effects of RFI is to turn the receivers off while the transmitters are on, but this approach forces the operators to choose between mission accomplishment and survivability [199]. Some scheduling and prioritization of the many functions is therefore necessary so that the RFI problem is managed and the finite power available supports the functions that ensure the survival and offensive roles of the platform.

A platform can perform defensive/offensive functions such as communications, surveillance, and tracking on the one hand, and missile guidance, threat illumination, and kill assessment on the other. In general terms the integrated system is threat-driven and adapts to the ever-changing dynamic environment. If one adopts a doctrinaire approach, the offensive functions are allotted a higher priority than those given to defensive functions. Such an approach, however, is not as successful as a heuristic energy management approach in scenarios where the radar operates at maximum load [200]. Such issues can be studied using an artificial intelligence (A)-based weapons system simulation as a system test bed and analytical tool [200]. Important as these system aspects are, we shall not pursue them further and we shall focus attention on system aspects closely related to the RF subsystems.

2.7.4.2. The Case for Shared Aperture Systems

If the total system RCS is to be reduced, it makes sense to decrease the number of antennas by consolidating the functionality of several systems into a single shared aperture antenna. The derived multifunction system performs its assigned functions either on a time sharing basis or by processing the derived information in parallel. As we shall see in section 2.7.4.4, both approaches have merits.

An important argument in support of shared aperture systems is that a receive planar array of N antenna elements can yield at least N independent and agile antenna beams. If only one antenna beam is derived from the array, the opportunity of utilizing the remaining $(N–1)$ beams is lost. An ESM system, for instance, can readily utilize all the derived beams to increase its POI in the spatial domain. The same argument can be used against the approach of using separate apertures to perform the required functions.

Are there any shortcomings related to reliability if all functions are performed by one system? Here we have assumed that the shared aperture system has graceful degradation, self-calibration, and self-healing features that allow it to operate at high efficiency under a diverse set of operational conditions.

Cost comparisons between shared aperture systems and an ensemble of independent systems are not firm as yet, but the indications are favorable for the former systems—see section 2.7.4.5.

2.7.4.3. The Case for Independent Systems

The dissenting views from the shared aperture concept [170,171,201] are well founded. The competing concept is to have several systems each performing a limited set of functions almost without any time constraints. Several radars coordinated by distributed computers can provide the same capability as shared aperture arrays at lower cost and with greater survivability, it is claimed [201].

For a start, the timelines for many critical functions can be long and are expected to be longer in the future [198, 201], so a multifunction radar operating on a time-sharing basis will not be able to meet current and future challenges. The other aspect is related to the detection of low-RCS targets in environments containing rain, chaff, and other sources of clutter. To perform this task the radar transmits high-PRF waveforms over long dwells: 'an almost insoluble problem when the multifunction approach is adopted' [171]. These views are based on comparisons between multifunction radars operating in the West and independent radars performing several functions in Russia [170,171].

Lastly, independent systems can have different array architectures to perform the radar and communications functions. Radar arrays can have conventional architectures, while self-focusing/self-scanning arrays are dedicated for communications.

2.7.4.4. The Ideal *Shared Aperture Array*

While there are many compelling arguments for the integration of many functions, no universal support for the shared aperture concept is apparent. A critical review of the many realization options is therefore warranted before an ideal shared aperture is arrived at. After the definition of the *ideal* shared aperture, we shall list the system and technological issues that have to be addressed before it can be realized.

Let us revisit the shared aperture that performs all functions in a time multiplexing mode, the realization approach widely used by many narrowband multifunction radars. The entire aperture is used to derive one inertialess beam and the total system bandwidth can be wide enough to accommodate the radar surveillance and tracking functions. The derived system has a long range and is effective in an environment defined by fair weather, a low number of targets, and the absence of jammers.

While time multiplexing is used extensively in communications, the same approach for radar-related functions does not satisfy the demands of modern weapon systems, because the timelines for these systems are generally long and ever-increasing [198, 201]. Time multiplexing may be useful for tactical communications systems, but system analysis and trade-offs are required [198].

Aperture segmentation is another realization option [198]. The real estate available for the aperture is segmented and each segment performs one function; in an alternative realization each segment performs two or more functions on a time-sharing basis. With this arrangement different arrays can have different architectures. This option yields a shared aperture system that is efficient because it has many antenna beams but its radar/communications ranges are short because the full aperture is not used.

The ideal shared aperture system uses a single common aperture to perform multiple functions simultaneously. This is possible because the aperture provides multiple, independently controlled beams simultaneously. In the terminology developed in this book, the nexus between the many functions of the system is completely broken. Practical aspects considered in the next section dictate the separation on the receive and transmit arrays when available components are used. While this approach of realizing the ideal shared aperture is potentially the most advantageous, it is fraught with significant instrumental challenges as discussed below. In what follows, we shall assume the availability of wideband antennas, polarizers, T/R modules and beamformers, the subsystems that we shall consider in the remaining chapters of this book.

In considering broadband systems, care should be taken that the broadband capability is not attained at the expense of performance. A narrowband high-power amplifier (HPA) may, for instance, have 2–3 times the power output and 2–4 times the PAE of a wideband HPA [199]. How narrow should the bandwidth be? Technological and economical issues considered, an octave bandwidth is considered as a compromise between several subsystems operating over substantially narrower bandwidths and one subsystem operating over a multioctave bandwidth. Low-loss multiplexers can therefore be used to divide a multioctave band into octave bandwidth bands for further processing. In what follows we shall outline several proposals and prototypes to illustrate the new degrees of freedom offered to the designer by the shared aperture paradigm. While proposals promote and sustain the innovation process, prototypes bridge the gap between proposals and production models and define problem realization issues.

Not all problems associated with shared aperture systems are solved, but considerable progress has been made on several fronts.

2.7.4.5. US Proposals and Prototypes

The US Navy's current AEW capability is based on the APS-139 radar installed aboard the E-2C aircraft. The radar operates at the UHF band in conjunction with a mechanically scanned antenna implemented within a rotodome configuration. Current and future threats, threats scenarios, and technological advances support the need for a wideband AEW radar system with an electronically agile beam. Operation at the UHF band offers advantages in detection over L-band against reduced RCS targets and device technology that offers higher powers. The new system that is at the proof-of-concept stage utilizes an active phased array that has the following characteristics [202,203]:

Frequency range	0.4 –1.4 GHz, a ratio of 3.5:1
Number of elements	153 (9 rows and 17 columns)
Surveillance volume	±60°
Antenna elements	Suspended circular disks
RMS sidelobe level	45 dB or better
Polarization	Provision for two orthogonal polarizations

Theoretical work is centered in the derivation of the array pattern and the elements' input impedance over the entire frequency range of operation and scan angle. The calculations involved the generation and subsequent solution of matrix equations with the number of unknowns of the order of 6000 at the high end of the frequency band.

By the use of the reflection and translation symmetries of the array as well as the use of computational shortcuts, the CPU time for individual runs per frequency was reduced to 2–4 hours! Preliminary results are in excellent agreement with measurements. The active array area is 1.81 m × 0.96 m and the physical area of the array is 2.67m × 1.39 m. The periphery around the active array is populated by antenna elements terminated into matched loads in order to dampen any surface waves. The T/R module designated for this array has a peak transmit power of 500 W and an instantaneous IF bandwidth of 2 MHz on receive. In order to achieve maximum power capability, the T/R module operates over two bands extending from 400 to 850 MHz and from 850 to 1400 MHz. Other system and subsystem details are outlined in references [202,203].

The Johns Hopkins Applied Physics Laboratory (APL) realized an experimental shared aperture array to allow its researchers to explore system-level issues and risks associated with shared aperture arrays [198]. Table 2.19 lists the salient characteristics of the receive-only array and a separate transmit array is envisaged in a realistic environment. This configuration provides some isolation between closely spaced frequency bands without a duplexer and filter. Typical duplexers and filters are too large to meet the space constraints imposed by the interelement spacing in the array. Improvements in the state of the art of filter/duplexer technology are necessary to support full duplex, multibeam operation in a single aperture; this is an important conclusion reached by the designers of the prototype.

The array elements are Vivaldi tapered notch (also known as tapered slotline) antennas with a wide bandwidth microstrip-to-slotline transition. The two orthogonal linear polarizations are combined to derive one circular polarization. Again, polarization agility will be required in a realistic environment.

The receive modules consist of several GaAs MMIC commercially available chips performing the microwave functions and silicon integrated circuits performing all of the control functions. While additional measurements are in progress, the in-house design, manufacturing, and testing of the antenna elements and receive modules constitute a major achievement.

Table 2.19 The essential characteristics of the APL prototype

- 8 × 8 antenna element of the receive-only array
- 50° maximum scan angle
- Triangular lattice geometry
- One circular polarization derived from two orthogonal polarizations.
- Dual-channel receive MMIC modules based on commercially available chips
- Two independently steered beams
- 7–10 GHz band, continuous coverage
- Module noise figure and gain, 3.5 and 23 dB
- Amplitude and phase errors, rms, are 0.5 dB and 7°
- $G/T = -5$ dB/K
- 5-bit digital phase-shifters

The detailed studies on which the Advanced Shared Aperture Program (ASAP) is based are outlined in a classified paper, cited as reference 1 of [204].

According to this, these studies have shown that shared apertures along with advanced avionics architectures can:

- Reduce the number of antennas on board the aircraft by up to 4 to 1 with a related reduction on aircraft RCS.
- Reduce overall avionics size, weight, power and cost.
- Improve functional availability and performance.
- Reduce the number of specialized aircraft required . . .
- Simplify functional changes and addition of new functions [204].

Table 2.20 lists the pertinent characteristics of the array in support of the ASAP. The knowledge base derived from the realization and testing of this array can be used for arrays operating in the UHF band. Although the technological achievements reported are of considerable import, the studies leading up to the foundations of the ASAP constitute the catalyst for the support of many future shared aperture programs.

Table 2.20 The pertinent characteristics of the array in support of the ASAP

- Continuous frequency coverage from upper C-band, X-band, and Ku-band
- Antenna element: Flared notch (tapered slotline)
- Dual T/R module to yield selectable linear/circular polarizations
- Amplitude and phased errors, rms: 0.7 dB and 7°→5°
- Module noise figure and gain: 6–7 dB and 12–18 dB
- Average power per module: 2.75 W or 4 W near band center
- Various beamforming options explored

Table 2.21 gives a summary of the pertinent characteristics of the wideband array system realized by the Hughes Aircraft Company under the sponsorship of the Advanced Research Projects Agency (ARPA). The array uses a photonics-based beamformer that operates over a wide total bandwidth. The photonics beamformer provides the long delays (5 bits) while the shorter delays are provided by conventional microstrip delays lines (6 bits).

Table 2.21 The Hughes photonic-based wideband array system [205]

Aperture definition	Conformal on a 3 m radius surface; ~1 × 2.7 m
Frequency range	850–1400 MHz
Bandwidth	50% at 1124 MHz
Antenna element	Balanced notch or 'bunny ear' (a version of tapered slotline) (v.s.w.r. of <2 bandwidth of 3:1)
No. of elements	4 × 24 or 96
No. of T/R modules	~24
Element spacing	10.7 centimeters in azimuth and 11.2 centimeters in elevation
Beamwidth	~5° in azimuth and 30° elevation, at mid-band
Scan range	±60° in azimuth and no scanning in elevation
Time shifters	5 bits photonic and 6 bits electronic
Peak sidelobe level	−13 dB in azimuth and elevation for both receive and transmit modes
Directivity	~23 dBi at midband
Radiated power	~24 W

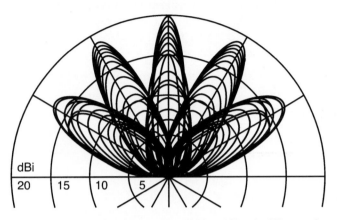

Figure 2.210 Beam patterns of broadband array at broadside, ±30° and ±60° over a 3-octave bandwidth from 0.2 to 1.6 GHz. (From [205]; © 1995, IEEE).

The attractive characteristics of the array are: (a) the array is conformal to a 3 m radius surface; (b) novel high quality wideband antenna elements (versions of tapered slotline antennas) are used; and (c) excellent scanning characteristics of ±60° taken in the frequency range 0.2–1.6 GHz have been verified. The measured array radiation patterns as shown in Figure 2.21 are polar plots taken at broadside, at ±30° and at ±60°. As can be seen, there are no beam squints in the measured polar diagrams.

The array provided a wealth of measurements related to the losses incurred when the wavelengths of operation are converted to optical wavelengths, the required delays are inserted, and another wavelength conversion takes place. The intrinsic conversion loss of the fiber optic link is 16.2 dB, but other losses such as the fan-out losses have to be offset by amplifiers, so that the beamformer is transparent to the overall system. Efforts to decrease these losses are in progress.

The Hughes array established once and for all the viability of arrays having large instantaneous and total bandwidths and excellent scanning characteristics.

The above prototype arrays are closely associated with military applications. The following proposal embraces military and civilian requirements. Unlike systems of the recent past, future radars designated for air traffic control and management should have very high reliability. Even 53 hours maintenance downtime per year for the airport surveillance radar is unacceptable [206].

The Department of Defense National Airspace System Modernization Program is part of the dual-use technology thrust [207] and includes civilian and military airports/airfields as well as the provision for mobility for the military systems; typically installation of the latter systems within an hour is a requirement for rapid deployment. The requirement for these highly reliable systems on the one hand and the decreasing costs of T/R modules on the other will lead the way for future air-traffic control functions to be performed by wideband multifunction phased array systems [207].

The following airport radars are considered for integration:

1. The airport surveillance radar (ASR) currently operating at S-band.
2. The precision approach radar (PAR) operating at X-band.
3. The terminal Doppler weather radar (TDWR) operating at S-, C-, or X-band [208].
4. The airport surface detection equipment (ASDE) operating at Ku-band.

The proposed wideband multifunction system is to operate in the X- to Ku-band and perform most of the system functions cited; if the frequency band of the ASR is shifted to C-band, all of the above system requirement can be performed by a system operating from C- to the Ku-band. The system will have the following attractive characteristics:

- Increased reliability owing to the graceful degradation property of the MMIC-based phased array.
- Affordable total cost (acquisition cost and LCCs).
- One or more inertialess beams will increase the radar range by as much as 50% when compared to conventional, mechanically scanned systems.
- Weather conditions and aircraft can be tracked simultaneously with parallel processing channels, saving the cost of another radar system. The performance advantage is obtained because the aircraft and weather conditions are seen on the same displays.

2.7.4.6. UK Research

The research reported in reference [209] is essentially a continuation of the United Kingdom's pioneering research into ship-based multifunction radars. The continuing R&D thrusts benefited from the knowledge base established from the highly successful multifunction electronically scanned adaptive radar (MESAR). Although the arrays under study are directed toward ship-based systems, the same systems can have wider uses.

The significant advantages that a radar operating at the optimal frequencies for the surveillance and tracking functions has over a radar operating at a compromise frequency are now appreciated [209]. Studies led to the selection of 1 and 10 GHz for the operating frequencies used for the surveillance and tracking functions. Since the frequency separation between the two bands is significant, two antenna elements—an open waveguide and a microstrip dipole—are selected for the low- and high-frequency bands. The decision to co-locate the two types of antenna elements on the same aperture is based on the need to maintain the alignment of the two arrays under the mechanical strain caused by ship flexure. Minimization of the total array area is also desirable when the array is to be located at the ship's highest mast. The available real-estate area of the aperture is therefore populated by the X-band elements interspersed by the L-band elements located at pseudo-random positions [209]. The proposed system will therefore perform these important radar functions at maximum efficiency and the research team is investigating the effects of coupling between similar and dissimilar antenna elements.

This approach to shared aperture systems is complementary to other R&D thrusts based on shared aperture arrays having large instantaneous bandwidths.

2.7.4.7. A Novel Approach

For ground-based systems the constraint to keep the total real-estate of a shared aperture system to a minimum is removed and the opportunity for novel systems is created. In a recent proposal [137] the receive array, occupying a substantial aperture area, is separated from 12 transmit arrays that are positioned in its periphery, as illustrated in Figure 2.22. Without any loss of generality, the proposal calls for 2 and 10 transmit arrays, shown shaded in the same figure, that are dedicated to the tracking and surveillance functions, respectively. Two or more transmit surveillance arrays (SAs) can, for instance, perform the horizon surveillance function exclusively and the other remaining arrays can perform other functions such as the surveillance of the remaining surveillance volume, target tracking and target illumination exclusively and as the need arises. The transmit SAs operate near 1 GHz while the transmit tracking arrays operate at 10 GHz.

The proposed system is based on the following core observations [137]:

- Theoretically phased arrays operating in a receive mode can generate many independent and simultaneous agile beams to meet challenges and requirements. If the receive array has 1000 or 5000 antenna elements, an equal number of simultaneous and independent beams can be generated. Instrumentally, the generation of these beams is facilitated by novel beamforming architectures and the availability of MMIC-based vector modulators.
- Future challenges imposed by multitarget and electronically hostile environments and by the requirement to perform several functions in parallel can be met by multibeam systems.

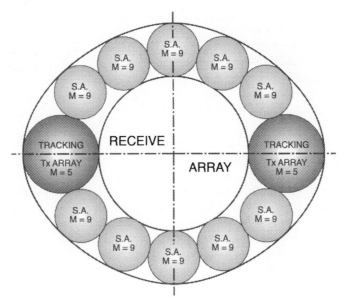

Figure 2.22 A shared aperture radar array accommodating one receive array, two tracking transmit arrays, and 10 surveillance arrays (SA). $M = 5$ for each of the tracking arrays while $M=9$ for each of the SAs.

- The realization of several simultaneous beams breaks the nexus between the many array functions and extends the timelines required for specific functions.
- The illuminated volume of space, at any time, is matched by the area occupied by the receive beams.
- Instrumental requirements are satisfied by subsystems such as MMIC-based vector modulators and solid-state transmitters that are readily available now. The technological risks involved in realizing the proposed system are therefore not high.

The area of each transmit array is either one-fifth or one-ninth of the area occupied by the receive array, so that approximately five or nine receive beams occupy the illuminated area. The transmit beam and its associated receive beams are scanned in unison toward any direction within the array surveillance volume. Given that the surveillance and tracking functions are preformed by nine and five beams respectively, the resulting system is robust. In the proposed system, 100 receive beams are processed simultaneously.

In order to maintain the same power–aperture product as that corresponding to a conventional phased array where the transmit and receive antennas are co-located, each transmit module for the novel array has to generate a power 25 ($M=5$) or 81 ($M=9$) times higher than that corresponding to the power of the T/R module of the conventional array.

When $M = 9$, the required power can be attained by combining the power of nine power amplifiers, each consisting of nine power amplifier chips, in a spatial field combiner.

As the SAs operate at the lower frequency band and the space normally occupied by the T/R switch and the receive module is not used, space is available for the extra power modules. The additional advantages of separating the receive and transmit functions are that (i) heat generated by the transmit modules is physically at some distance from the LNAs of the receive array; and (ii) the design guidelines derived by the researchers of the APL prototype are followed.

The original proposal called for a receive array populated either by wideband antenna elements operating over two to three octaves or by stacked patches resonating at the optimal frequencies designated for the surveillance and tracking function [137]. The receive array will therefore have either a continuous coverage or will operate at two bands centered at the two spot frequencies.

In Chapter 3 we shall consider wideband antenna elements in some detail, but it suffices here to mention that high-quality antenna elements operating over a 3:1 frequency range have been realized and thoroughly tested. The high-quality, wideband microstrip antennas are variants of the tapered slotline antennas and are commonly known as 'bunny ears'.

If a 9:1 frequency coverage is required, two sets of antenna elements operating over a 3:1 range randomly dispersed over the aperture area assigned for the receive array can meet requirements. The lowest frequency of the receive array can, for instance, be 1.11 GHz and the highest frequency 10 GHz. Continuous frequency coverage is thus attained with antenna elements already developed [205]. In the second option the receive array is populated by stacked patch antennas, operating at 1 and 10 GHz, and the array does not have a continuous frequency coverage.

The proposal illustrates the options available to the array designer. Other designs having more than two transmit arrays dedicated to the tracking function will meet another set of requirements. Similarly, horizon surveillance can be performed at 10 GHz to marginalize the multipath effects. The ECCM capabilities of the proposed array depend on the total bandwidth over which the array can operate; by contrast the ECCM capabilities of conventional narrowband multifunction arrays are poorer.

There is no doubt that shared aperture systems present the maximum number of challenges to the array designer.

2.8. CONCLUDING REMARKS

In this chapter we have erected the theoretical infrastructure that enabled us to explore a plethora of issues related to working arrays. Apart from having the required power–aperture product and spatial resolution, radar arrays should have no grating lobes in the array's surveillance volume and a prescribed array radiation pattern topology, i.e. minimum sidelobe level toward a range of directions and minimal sidelobe levels toward other directions.

The migration from analysis of arrays to the synthesis of arrays having a prescribed sidelobe level over the array surveillance volume is discernible. Adequate and user-friendly aids are available to allow the busy scientist or engineer to utilize the Taylor/Elliott/Bayliss/Hansen methodologies and define the complex weights on each antenna element of an array that meets the prescribed sidelobe levels over the array's surveillance volume.

The recent trend toward the synthesis of arrays having prescribed array radiation pattern topologies is based on powerful mathematical methodologies. Genetic algorithms suitable for the synthesis of large arrays are singled out because of their importance and means of speeding their execution time will render them even more useful. User-friendly software based on these methodologies is urgently required before these novel synthesis approaches are widely accepted by practicing engineers and scientists.

Irrespective of how the theoretical complex weights are derived, the designer has to realize the required array by using current affordable T/R modules, which unfortunately allow only a staircase approximation to the theoretical values. Several approaches have therefore been proposed to implement this approximation.

The following significant array topics have been considered:

- The definition of the conditions under which the array designer can use programmable phase-shifters instead of time-delays to steer the resulting beam to any direction within the array surveillance volume.
- The derivations of important parameters of working phased arrays, e.g. their G/T ratio, beamwidth, bandwidth, dynamic ranges, and noise figure.
- Approaches used to correct the resulting beam of working thinned arrays operating in the image synthesis mode before it can be used for imaging.
- Approaches used to manage the heat generated by the transmitters of working phased arrays operating at centimeter and millimeter wavelengths.
- The evolution of novel power supplies that meet a raft of hard to attain requirements. Considerable scope for improvements exists if the array designer collaborates with the specialists in this area.

- The quantitative assessment of the impact of random and systematic errors occurring in the RF subsystems on the array performance.
- The delineation of the effects of mutual coupling between the elements of arrays in general and approaches that marginalize their influence on the radiation pattern of small arrays. For small arrays the effects of mutual coupling between antenna elements have to be marginalized and the tolerances on the phase-shifters used have to be tight if low sidelobe levels are required. Phase-shifters operating at the IF of the system can meet this requirement.

On the basis of the above work, useful design guidelines have been deduced and trade-offs between the many array parameters can be considered to meet a wide range of specifications.

Realization aspects of proposals related to bistatic radars utilizing transmitters of opportunity are examined in detail. While some systems using FM/TV transmissions are now available, long development times are required to utilize the transmissions from current or future LEO/MEO constellations of satellites as illuminators of opportunity. Similarly, long development times are required to realize a pseudo-satellite that can be used for communications and surveillance functions over an area on the Earth's surface 1000 km in diameter.

The advantages of the active arrays over the 'one-bottle passive' array are overwhelming, but recent developments have to be taken into account. The migration from the 'one-bottle-passive' phased array to arrays utilizing several mini-TWTs or MPMs presents a serious threat to MMIC-based active phased arrays that went unchallenged for a long time. This threat is especially serious at millimeter wavelengths where MMIC transmitters yield minuscule powers.

While different approaches for cost reduction of conventional arrays will result in affordable arrays, other radical approaches for the realization of low-cost arrays have been flagged. The approaches considered in detail in Chapters 3 and 5 include the use of lenses, and the partial or complete elimination of phase-shifters and their control subsystems. These are the 'intelligent' or self-focusing or self-coherent low-cost arrays that we shall consider in Chapter 5.

The end of the cold war allowed western scientists and engineers to compare radar design approaches developed in the West with those developed in the former Soviet Union. Many low-cost approaches adopted by the Russian designers of radars provide useful comparisons, and significant hardware components like the ESAs are considered in detail in Chapter 4.

Multifunction arrays operating over wide bandwidths represent a raft of challenging systems problems. RFI problems have to be dealt with and it is now widely accepted that the multiplexing of the many functions required to protect a platform will not provide an optimal solution. Whenever possible, array functions have to be performed in parallel by multiple independent beam systems. The instrumental issues of realizing wideband antenna elements, T/R modules, and MMIC-based multiple beam beamformers are important and are considered in detail in the subsequent chapters of this book.

The many proposals and prototypes considered in some detail illustrate the many approaches studied to realize truly wideband multifunction systems. While there is a general thrust toward shared aperure arrays, the dissenting views have been adequately presented in this chapter. The requirement for parallel processing of the information derived from shared aperture arrays, for instance, originated from considerations of the dissenting views. R&D related to shared aperture arrays is indeed very important and challenging.

The quantitative assessment of the impact of the non-linearities of the FPAs on the resulting transmit array radiation pattern when linear wideband arrays are used to transmit one beam at one frequency, one beam at two frequencies, or two beams at different frequencies has been considered. The preliminary work reported is to be extended to planar arrays, and experimental verification of the pointing angles of the intermodulation beams for active arrays would be useful.

The requirements of phased array-based radiotelescopes are:

- The maximization of the available baselines between antenna elements.
- The attainment of the narrowest beamwidths from a given aperture.
- Prescribed spatial resolution and sensitivity attained at several frequency bands.

While the early radiotelescopes aimed toward maximum spatial resolution with adequate sensitivity, the current trend is toward high sensitivity with adequate resolution. Naturally there are other possible boundary cases. The other discernible trends are toward observations taken in several instantaneous fields of view and at several frequency bands. Put differently, the radio equivalent is sought of the optical Schmidt telescope operating over many frequency bands instantaneously. Some proposals meeting these new requirements have been considered together with a proposal for a sensitive compact array operating at millimeter wavelengths.

The transmissions from several LEO/MEO constellations of satellites pose a threat for radioastronomy observations, and the excision of these unintentional 'jammers' in the spatial and frequency domains presents a new set of challenges for the designer of modern phased array-based radiotelescopes.

Considering the diverging aims of radar and radioastronomy systems, the discernible trends for future radar systems and radiotelescopes are not very different.

REFERENCES

[1] Sweetman B. Leading-edge technology for Swedish AEW. *Int. Defense Rev.* **2** (3), 277 (1988).
[2] Kronhamn T. R. AEW performance improvements with the ERIEYE phased array radar. *1993 IEEE Natl. Radar Conf.*, pp. 34–39 (1993).
[3] Christiansen W. N. and Mathewson D. S. Scanning the sun with a highly directional antenna. *Proc. IRE* **46**(1), 127 (Jan. 1958).
[4] Hemmi C. Bandwidth of the array factor for phased-steered arrays, in 'Hal Schrank's Antenna Designer's Notebook' *IEEE Antennas Propag. Mag.* **AP-35**(1), 72 (1993).
[5] Elliott R. S. Beamwidth and directivity of large scanning arrays. *Microwave J. Part I*, (Dec.), 53 (1963).
[6] Fletcher P. N. and Dean M. Derivation of orthogonal beams and their application to beamforming in small phased arrays. *IEE Proc. Microwave Antennas Propag.* **AP-143**(4) 304–308 (Aug. 1996).
[7] Hansen R. C. *Microwave Scanning Antennas*, vol. II, ch. 4. Academic Press, New York (1986).
[8] Fletcher P. N., Wicks A. E. and Dean M. Improvement of coupling corrected difference beams in small phased arrays. *Electron. Lett.*, **33**(5), 352–353 (27 Feb. 1997).

[9]   Steyskal I. L. and Herd J. S. Mutual coupling in small array antennas. *IEEE Trans. Antennas Propag.* **AP-38**(12) 1971–1975 (1990).

[10] Fletcher P. N. and Dean M. Application of retrodirective beams to achieve low sidelobe levels in small phased arrays. *Electron. Lett.* **32**(6), 506–508 (14 Mar. 1996).

[11] Steyskal H. Digital beamforming at Rome laboratory. *Microwave J.* (Feb.) 100–124 (1996).

[12] Harrington R. F. Sidelobe reduction by nonuniform element spacing. *IRE Trans. Antennas Propag.* (Mar.), 187–192 (1961).

[13] Hodjat F. and Hovanessian S. A. Nonuniformly spaced linear and planar array antennas for sidelobe reduction. *IEEE Trans. Antennas Propag.* **AP-26**(2), p 198–204 (Mar.1978).

[14] Tomiyasu K. Combined equal and unequal element spacing for low sidelobe pattern of a symmetrical array with equal-amplitude elements. *IEEE Trans. Antennas Propag.* **AP-39**(2), 265–266 (Feb. 1991).

[15] Schuman H. K. and Strait B. J. On the design of unequally spaced arrays with nearly equal sidelobes. *IEEE Trans. Antennas Propag.* **AP-16**, 493–494 (July 1968).

[16] Sandrin W. A., Glatt C. R. and Hague D. S. Design of arrays with unequal spacing and partially uniform amplitude taper. *IEEE Trans. Antennas Propag.* **AP-17**, 642–644 (Sep. 1969).

[17] Lo Y. T. A mathematical theory of antenna arrays with randomly spaced elements. *IRE Trans. Antennas Propag.* **AP-12**, 257 (May 1964).

[18] Steinberg B. D. *Principles of Aperture and Array Design*. Wiley, New York (1976).

[19] Steinberg B. D. The peak sidelobes of the phased array having randomly located elements. *IEEE Trans. Antennas Propag.* **AP-20**, 129 (Mar. 1972).

[20] Leeper D. C. Thinned aperiodic antenna arrays with improved peak sidelobe level control. US Patent 4,071,848 (31 Jan. 1978).

[21] Arsac J. Nouveau reseau pour l'observation radioastronomique de le brillance sur le soleil a 9350 Mc/s. *C. R. Acad. Sci. (Paris)* **240**, 942–945 (1955).

[22] Moffet A. T. Minimum redundancy linear arrays. *IEEE Trans. Antennas Propag.* **AP-16**, 172 (1968).

[23] Bracewell R. N. *et al.* The Stanford five element radiotelescope. *Proc. IEEE* **61**, 1249 (1973).

[24] Napier P. J., Thompson A. R. and Ekers R. D. The very large array: design and performance of a modem synthesis radio telescope. *Proc. IEEE* **71**(11) 1295 (1983).

[25] Milman A. S. Sparse-aperture microwave radiometers for Earth remote sensing. *Radio Sci.* **23**(2), 193 (1988).

[26] Ruf C. S. *et al.* Interferometric synthetic aperture microwave radiometry for remote sensing of the Earth. *IEEE Trans. Geoscience Remote Sensing* **26**(5), 596 (1988).

[27] Swift C. T., Le Vine D. M. and Ruf C. S. Aperture synthesis concepts in microwave remote sensing of the Earth. *IEEE Trans. Microwave Theory Tech.* **39**(12, 1931 (1991).

[28] Le Vine D. M. *et al.* Initial results of the development of a synthetic aperture microwave radiometer. *IEEE Trans. Geoscience Remote Sensing* **28**(4), 614 (1990).

[29] Lee Y. and Pillai S. U. An algorithm for optimal placement of sensor elements. *Int. Conf. Acoust., Speech, Signal Process.* vol. 5, p. 2674 (1988).

[30] Chen W.-L. and. Yeheskel B.-N. Minimum redundancy array structure for interference cancellation. *IEEE-AP-S Int. Symp.*, London, Ontario, Canada, p. 121 (1991).

[31] Huang X., Reilly J. P. and Wong M. Optimal design of linear array of sensors. *Proc. Int. Conf. Acoust., Speech, Signal Process.* vol. 2, p. 1405 (1991).

[32] Jorgenson M. B., Fattouche M. and Nichols S. T. Applications of minimum redundancy arrays in adaptive beamforming. *IEE Proc., Part H, Microwaves, Antennas Propag.* **AP-138**(5), 441 (1991).

[33] Felli M. and Pampaloni P. The information of a minimum redundancy linear antenna array. *Alta Freq.* **44**(5), 240 (1975).

[34] Blanton Y. A. and. McClellan J. H. New search algorithm for minimum redundancy linear arrays. *Int. Conf. Acoust., Speech, Signal Process.*, vol. 2, p. 1361 (1991).

[35] Linebarger D. A., Sudbough I. H. and Tollis I. G. A unified approach to design of minimum redundancy arrays. *24th Asilomar Conf. Signals, Syst. Comput.* vol. 1, p. 143 (1990).

[36] Ruf C. S. Numerical annealing of low-redundancy linear arrays. *IEEE Trans. Antennas Propag.* **AP-41**(1), 85 (1993).

[37] Special Issue: The Australia Telescope. *J. Elect. Electron. Eng. Aust.* **12** (2), (June 1992).

[38] Tseng C.-Y. and Griffiths L. J. Sidelobe suppression in minimum redundancy linear arrays. *Proc. IEEE 6th SP Workshop on Statistical Signal and Array, Processing Conference,* Victoria BC, Canada, pp. 288–291 (1992).

[39] Tseng C.-Y. and Griffiths L. J. A simple algorithm to achieve desired patterns for arbitrary arrays. *IEEE Trans. Signal Processing,* **40**(11), 2737– 2746 (Nov. 1992).

[40] Rossouw M. J., Joubert J. and McNamara D. A. Thinned arrays using a modified minimum redundancy synthesis technique. *Electron. Lett.* **33**(10), 826–827 (8 May 1997).

[41] Vertatschitsch E. and Haykin S. Nonredundant arrays. *Proc. IEEE* **74**(1), 217 (Jan. 1986).

[42] Linebarger D. A., Sudborough I. H. and Tollis I. G. Difference bases and sparse sensor arrays. *IEEE Trans. Information Theory* **39**(2), 716–721 (Mar. 1993).

[43] Hansen R. C. Array pattern control and synthesis. *Proc. IEEE* **80**(1), 141–151 (Jan. 1992).

[44] Stone J. S. US Patents 1,643,323 and 1,715,433.

[45] Balanis C. A. *Antenna Theory—Analysis and Design.* Harper & Row, New York (1982).

[46] Dolph C. L. A current distribution for broadside arrays which optimizes the relationship between beam width and side-lobe level. *Proc. IRE Waves Electrons* **34**, 335 (June 1946).

[47] Riblet H. J. Discussion on 'A current distribution for broadside arrays which optimizes the relationship between beam width and side-lobe level' by C. L. Dolph. *Proc. IRE* **35**, 489 (1947).

[48] Barbiere D. A method for calculating the current distribution of Tschebyscheff arrays. *Proc. IRE* **40**, 78 (Jan. 1952).

[49] Stegen J. Excitation coefficients and beamwidths of Tschebyscheff arrays. *Proc. IRE* **41**, 1671 (Nov. 1953).

[50] Drane, J. Jr., Useful approximations for the directivity and beamwidth of large scanning Dolph–Chebyshev arrays. *Proc. IEEE* **56**, 1779 (Nov. 1968).

[51] Kraus J. D. *Antennas.* McGraw-Hill, New York (1988).

[52] Shelkunoff S. A. A mathematical theory of linear arrays. *Bell Syst. Tech. J.* **22**, 80 (1943).

[53] Blanton J. L. Approximations for computing the weighting parameters for one-parameter Taylor and Hansen aperture distributions. Antenna Designer's Notebook. *IEEE Antennas Propag. Mag.* **AP-34**(4) 34–35 (1992).

[54] Elliott R. S. Improved pattern synthesis for equispaced linear arrays. *Alta Freq.* **51**(6), 296 (1982).

[55] Taylor T. T. The design of line-source antennas for narrow beamwidth and low sidelobes. *IRE Trans. Antennas Propag.* **AP-3** (1), 16 (1955).

[56] Bayliss E. T. Design of monopulse antenna difference patterns with low sidelobes. *Bell Syst. Tech. J.* **47**, 623 (1968).

[57] Elliott R. S. Design of line source antennas for difference patterns with sidelobes of individually arbitrary heights. *IEEE Trans. Antennas Propag.* (May), 310–316 (1976).

[58] Elliott R. S. Pattern synthesis for antenna arrays. *IEEE Lecture Notes* (1988).

[59] Elliott R. S. *Antenna Theory and Design.* Prentice-Hall, Englewood Cliffs, NJ (1981).

[60] Elliott, R. S. Improved pattern synthesis for equispaced linear arrays. *Alta Freq.* **51**(6), 296 (1982).

[61] Kim Y. U. and Elliott R. S. Shaped-pattern synthesis using pure real distributions. *IEEE Trans. Antennas Propag.* **AP-36**(11), 1645–1649 (Nov. 1988).

[62] Orchard H. J, Elliott R. S. and Stern G J. Optimizing the synthesis of shaped beam antenna patterns. *Proc IEE Part H* **132**, 63–68 (1985).

[63] Ng B. P., Er M. H. and Kot C. Linear array geometry synthesis with minimum sidelobe level and null control. *IEE Proc. Microwwave Antennas Propag.* **141**(3), 162–166 (June 1994).

[64] Nagesh S. R. and Vedavathy T. S. A procedure for synthesizing a specified sidelobe topography using an arbitrary array. *IEEE Trans. Antennas Propag.* **43**(7), 742–745 (July 1995).

[65] Holland J. H. Genetic algorithms. *Sci. Am.,* **267** 44–50 (1992).

[66] Alander J. T. *An indexed bibliography of genetic algorithms: Years 1957–1993.* Dept Information Technology, Production Econ, University of Vassa, Finland, Rep Ser 94–1 (Feb 1994).

[67] Holland J. H. *Adaptation in Natural and Artificial Systems.* MIT Press, Cambridge, MA (1975).

[68] Man K. F., Tang K. S. and Kwong S. Genetic algorithms. *IEEE Trans. Ind. Electron.* **43**(5) 519–534 (Oct. 1996).

[69] Mohammed O. A. Practical issues in the application of genetic algorithms to optimal design problems in electromagnetics. *IEEE* South East Con '96, 634–640 (1996).

[70] Haupt R. L. Genetic algorithms design of antenna arrays. *IEEE Aerospace Applications Conf.* pp. 103–109 (1996).

[71] Haupt R. L. Thinned arrays using genetic algorithms. *IEEE Trans. Antennas and Propag.* **AP-42**(7), 903–906 (July 1994).

[72] Haupt R. L. Optimum quantized low sidelobe phase tapers for arrays. *Electron. Lett.*, **31**(14), 1117–1118 (6 July 1995).

[73] Mitchell R. J., Chambers B. and Anderson A. P. Array pattern synthesis in the complex plane optimised by a genetic algorithm. *Electron. Lett.* **32**(20), 1843–1845 (26 Sept. 1996).

[74] Haupt R. L. Phase-only adaptive nulling with a generic algorithm. *IEEE Trans. Antennas Propag.* **AP-45**(6), 1009–1015 (June 1997).

[75] Mandelbrot B. B. *The Fractal Geometry of Nature.* WH Freeman, New York (1983).

[76] Barnsley M. F. *Fractals Everywhere.* Academic Press, Boston, MA (1993).

[77] Cvitanovic P. *Universality in Chaos.* Adam Hilger, Bristol, UK (1984).

[78] Gleick J. *Chaos.* Penguin Books (1987).

[79] Waldrop M. M. *Complexity.* Viking (1992).

[80] Jaggard D. L. Prolog to Special section on fractals in electrical engineering. *Proc. IEEE*, **81**(10), 1423–1427 (Oct. 1993).

[81] Kim Y. and Jaggard D. L. The fractal random array. *Proc. IEEE* **74**(9), 1278–1280 (Sept. 1986).

[82] Puente-Baliarda C. and Pous R. Fractal design of multiband and low-sidelobe arrays. *IEEE Trans. Antennas Propag.* **AP-44**(5), 730–739 (May 1996).

[83] Werner D. H. and Werner P. L. Frequency-independent features of self-similar fractal arrays. *Radio Science* **31**(6), 1331–1343 (Nov./Dec. 1996).

[84] Werner D. H. and Werner P. L. On the synthesis of fractal radiation patterns. *Radio Science* **30**(1), 29–45 (Jan./Feb. 1995).

[85] Kopilovich L. E. and Sodin L. G. Two-dimensional aperiodic antenna arrays with a low sidelobe level. *IEE Proc, Part H*, **138**(3), 333–337 (June 1991).

[86] Kopilovich L. E. and Sodin L. G. Aperture optimization of telescopes and interferometers: a combinational approach. *Astron. Astrophys. Suppl Series* **116**, 177–185 (1996).

[87] Calabro D. and Wolf J. K. On the synthesis of two-dimensional arrays with desirable properties. *Information Control* **11**(5–6) 537–560 (Nov–Dec, 1967).

[88] Wild P. Infinite families of perfect binary arrays. *Electron. Lett.* **24**(14), 845–847 (7 July 1988).

[89] Bomer L. and Antweiler M. Perfect binary arrays with 36 elements. *Electron. Lett.* **23**, 730–732 (1987).

[90] Kopilovich L. E. On perfect binary arrays. *Electron. Lett.* **24**(9), 566–567 (28 Apr. 1988).

[91] Sverdlik M. B. and Meleshevich A. N. Synthesis of optimum pulsed sequences having the property of "No more than one coincidence". *Radio Eng. Electron. Phys.* **19**(4), 46–54 (Apr. 1974).

[92] Kopilovich L. E. On signals with the minimal ambiguity function for simultaneous determination of range and velocity of objects. *Radiotekh & Elektron.* **32**, 1544–1547 (1987) (in Russian).

[93] Gardner M. Mathematical games. *Sci. Am.* **226**, 114–118 (1972).

[94] Dewdney A. K. Mathematical games. *Sci. Am.* **253**, 16–20 (1985).

[95] Dewdney A. K. Mathematical games. *Sci. Am.* **254**, 8–13 (1986).

[96] Von Aulock W. H. Properties of phased arrays. *IRE Trans. Antennas Propag.* **AP-9**, 1715 (1960).

[97] Cheston T. C. and Frank J. Phased array radar antennas. In *Radar Handbook* (M. Skolnik, ed.), ch. 7. McGraw-Hill, New York (1990).

[98] Tseng F.-I. and Cheng D. K. Optimum scannable planar arrays with an invariant sidelobe level. *Proc IEEE* **56**(11), 1771–1778 (Nov. 1968).

[99] Elliott R. S. Beamwidth and directivity of large scanning arrays. *Microwave J. Part II* **7**, 74 (1964).

[100] Labrum N. R. *et al.* A compound interferometer with a 1.5 minute of arc fan beam. *Proc. IRE, Aust.* **24**(2), 148 (Feb. 1963).

[101] Convington A. E. A compound interferometer. *J. Royal Astron. Soc. Canada* **54**, 67 (Feb. 1960).

[102] Hansen R. C. A one-parameter circular aperture distribution with narrow beamwidth and low sidelobe. *IEEE Trans. Antennas Propag.* **AP-14**, 477 (1976).

[103] Taylor T. T. Design of circular apertures for narrow beamwidth and low sidelobes. *IRE Trans. Antennas Propag.* **AP-18**, 17 (1960).

[104] Hansen R. C. Tables of Taylor distribution for circular aperture antennas. *IRE Trans. Antennas Propag.* **AP-8**, 23–36 (1960).

[105] Davies D. E. N. Circular arrays. In *The Handbook of Antenna Design*, vol. 2. A. W. Rudge, K. Milne, A. D. Olver and P. Knight, eds. Peter Peregrinus Ltd. (1983).

[106] Rehnmark S. Broadband antennas and antenna feed networks for precision ESM. *Antenna '91*, Faro, Sweden, pp. 1–23 (1991).

[107] Chow Y. L. Comparisons of some correlation array configurations for radio astronomy. *IEEE Trans. Antennae Propag.* **AP-18**(4), 567–569 (July 1970).

[108] Wild J. P. A new method of image formation with annular aperture and an application in radio astronomy. *Proc. Roy. Soc. London, Ser. A* **286**, 499 (1965).

[109] Wild J. P. The Culgoora radioheliograph. 1. Specification and general design. *Proc. IREE, Aust.* **28**(9), 279 (1967).

[110] Goto N. and Tsunoda Y. Sidelobe reduction of circular arrays with a constant excitation amplitude. *IEEE Trans. Antennas Propag.* **AP-25**(6), 890 (1977).

[111] *The Millimeter Array.* National Radio Astronomy Observatory flyer.

[112] Harper J. C., Stangel J. J. and Valentino P. A. An octave band electronic scan cylindrical array. *IEEE Int. Symp. Antennas Propag.* pp. 670–673 (1993).

[113] Harper J. C., Stangel J. J. and Valentino P. A. Cylindrical array antenna design considerations and configurations. *IEEE Int. Symp. Antennas Propag.* pp. 222–225 (1993).

[114] Kiuchi E. and Ueda I. Tactical cylindrical active phased array radar. *1996 IEEE Int. Symp. Phased Array Systems and Technology*, Boston MA., pp. 222–225 (1996).

[115] Budsinsky Yu. A. and Kantyuk S. P. A new class of self-protecting low-noise microwave amplifiers. *IEEE-MTT-S Dig.* pp. 1123–1125 (1993).

[116] Lee J. J. G/T and noise figure of active array antennas. *IEEE Trans. Antennas Propag.* **41**(2), 241–244 (Feb. 1993).

[117] Mailloux R. J. Antenna array architecture. *Proc. IEEE* **80**(1), 163 (1992).

[118] Brookner E. (ed.) *Aspects of Modem Radar*, ch. 2, p. 28. Artech House, Dedham, MA (1988).

[119] Cohen E. D. Active electronically scanned arrays. *IEEE Nail. Telesyst. Conf.*, p. 3 (1994).

[120] Edward B. and Rees D. A broadband printed dipole with integrated balun. *Microwave J.* **30**, 339 (1987).

[121] Fourikis N., Lioutas N. and Shuley N. V. Parametric study of the co- and cross-polarisation of tapered planar and antipodal slotline antennas. *IEE Proc., Part H: Microwave Opt. Antennas* **140**(1), 17 (1993).

[122] Langley L. D. S., Hall P. S. and Newham P. Novel ultrawide-bandwidth Vivaldi antenna with low cross polarisation. *Electron. Lett.* **19**(23), 2005 (1993).

[123] Povinelli M. J. and Johnson J. A. Design and performance of wideband dual-polarized stripline notch arrays. *Proc. IEEE AP-S Int. Symp.*, vol. 1, p. 200 (1988).

[124] Newberg I. L. and Wooldridge J. J. An affordable low-profile multifunction structure (ALMS) for an optoelectronic (OE) active array. *IEEE MTT-S Int. Microwave Symp. Dig.*, p. 509 (1993).

[125] Newberg I. L. and Wooldridge J. J. Revolutionary active array radar using solid state 'modules' and fiber optics. *Rec. 1993 IEEE Natl. Radar Conf.*, p. 88 (1993).

[126] Scalsi G. J., Turtle J. P. and Carr P. H. MMICs for airborne phased arrays. *Monolithic Microwave Integ. Circuits Sensors, Radar Commun. Syst., Proc. SPIE* **1475**, 2 (1991).

[127] Kinzel J. A., Edward B. J. and Rees D. V-band, space-based phased arrays. *Microwave J.* **30**, 89 (1987).

[128] Carey D. R. and Evens W. The Patriot radar in tactical air defense. *Microwave J.* **31**, 325 (1988).

[129] Daryoush A. S., Ackerman E., Samant N. R., Wanuga S. and Kasemset D. Interfaces for high-speed fiber-optic links: analysis and experiment. *IEEE Microwave Theory Tech.*, **39**(12), 2031–2044 (1991).

[130] Lee J. J. Sidelobes control of solid-state array antennas. *IEEE Trans. Antennas Propag.* **AP-36**(3), 339 (1988).

[131] Mailloux R. J. and Cohen E. Statistically thinned arrays with quantized element weights. *IEEE Trans. Antennas Propag.* **AP-39**(4), 4 (1991).

[132] Willey R. E. Space tapering of linear and planar arrays. *IRE Trans. Antennas Propag.* **J-AP62**, 369 (1962).

[133] Hacker P. S. In Hal Schrank's Antenna Designer's Notebook, Thinned Arrays: some fundamental considerations. *IEEE Antennas Propag. Mag.* **34**(3), 43 (1992).

[134] Skolnik M. *et al.* Statistically designed density-tapered arrays. *IEEE Trans. Antennas Propag.* **AP-12**, 408 (1964).

[135] Frank J. and O'Haver K. W. Phased array antenna development at the Applied Physics Laboratory. *Johns Hopkins APL Tech. Dig.* **14**(4), 339 (1993).

[136] Frank J. and Coffman R. Hybrid active arrays. *IEEE Proc. Natl. Telesyst. Conf.*, p. 19 (1994).

[137] Fourikis N. Novel shared-aperture phased arrays. *Microwave Optical Tech. Lett.*, (20 Feb.), 189–192 (1998).

[138] Mallavarpu R. and MacMaster G. 100 W peak/30 W average broadband X-band solid-state amplifier. *Conf. Proc. Military Microwaves '86*, pp. 354–359 (1986).

[139] Ruze J. The effect of operature errors on the antenna radiation pattern. *Nuovo Cimento, Suppl.* **9**(3), 361 (1992).

[140] Elliott R. E. Mechanical and electrical tolerances for two-dimensional scanning antenna arrays. *IRE Trans. Antennas Propag.* **AP-6**, 114 (1958).

[141] Allen J. L. The theory of array antennas. *MIT Lincoln Lab. Tech. Rep. No. 23*. Massachusetts Institute of Technology, Cambridge, MA (1963).

[142] Skolnik M. Non-uniform arrays. In *Antenna Theory* (R. E. Collin and F. J. Zucker, eds.), ch. 6. McGraw-Hill, New York (1969).

[143] Moody J. A survey of array theory and techniques. *Res. Labs Rep. No. 501.3*. RCA Victor Co. (1963).

[144] Cheston C. Effect of random errors on sidelobes of phased arrays. *IEEE APS Newsl.—Antennas Des. Notebook* (Apr. 1985).

[145] Borkowski M. T. and Leighton D. G. Decreasing cost of GaAs MMIC modules is opening up new areas of application. *Electron. Prog.* **29**(2), 32 (1989).

[146] Rice S. O. Mathematical analysis of random noise. In *Selected Papers on Noise and Stochastic Processes* (N. Wax, ed.). Dover, New York (1954).

[147] Hsiao J. K. Array sidelobes, error tolerances, gain and beamwidth. *RL Rep. 8841*. Naval Research Laboratory, Washington, DC (1984).

[148] Hsiao J. K. Design of error tolerance of a phased array. *Electron. Lett.* **21**(19), 834 (1985).

[149] Kaplan R. D. Predicting antenna sidelobe performance. *Microwave J.*, **29**, 201 (1986).

[150] Mailloux R. J. *Phased Array Handbook*. Artech House, Norwood, MA (1994).

[151] Mailloux R. J. Array grating lobes due to periodic phase, amplitude and time delay quantization. *IEEE Trans. Antennas Propag.* **AP-32**(12), 1364 (1984).

[152] Schlesak J. J., Alden A. and Ohno T. A microwave powered high altitude platform. *IEEE MTT-S Dig.* pp. 283–286 (1988).

[153] Alden A. and Ohno T. A power reception and conversion system for remotely-powered vehicles. *1989 IEE Int. Conf. Antennas and Propag.*, University of Warwick, UK, pp. 535–538 (1989).

[154] Pinhasi Y., Yakover I. M., Eichenbaum A. L. and Gover A. Efficient electrostatic-accelerator free-electron masers for atmospheric power beaming. *IEEE Trans. Plasma Sci.* **24**(3), 1050–1057 (June 1996).

[155] Koert P. and Cha J. T. Millimeter wave technology for space power beaming. *IEEE Trans.* **MTT-40**(6), pp. 1251–1258 (June 1992).

[156] East T. W. R. A self-steering array for the SHARP microwave-powered aircraft. *IEEE Trans. Antennas Propag.* **40**(12), 1565–1567 (Dec. 1992).

[157] Taylor A. H., Young L. C. and Hyland L. A. System for detecting objects by radio. US Patent 1,981, 884 (Nov. 1934).

[158] Griffiths H. D. and Long N. R. W. Television-based bistatic radar. *IEE Proc., Part F, Commun. Radar and Signal Processing*, **133**, 649–657 (1986).

[159] Fulghum D. A. Passive system hints at stealth detection *Avian Week & Space Tech*. (30 Nov.), 1998.
       Nordwall B. D. 'Silent Sentry' A new type of radar. *Aviation Week & Space Tech*, (30 Nov.) 1998.
[160] Koch V. and Westphal R. A new approach to a multistatic passive sensor for air defense. *IEEE Int. Radar Conf.* pp. 22–28 (1995).
[161] Soule L. High performance thermal systems. *Powertechnics Magazine*, **7**(2), 29–31 (Feb. 1991).
[162] Lingle D. E., Mikszan D. P. and Mukai D. Advanced technology ultrareliable radar. *Proc. IEEE Natl. Radar Conf.*, p. 1 (1989).
[163] Ramalingam M. L., Donovan B. D. and Beam J. E. Cryogenic refrigeration thermodynamics for a power conditioning electronic component [for radar systems]. *IECEC '96. Proceedings of the 31st Intersociety Energy Conversion Eng. Conf.*, vol. 2, pp. 1390–1395.
[164] Lockerd R. M. and Crain G. E. Airborne active element array radars come of age. *Microwave J.*, **33**, 101 (1990).
[165] Skolnik M. *Introduction to Radar Systems*. McGraw-Hill, New York (1980).
[166] Gardenghi R. A. and Moulne R. C. Power supply considerations for pulsed solid-state radar. *IEEE Conf. Rec. 1990 Ninteenth Power Modulator Symp.*, pp. 145–152 (1990).
[167] Cathell F. and Strickland E. High voltage capacitor charging power supplies with high regulation. *PCIM '93. Power Conversion and Intelligent Motion USA*. pp. 166–180, 492 (1993).
[168] Abrams R. H. Jr. and Parker R. K. Introduction to the MPM: what it is and where it might fit. *IEEE MTT-S Int. Microwave Symp. Dig.*, p. 107 (1993).
[169] Christensen J. A. *et al.* MPM technology developments: An industry perspective. *IEEE MTT-S Int. Microwave Symp. Dig.*, p. 115 (1993).
[170] Skolnik M. The radar antenna—circa 1995. *J. Franklin Institute*, **332B**(5) 503–519 (1995).
[171] Barton D. K. The 1993 Moscow Air Show. *Microwave J.* **37**, 34–39 (1994).
[172] Trunk G. V. and Patel D. P. Optimal number of phased array faces for horizon surveillance. *1996 IEEE Int. Symp. Phased Array Systems and Technology*, Boston MA, pp. 214–216 (1996).
[173] Keizer W. P. M. N. New active phased array configurations. *Conf. Proc. Mil. Microwaves*, p. 564 (1990).
[174] Headrick J. M. Looking over the horizon. *IEEE Spectrum* (July), 36–39 (1990).
[175] Georges T. M. and Harlan J. A. The first large-scale map of ocean surface currents made with dual over-the-horizon radars. *OCEANS '96 MTS/IEEE Conf. Proc. The Coastal Ocean—Prospects for the 21st Century*, vol. 3, pp. 1485–1487 (1996).
[176] Georges T. M., Harlan J. A. and Lematta R. A. Large-scale mapping of ocean-surface currents with dual over-the-horizon radars. *Nature* **379**, 434–436 (1996).
[177] McNally R. *Jindalee Operational Radar Network Project*. Department of Defence. The Auditor-General performance audit. Audit Report No 28. 1995–96. Australian Government Publishing Service (1998).
[178] *Jindalee Operational Radar Network Project*. Joint Committee of Public Accounts and Audit, Report 357 (Mar. 1998).
[179] Barrick D. E., Evans M. W. and Weber B. Ocean surface currents mapped by radar. *Science* **198**, 138–144 (1977).
[180] Harlan J. A. and Georges T M. Observations of Hurricane Hortense with two over-the-horizon radars. *Geophys. Res. Lett.*, **24**(24), 3241–3144 (Dec. 1997).
[181] Cottony H. V. Wide-frequency band array system. *IEEE Trans. Antennas Propag.* **AP-18**, 774 (1970).
[182] Cohen M. N. A tabular synthesis technique for broadband/thinning linear phased arrays. *Proc. 16th Int. Conf. Infrared Millimeter Waves*, Lausanne, Switzerland, p. 501 (1991).
[183] Laughlin G. J., Byron E. V. and Cheston T. C. Very wideband phased-array antenna. *IEEE Trans. Antennas Propag.* **AP-20**, 699 (1972).
[184] Boyns J. E. and Provencher J. H., Experimental results of a multifrequency array antenna. *IEEE Trans. Antennas Propag.* **AP-20**, 106 (1972).
[185] Shively D. G. and Stutzman W. L. Wideband arrays with variable element sizes. *IEE Proc., Part H: Microwaves, Optical Antennas* **137**(4), 138 (1990).

[186] Stutzman W. T. Wide bandwidth antenna array design. *Proc. IEEE Southeast. Reg. Meet.*, Raleigh, NC, p. 97 (1985).

[187] Simpson M. High ERP phased array ECM systems. *J. Electron. Defense* (Mar.) 41 (1982).

[188] Curtis R. D. On overview of surface navy ESM/ECM developments. *J. Electron. Defense* (Mar.) 37 (1982).

[189] Rossi V. and Damen G. Solid state jamming antenna. *Conf Proc Military Electronics*, 1980, p. 446, (1990).

[190] Moynihan R. L. Phased arrays for airborne ECM- The rest of the story. *Microwave* J. January, p. 34 (1987).

[191] Rehnmark S. DF Measurements for precision ESM. *Antenna '94*, Eskilstuna, Sweden, pp. 1–10, 1994.

[192] Goodwin P. L. Ambiguity resistant three- and four-channel interferometers. *NRL Report 8005.* Naval Research Laboratory, Washington DC. (1976).

[193] Hemmi C. Pattern characteristics of harmonic and inter-modulation products in broad band active transmit arrays. Private communication (Aug. 1998).[In preparation for publication.]

[194] Field G. and Goldsmith D. *The Space Telescope: Eyes Above the Atmosphere.* Contemporary Books, Chicago, IL (1989).

[195] Fourikis N. Novel radiometric phased array systems. *Microwave Opt. Tech. Lett.* **18**(2), 100–104 (5 June 1998).

[196] Pozar D. M. and Duffy S. M. A dual-band circularly polarized aperture-coupled stacked microstrip antenna for global positioning satellite. *IEEE Trans. Antennas Propag.*, **AP-45**(11), 1618–1625 (Nov. 1997).

[197] Miller B. Satellites free the mobile phone. *IEEE Spectrum* (Mar.), 26–35 (1998).

[198] Axness T. A., Coffman R. V., Kopp B. A. and O'Haver K. W. Shared aperture technology development. *Johns Hopkins APL Tech. Dig.*, **17**(3), 285–294 (1996).

[199] Burke M. A. Multifunction/shared aperture systems or smart skins now. *J. Electron. Defense* (Jan.), 29–32 (1991).

[200] Stoffel A. P. Heuristic energy management for active multifunction radars. *IEEE Natl. Telesystems Conf.*, pp. 71–74 (1994).

[201] Symons R. S. Tubes: still vital after all these years. *IEEE Spectrum* (Apr.), 52–63 (1998).

[202] Teti J. G. *et al.* Wideband airborne early warning (AEW) radar. *Rec. IEEE Natl. Radar Conf.*, p. 239 (1993).

[203] Kalbasi K., Plumb R. and Pope R. An analysis and design tool for a broadband dual feed circles array antenna. *IEEE AP-S Int. Symp.*, vol. 4, p. 2085 (1992).

[204] Hemmi C., Dover T., Vespa A. and Fenton M.-W. Advanced shared aperture program (ASAP) array design. *1996 IEEE Int. Symp. Phased Array Systems and Technology. Revolutionary Developments in Phased Arrays*, Boston MA., pp. 278–282 (1996).

[205] Lee J. J., Loo R Y., Livingston S., Jones V. I., Lewis J. B., Yen H.-W., Tangonan G. L. and Wechsberg M. Photonic wideband array antennas. *IEEE Trans. Antennas Propag.* **AP-43**(9), 966–982 (Sep. 1995).

[206] Brukiewa T. F. Active array radar systems applied to air-traffic control. *IEEE Natl. Telesyst. Conf.*, p. 27 (1994).

[207] Bowies R. J. and Goodson J. Dual-use technology and the national airspace system. *IEEE Natl. Telesyst. Conf.*, p. 1A–4 (1994).

[208] Michelson M., Shrader W. W. and Wieler J. C. Terminal Doppler weather radar. *Microwave J.* **33**, 139 (1990).

[209] Moore S. A. W. and Moore A. R. Dual frequency multifunction radar antenna research. *10th ICAP, IEE Conf. Pub.* **436**, pp. 1522–1526 (1997).

3

Array Antenna Elements

It's better to know some of the questions than all of the answers

James Thurber, 1894–1961

In the first two chapters we considered current and future radar, radar-related, communications, and radiometric systems as well as the associated applications. In exploring future systems we accumulated a seemingly endless wish list of instrumental requirements related to systems ranging from narrowband to wideband multifunction systems. In the next three chapters we shall demonstrate that our wish list is firmly embedded in the realms of what is available now and in the not too distant future. Hopefully the review of technological offerings and opportunities will go some way toward the definition of future systems that have better performance than the systems we have considered without any regard to technological issues. The scope of our coverage is, however, restricted to the RF front-end of the systems, namely the antenna elements, polarizers, T/R modules, and beamformers. While array antenna elements and polarizers are considered in this chapter, T/R modules, and beamformers are treated in Chapters 4 and 5, respectively.

In this chapter we shall consider the widely used reflector antennas. Both phased arrays and reflectors can yield several independent and simultaneous beams and yet the two apertures have the following distinct characteristics:

- Phased arrays can yield extremely narrow, agile/staring antenna beams; with present technology it is mechanically impossible to realize reflectors yielding antenna beams that have comparable spatial resolutions.
- Even when reflectors and phased arrays have the same geometric area, the phased arrays offer the user several degrees of freedom, ranging from the interbeam spacing to the resulting beam shapes, but are relatively expensive to realize. Reflectors, on the other hand, offer fewer degrees of freedom but are less expensive.

Clearly, reflector antennas and phased arrays are complementary systems. Reflectors are, for instance, the building blocks of radioastronomy arrays that yield extremely narrow beams, and reflectors having a phased array at or near their focal plane constitute affordable and sensitive imagers and radar systems.

As reflectors yielding one antenna beam are adequately treated in basic antenna treatises and reference books, multiple-beam reflectors that can be used either as standalone imagers or in conjunction with phased arrays are considered in some detail in this chapter. In a similar vein, we shall consider approaches that allow simultaneous operation over two or more frequency bands.

Beam forming lenses are low-cost solutions often used in wideband, wide scan angle applications where a relatively small number of well-isolated staring beams are required. As some of these lenses perform the aperture function also, we shall cover these beamforming lenses in this chapter.

3.1. A REVIEW OF TRENDS AND REQUIREMENTS

Early radar array models were heavy structures that used metal horns as antenna elements followed by metal waveguides. The migration to lightweight structures that consist of dipoles or microstrip antennas and transmission lines is pronounced. Recently, microstrip transmission lines have been replaced with optical fibers and a further decrease in array weight is in the offing.

Apart from weight reduction, microstrip antennas are inexpensive to produce and can be conformal to the platform's skin. The migration from an aerodynamically unattractive filled aperture housed in an aerodynamically attractive but lossy radome to a conformal aerodynamically attractive array is a significant trend.

The third attractor of microstrip antenna elements is their bandwidth. Recent developments indicate that these antenna elements can meet narrowband, moderate band and wideband requirements. High-quality performance, for instance, can be expected from microstrip antennas operating over a 3:1 frequency range. Our coverage of microstrip antenna elements is therefore extensive and our scope does not include slotted waveguide antennas because these antennas are basically narrowband structures and have been adequately covered in other books and references.

In general terms we are interested in antenna systems that are highly efficient over different instantaneous and total system bandwidths; here the efficiency of an antenna is related to well-behaved radiation patterns, adequate isolation between the two polarizations, and a good match between the antenna and the LNA/FPA. For a receive antenna a voltage standing-wave ratio (VSWR) equal to or less than 2 is acceptable. For antennas operating in the HF band or over a narrow bandwidth the antenna input VSWR is considerably tighter. The input VSWR for a transmit antenna is preferably much less than 2.

The first three entries in Table 3.1 summarize the instantaneous fractional bandwidths (IFBs) and total system bandwidths of current systems. Systems in category I represent current multifunction radar systems having an IFB that is equal to or less than 10% centered at f_1 and the total system bandwidth is equal to about 10% of f_1. Requirements or the array bandwidth constraints determine the exact IFB when phase-shifters are used to form and steer the resulting beam. Current compact array radiotelescopes operating at centimeter wavelengths are represented in category II and have an IFB of 10% or less centered at different frequencies $f_1, f_2, \ldots, f_N$. The exact IFB is determined either by requirements or by the array bandwidth constraints when phase-shifters are used to form and steer the resulting beam. These bands can be contiguous and are not available simultaneously. The total system bandwidth can be at least 10:1. Compact arrays operating at millimeter wavelengths operate over one or more waveguide bands Current OTHRs represented in category III are tunable over a 30:1 range and their IFB is less than 10%.

The remaining entries, categories IV to VII, represent proposed systems under development. Radar shared aperture systems in category IV can operate at two frequencies f_1 and f_2 simultaneously and the IFB is equal to or less than 10%—see section 2.7.4.6. Category V systems represent radar systems operating at many frequency bands and the number of bands and total bandwidth are defined by requirements, the available budget, and technology. While two bands are required to perform the essential radar functions, the remaining bands can be used to perform EW/communications or radiometric functions essential for the platform's survival. The many bands can also be used to attain ECCM capabilities for the radar system.

Table 3.1 The instantaneous fractional bandwidth (IFB) and total system bandwidth for several current and proposed systems

Systems	IFB centered at f_1	IFB centered at f_2	$f_3, \ldots, f_{N-1}$	IFB centered at f_N	Total system bandwidth
I. Current canonical multifunction radars	$\leq 10^a\%$				$10\%f_1$
II. Current radiotelescopes	$\leq 10\%^b$, not simultaneous	$\leq 10\%^b$, not simultaneous	....	$\leq 10\%^b$, not simultaneous	At least 10:1 in frequency
III. Current OTHRs	$\leq 10\%$– Tunable	$\leq 10\%$– Tunable	....	$\leq 10\%$– Tunable	At least 10:1 in frequency
IV. Proposal. Shared aperture radar systems	$\leq 10\%$ simultaneous	$\leq 10\%$ simultaneous	———	———	$f_2/f_1 \approx 10$
V. Proposal. Shared aperture radar systems	$\leq 10\%$ simultaneous^b	$\leq 10\%$ simultaneous^b	....	$\leq 10\%$ simultaneous^b	Number of bands and total bandwidth is defined by many factors^c
VI. Proposal. Radiotelescopes	$\leq 10\%$ simultaneous^b	$\leq 10\%$ simultaneous^b	....	$\leq 10\%$ simultaneous^b	Number of bands and total bandwidth is defined by many factors^c
VII. Proposal. Shared aperture radar systems	Equal to the total bandwidth^d				2–3 octaves

[a] The IFB is defined by system requirements or the array bandwidth constraints when phase-shifters are used to form and steer the resulting beam.
[b] The IFB is available within an octave bandwidth and is defined by system requirements and array bandwidth constraints when phase-shifters are used to form and steer the resulting beam.
[c] Some of these factors are requirements, available budget, and current technology.
[d] The bandwidth has to be partitioned to meet the array bandwidth constraints when phase-shifters are used. Alternatively, delay lines and phase-shifters have to be used. The technological challenges involved to realize these systems are very significant.

Radiotelescopes in category VI have similar characteristics to the category V systems. Lastly, the systems of category VII have large instantaneous bandwidths. The bandwidth has to be partitioned to meet the array bandwidth constraints when phase-shifters are used. Alternatively, delay lines and phased shifters have to be used. The technological challenges involved in the realization of these systems are very significant.

Another system requirement is the utilization of two orthogonal linear polarizations or two opposite circular polarizations. We already concluded, in Chapter 1, that polarization isolation is application dependent but usually ranges between 25 and 40 dB. Several realizations of polarizers are explored in this chapter.

Subarrays reduce the total array cost and marginalize the occurrence of array scan blindness. However, the price the designer pays for these advantages is the occurrence of amplitude quantization (AQ) lobes that can be marginalized by the approaches we shall explore in section 3.8.1.

The other antenna elements for phased arrays that we shall consider are conventional monopoles and printed board dipoles. As can be expected, our focus will be on the novel aspects of widely used antennas and on antenna elements not usually covered in other books.

While meeting any of the above requirements, e.g. cross-polarization level or bandwidth of operation, is relatively straightforward, meeting a set of these requirements is challenging.

3.2. MONOPOLES/DIPOLES

Conventional monopoles/dipoles have been used extensively in many narrowband systems. While they are eminently suitable for narrowband operation, numerous efforts have been made to increase their bandwidth of operation. In section 3.6.3.2 we consider wideband printed board dipoles and in this section we shall consider one approach to widen the useful bandwidth of conventional monopoles. Monopoles are widely used in OTHR systems that have to operate over a bandwidth of at least 10:1.

3.2.1. The Tuning of Monopoles

The gain of an electrically short monopole, $G(\theta,\phi)$, over an infinite plane, is given by

$$G(\theta,\phi) = \eta_r \eta_m D(\theta,\phi) \tag{3.1}$$

where $D(\theta,\phi)$ is the antenna directivity, and η_r, η_m are the radiation and mismatch efficiencies given by

$$\eta_r = \frac{P_r}{P_r + P_L} \tag{3.2}$$

and
$$\eta_m = 1 - |\Gamma|^2 \tag{3.3}$$

respectively; P_r is the radiated power, P_L is the power dissipated by the loss resistance, and Γ is the antenna input voltage reflection coefficient. For a monopole, η_r represents a fundamental constraint that is related to its length, ℓ, and varies slowly with frequency. η_m, on the other hand, varies rapidly as a function of frequency.

The input impedance of an electrically small monopole antenna over an infinite perfectly conducting ground plane is given by [1]

$$Z = R + jX$$

where

$$R = 40\left(\frac{\pi\ell}{\lambda}\right)^2 \quad \text{and} \quad X = -60\left(\ln\frac{2\ell}{d} - 1\right)\cot\left(\frac{2\pi\ell}{\lambda}\right) \tag{3.4}$$

λ is the wavelength of operation and d is the diameter of the dipole. As can be seen, the input impedance of the monopole can be matched to the input of the LNA/FPA by appropriate tuning over a narrow bandwidth centered at the wavelength of operation.

The maximization of η_m by appropriate tuning has been reported in early references [2,3] and more recently in references [4] and [5]. The tuning of monopoles in the VHF band utilized by frequency hopping systems was implemented with the aid of PIN-diode switches and appropriate high-Q passive components at the rate of 1000 Hz [4]. In reference [5] the tuning of a monopole resonant at 180 MHz over the frequency band 30–90 MHz with the aid of PIN-diode switches, low-loss passive components, and a microprocessor is reported. A 10 dB improvement in gain was measured over the band of operation and the comparison is here made with respect to the untuned monopole. The issues of importance are the use of high-Q tuning circuits and low-loss PIN-diode switches.

Comparable improvements in the gain of monopoles used in the receive arrays of OTHRs can be expected over the useful frequency range. Tuning of microstrip dipoles at centimeter wavelengths is more challenging because losses have to be kept to a minimum.

Tuning also applies to other types of microstrip antennas such as patches.

3.3. REFLECTOR SYSTEMS

Reflector systems are widely used in many applications such as communications and radioastronomy installations. The prime focus reflector shown in Figure 3.1a consists of a symmetrical parabolic reflector and a focus cabin placed at the focal plane of the reflector. The cabin houses a high-efficiency feed placed at the focus of the paraboloid F_1, and a polarizer followed by the receiving/transmitting system. The reflector and its cabin are on a mount that points the antenna toward any azimuth/elevation angle. N simultaneous and independent antenna beams can be attained if an equal number of horns/polarizers and receiving/transmitting systems are placed at the focal plane of the reflector.

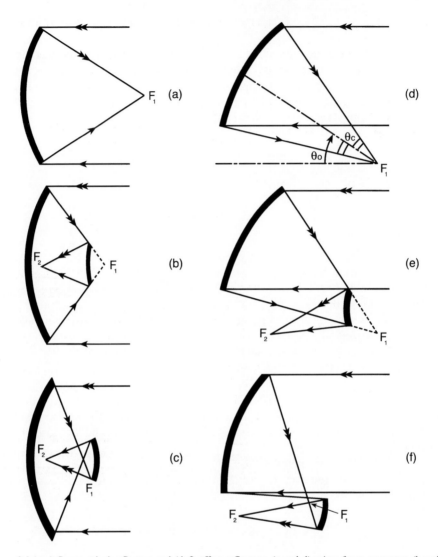

Figure 3.1 (a–c) Symmetrical reflectors and (d–f) offset reflectors. (a and d) prime-focus geometry, (b and e) Cassegrain, and (c and f) Gregorian systems.

In Figures 3.1b and c the Cassegrain and Gregorian systems of reflectors are shown. The main reflectors for both systems are again symmetrical paraboloids, but the sub-reflectors are hyperboloid or ellipsoid, respectively. The subreflectors reflect the incoming radiation from the primary focus F_1 to the secondary focus F_2. All three versions of symmetric reflectors are widely used.

The attractors of symmetrical reflector systems are:

- Frequency-independent operation within the frequency band f_H–f_L.
- The attainment of high gain at reasonable costs.
- The systems have no grating lobes and are volumetrically attractive.

- To a first approximation, the resulting beamwidth is invariant (with elevation/ azimuth angles).
- N independent and simultaneous antenna beams can be attained by placing N feed horns/polarizers and receiving/transmitting systems in the focal plane of the reflector system.

For a prime focus reflector, f_H is determined either by the random phase errors of the reflector's surface from a perfect paraboloid [6,7], or by the systematic phase errors on the reflector's surface caused by gravity forces, thermal distortions, initial rigging, or adjustment bias on the antenna structure [8]. If both kinds of errors are taken into account, the shortest wavelength of operation, λ_{min}, of a reflector antenna of diameter D is given by [9]

$$\lambda_{min} = 5(\text{mm}) \ \frac{D}{100} \tag{3.5}$$

where D is in meters. f_L, on the other hand, is defined by considerations related to the generation of cross-polarization fields due to diffraction; if the smallest reflector diameter used in a system of reflectors is greater than 25λ at the lowest frequency of operation, the cross-polarization fields can be neglected [10].

In its simplest form, the maximum gain, G, of a reflector having a geometric area, A, at a wavelength, λ, is given by

$$G = \frac{4\pi A}{\lambda^2} \ \eta_{so}\eta_b\eta_l\eta_{sc} \tag{3.6}$$

where η_{so} is the spillover efficiency, η_b is the blocking efficiency, η_l is the illumination efficiency, and η_{sc} is the scattering efficiency. For prime reflectors operating in a transmit mode, a spillover efficiency of unity is attained when all the power radiated from the feed horn illuminates the primary reflector; similarly, the blocking and illumination efficiencies are equal to unity when the transmitted radiation is not blocked and when the reflector is uniformly illuminated. Any blockages decrease η_b and raise the sidelobe level of the reflector [11]. For a prime focus reflector, the radiation is blocked by the focus cabin and its three or four supports. For Cassegrain and Gregorian systems the subreflectors used and their supports block the received/transmitted energy. Lastly, η_{sc} is maximized when the reflector's surface approximates a perfect paraboloid that is adequately supported so that systematic distortions are minimized.

If we attempt to maximize η_l by uniformly illuminating the main reflector, the resulting first two sidelobes will be -13.3 and -17.6 dB below the maximum main lobe, if the reflector is square or circular, respectively. As this level is too high, some amplitude taper is used to decrease the sidelobes at a slight decrease of the illumination efficiency. Standard antenna books such as [12] and papers [13] and [14] relate the types of amplitude taper with the level of the first and subsequent sidelobes. When uniform illumination is used, the resulting beamwidths are minimum. For a square aperture of side A or a circular aperture of diameter D, the resulting half-power beamwidths (HPBWs) are $0.89\lambda/A$ and $1.02\lambda/D$, respectively. As the illumination is tapered, the resulting sidelobes decrease but the beamwidths increase.

The maximization of η_{so} and η_{sc} call for efficient feed horns and mechanical engineering considerations, respectively. While the blockage due to the subreflector/ focus cabin is tolerable when only one beam is derived from the reflector, the blockage increases when many beams are required.

The elimination of any structures blocking the received/transmitted radiation has been pursued over many decades with considerable success, and offset antennas are now commonly used [15–19].

3.3.1. Offset reflectors

The antennas shown in Figures 3.1d to f are the offset equivalents of their symmetric counterparts shown in Figures 3.1a to c. As can be seen, there is no blockage of the incoming radiation.

The offset reflectors share many of the attractors attributed to symmetrical reflectors but have the following additional characteristics:

*   Maximization of η_b.
*   The near and far sidelobes are solely determined by the illumination function and can be at acceptably low levels; thus, higher isolation between beams and lower antenna temperatures result [20]. The comparisons are made here with respect to their symmetrical counterparts.

Offset antennas have been used extensively in satellite multiple-beam applications and point-to-point communication systems [20]. The new 100 m Green Bank (West Virginia, USA) radiotelescope operated by the National Radioastronomy Observatory has an offset geometry [21].

Small reflectors, symmetrical or offset, are inexpensive because they can be stamped out of metal. Compared to their symmetrical counterparts, large offset antennas are more expensive to realize owing to their asymmetry. Apart from increased costs, the widespread use of offset antennas was also delayed because the antennas give rise to cross-polarization fields, γ, when they are illuminated by linear polarization [22]. If an offset reflector is illuminated by circular polarization, the resulting beams corresponding to the two polarizations are slightly displaced by an angular distance equal to δ [22]; some authors use the term 'beam squint' for the beam displacement observed when the offset reflector is illuminated by circular polarization.

The magnitudes of γ and δ are proportional to the angles θ_0 and θ_c, shown in Figure 3.1c. The minimization of θ_0 and/or θ_c therefore results in the minimization of γ and δ. For prime focus offset reflectors, γ and δ are negligible if the ratio of the focal length to its diameter, or f/D ratio, is equal to or greater than unity [22].

A frequency-independent design procedure that renders γ and δ of Cassegrain and Gregorian systems of offset antennas negligible is outlined in reference [23]. The assumptions made in formulating the procedure are:

*   The feedhorns used to illuminate the reflectors are ideal or have negligible cross polarization over the bandwidth of interest.
*   The size of the subreflectors used is greater than 25 wavelengths, so cross polarization fields due to diffraction are not generated [10].

Pre-1988 work on the minimization of γ and δ is outlined in reference [23]. Reference [24] (1993) outlines a design procedure for the derivation of offset parabolic reflectors having low cross-polarization and low sidelobes.

3.3.2. Focal Plane Imagers

Focal plane imagers consist of a system of reflectors capable of yielding several simultaneous and independent antenna beams and an equal number of receiving/transmitting systems, each accommodating one antenna beam. The arrangement under consideration represents the radio equivalent of the optical camera in the sense that each pixel on a photographic film is substituted by a feed horn and receiver/transceiver combination. Offset reflectors are ideal for the generation of multiple independent and simultaneous beams.

While the main beam, accommodated in the feed horn positioned at the focus of the reflector, will have sidelobes determined by the illumination function (no blocking is assumed), the antenna beams accommodated in the neighboring feeds, directed toward θ_1, θ_2, θ_3 etc. with respect to the boresight axis, will have their first sidelobe, also known as the coma lobe, raised because Abbe's sine condition (modified for reflectors), is not satisfied. Apart from the raising of the coma lobe, the gain of the resulting beams decreases as the angular distance from boresight increases. For commercial applications the quality of the resulting beams is not as high as that corresponding to imagers used for radioastronomy applications.

The total number of beams on the sky, N_{TOT}, is given by [25–27]

$$N_{TOT} = 1520(f/D)^4 \tag{3.7}$$

It is here assumed that the beam deviation factor is 22, the coma level is –10 dB (with respect to the main beam), the gain loss is 1 dB and the illumination taper is 10 dB [26,27]. The quality of the resulting beams is acceptable for many commercial and applied science applications. By contrast, the total number of high-quality beams on the sky, N_{TOT}, is given by [25–27]

$$N_{TOT} = 28(f/D)^4 \tag{3.8}$$

The assumptions made are that the beam deviation factor is 3, the coma level is –20 dB, the gain loss is 0.05 dB and the illumination taper is 10 dB [26,27]. As can be seen:

- Substantial numbers of antenna beams cannot be derived from structurally attractive prime focus reflectors having f/D ratios less than 1.
- The total number of derived beams from a reflector is inversely proportional to the quality of the beams.

Table 3.2 shows the dependence of N_{TOT} on the antenna f/D ratio; in the same table is listed θ_c (see Figure 3.1d) of the reflector having the same range of values of f/D. As can be seen, a considerable number of high-quality beams can be generated theoretically when the f/D ratio is higher than 1.

The diameters $2a$, and slant lengths, R, of the hybrid mode feeds required to illuminate the reflectors having different f/D ratios are also listed in the same Table 3.2 [27].

Table 3.2 The dependence of N_{TOT} on the antenna f/D ratio and θ_c, $2a$, and R of the feed horn[a] [27]

f/D	N_{TOT}	θ_c (deg)	$2a$ (λ)	R (λ)
0.8	11	33.7	2	5
0.9	18	30.5	2.2	7
1	28	27.7	2.6	8
1.2	58	23.3	3.2	13
1.4	108	20	3.6	15
1.5	143	19.3	4	17
2	452	14.2	4.8	28
3	2290	9.5	8	70

[a] The beam deviation factor is assumed to be 3.

A hybrid mode feed is shown in Figure 3.2a and c, while a scalar feed is shown in Figure 3.2b. From Table 3.2 we can see that as the f/D ratio increases, θ_c decreases and $2a$ and R increase. The graphs relating θ_c to the corresponding hybrid mode horn radius, a, and slant length, R, are shown in Figure 3.2b when the taper is 10 dB. As can be seen, the radii and slant lengths of the feeds increase exponentially when the f/D ratio of the reflector is higher than about 1.4. Realistically then, the lower and higher bounds of N_{TOT} are about 10 and 100, respectively, with conventional offset reflectors [27].

In order to increase the packing density of the hybrid mode feeds, the assumption is made that the corrugations of the hybrid horns can be cut away at the aperture of the horns in order to pack them tightly in the focal plane without a penalty in increased beamwidth or sidelobes [28]. With this arrangement, the requirements for (a) low

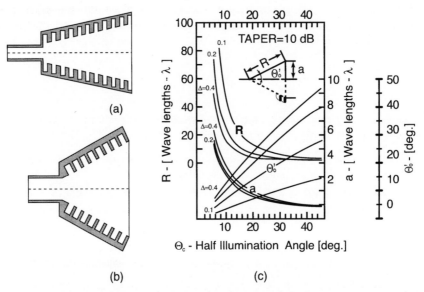

Figure 3.2 High-quality horns. (a) The hybrid mode horn. (b) The scalar horn. (c) The dimensions of hybrid mode horns as a function of the half illumination angle θ_c and when the amplitude taper is –10 dB [39,40].

cross-polarization radiation, (b) low sidelobes, (c) low spillover losses, and (d) low mutual coupling between feeds over a waveguide bandwidth are met, but optimal sampling of the focal plane is not attained [27,29–31]. This approach is acceptable for radioastronomical observations since a full image of an extended source can be assembled by interleaving two or more sets of observations obtained when the array is pointed at slightly different directions [25,31].

For imaging systems designated for commercial applications, the requirement for optimal sampling of the focal plane is of paramount importance and can be approximately met if tapered slotline antennas are used [29] instead of hybrid mode horns. The whole question of generating high-quality independent and simultaneous antenna beams and of sampling the focal plane optimally is problematic with the feeds presently known [31]. The annular synthesis antennas do not share this problem—see section 3.5.

3.3.2.1. Proposals for Schmidt Radiotelescopes

Current radiotelescopes have excellent spatial resolutions and adequate sensitivities; indeed their spatial resolutions often match or exceed those attained by optical telescopes. The other defining characteristics of current radiotelescopes are:

- One narrow FOV is observed at a time.
- Dual-polarization capability.
- Operation over one narrow frequency band at a time.
- A range of short spacings between adjacent antenna elements is often missing.

The requirements for future radiotelescopes are:

- High sensitivity and moderate spatial resolution. More explicitly, the required sensitivity is equal to that attained by a radiotelescope fully populating an area of one square kilometer [32,33]. Various realizations of the one square kilometer array (SKA) are now seriously considered in many countries—see also section 3.10.
- The availability of one or more wide fields of view simultaneously [31,32].
- Inexpensive realizations are preferred.
- Operation at many frequency bands simultaneously [27].

The last requirement not only minimizes the observation time required to attain maps of a celestial source at many frequency bands but also meets the requirement for observations of variable sources at many frequency bands simultaneously. The same radiotelescopes can also be used for the search for extraterrestrial intelligence (SETI). In section 3.10 we shall expound on flexibility as an emerging important requirement for future radiotelescopes.

As offset systems of reflectors are eminently suited for multiple-beam operation, they meet most of the above requirements. Dipoles raised above ground and other antenna elements are also being considered for future radioastronomy arrays [32,33]. Dipoles having a hemispherical FOV are inexpensive antenna element realizations and are particularly suitable for operation below 1 GHz. Their bandwidth is narrow, but tunable dipoles can meet all but the last requirement.

Regardless of which antenna elements are used, future radiotelescopes will have a FOV that is considerably wider than that of conventional radiotelescopes; thus the radio equivalent of the Schmidt optical telescope will be realized. The promise of ever-increasing computing power at affordable costs is assumed.

Figure 3.3 illustrates an artist's impression of a proposed synthesis radiotelescope operating at centimeter/millimeter wavelengths [27]; it consists of a number of offset reflectors arranged so as to meet the spatial resolution and sensitivity requirements, and 10 or 100 non-contiguous fields of view can be observed simultaneously. If the projected diameter of each offset antenna is small, each of the derived fields of view can be wide. As the altitude over azimuth mounts of the offset antennas have the azimuth tracks on the ground, the receiving systems are easily accessible. Other realizations based on inexpensive offset dishes normally used for the reception of TV images are also promising.

The provision of suitable, low-loss frequency multiplexers following the polarizers will enable the array to operate at several frequency bands within the band of the hybrid mode or dielectrically loaded feeds simultaneously (see section 3.10 on the capabilities of dielectrically loaded feeds). Given that most radioastronomy arrays utilize cryogenically cooled front-ends, it is natural to include low-loss multiplexers utilizing superconductors [34] to separate the radioastronomy bands, before low noise amplification, in the same cryodyne. With this arrangement, the losses between the polarizer and the LNA are minimized.

If the subreflector of a Cassegrain antenna is a frequency-selective surface (FSS) the systems can yield as many staring beams at another waveguide band—see Section 3.3.4. The derivation of simulaneous multiple beams at several bands meets the important requirements for radiotelescopes for the next millennium.

A cost-effective engineering approach will therefore be to realize the proposed second-generation radiotelescopes and use a substantial number of staring beams economically, in the first instance; as the costs of computing power decline with the passage of time, the users can implement a progressively higher number of staring beams [27]. If a higher number of independent and simultaneous beams is required, the reflectors proposed in references [35] and [36] can be considered.

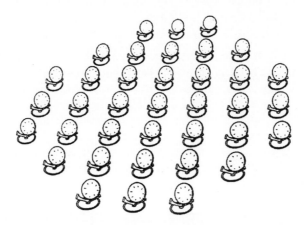

Figure 3.3 Artist's impression of a proposed radiotelescope; each antenna element is an offset system of reflectors. The array geometric area and the degree of thinning is derived from considerations related to the required spatial resolution and sensitivity. (From [27]; ©1998, Wiley.)

The data rate derived from the novel phased arrays having multiple fields of view and capable of operating at several frequency bands simultaneously will be prodigious [27]; the processing of the data derived will therefore be costly. However, the cost of computing power is expected to decrease dramatically over time [37].

The annular synthesis antenna (ASA) [38–40], which consist of a circular array of antennas mounted on a fully steerable structure, offers the designer the following attractors:

- A range of short spacings between antenna elements
- An economic realization, since one mount can accommodate a number (eight or more) of small and inexpensive antennas. The antenna elements of the circular array can be offset reflectors commonly used for direct television reception. As these reflectors are small in size, this realization meets the important requirements of low cost and a wide instantaneous FOV. Annular synthesis antennas are further considered in section 3.5.

We have already noted in the previous two chapters that the excision of unintentional jammers is important for sensitive radioastronomy observations. Apart from the terrestrial emissions used by many services, the transmissions of the constellation of navigational and communications satellites operating at S-band and other bands (listed in Table 1.16), are significant sources of interference for all radiotelescopes regardless of their position on Earth. Recent approaches to mitigate the effects of radiofrequency interference are outlined in section 3.10.

3.3.2.2. Other Applications

Hybrid systems employ one or more reflectors/lenses and a phased array placed near or at the focal plane of the aperture. Reference [41] reviews focal plane imaging systems and hybrid systems, while reference [42] is a treatise on hybrid systems. In this section we shall delineate a variety of focal plane imagers and some hybrid systems. Reference [43] is a recent and useful review of applications.

In the radioastronomy context, the first step toward increasing the efficiency of observations is taken by converting the existing reflectors, designed to yield one antenna beam into multiple beam reflectors. An example will illustrate the rationale upon which the requirement for imagers operating at millimeter wavelengths is based. We have already noted that carbon monoxide molecules are abundant in the cosmos and the CO molecule has transitions at about 115.3 GHz and multiples thereof. The mapping of molecular clouds contributes to our knowledge of cloud kinematics and molecular formation processes. As the spatial resolutions attainable by moderate size antennas, e.g. 14 m are of the order of 50 arc seconds at 115.3 GHz [43], the time taken to map a molecular cloud of angular extent of 2.5° by 3° would be far too long with a conventional single beam aperture. It is recalled here that meteorological conditions further constrain the observation times available for radiotelescopes operating at millimeter wavelengths.

Reference [44] collates important focal plane imaging systems. Of interest are the following systems:

(a)   An 8-element imager operating in the frequency band 220–230 GHz in conjunction with the 12 m radiotelescope operated by the NRAO (National Radio Astronomy Observatory).

(b)   A 15-element imager operating in the frequency band 86–115 GHz in conjunction with the 14 m radiotelescope operated by the FCRAO (Five Colleges Radio Astronomy Observatory) [45]. The array consists of suitably modified corrugated feed horns and polarization interleaving is used.

Both systems utilize front-end mixers, cooled to 15 and 20 K, respectively, to downconvert the incoming signals to a suitable IF. Other developments taking place at different observatories are also outlined in the same reference.

Passive millimeter-wave radiometric imagers are used widely. Imaging systems operating at 140 GHz are used for plasma diagnostics, which are essential for the understanding of energy loss processes [44]. Other imaging systems are used for the detection of concealed weapons and explosives in airports and imagers that enable aircraft to land autonomously [46]. Radiometric imagers operating at millimeter wavelengths have adequate spatial resolutions, are compact, and offer an effective aid, or an alternative to existing technology for aircraft landing and surface operations under inclement weather conditions. The impact of this technology on air traffic can be significant in terms of reducing weather-related accidents and delays as well as affecting the economics of airlines, mail and parcel carriers, and general aviation. The last applications deserve considerable attention because of the recent technological breakthroughs reported [46,47].

Focal plane imagers have adequate sensitivity even when images are obtained at frame rates of 10–30 frames a second. Several imagers operating at 94 GHz have been realized [46,47]. While current systems typically have a few tens of elements, future systems having 10^4 antenna elements are seriously being considered [43]. When the number of pixels is 10^4, direct low-noise amplification without any heterodyning is preferred as a means of realizing simpler systems [44,46].

As space for commercial applications is limited, lenses having an f/D ratio ≤ 1.25 [43], rather than reflectors, are typically used. The insertion loss of lenses operating at millimeter wavelengths can be minimized if Fresnel zone plates [48] are used.

While the sensitivity of millimeter wave radiometric systems is adequate, increased spatial resolution and sensitivity are desirable, particularly for airborne imaging applications. The applications envisaged are for airborne systems performing the surveillance and reconnaissance functions under all weather conditions. We have already considered systems or proposals for systems operating at centimeter and millimeter wavelengths in section 1.7.5.2.

As can be expected, hybrid systems have considerable collecting areas and are less expensive than fully populated phased arrays. Their scan angle, however, is limited to say ±10° [49, 50]. As the angle subtended by a satellite located at the geostationary orbit is only 17.34°, hybrid systems have been used. The transmit antenna of the INTELSAT VI (4/6 GHz) global system consists of a 3.2 m offset reflector fed by an array of 146 dual-polarized feed horns; groups of feeds are excited together to produce the required beam patterns and to cancel sidelobes [51].

Trends in satellite systems toward higher capacity resulted in stringent demands on the antenna design to provide more reuse of the spectrum. These requirements have driven conventional reflector systems to their limits defined by volumetric constraints and mass [52,53]. Phased arrays are eminently suited for satellite antenna applications owing to their ability to form multiple beams, to provide power-sharing among beams through distributed amplification, and to rapidly reconfigure and re-point beams. However, their adoption in communication satellites has been delayed because of concerns related to overall mass, DC-to-RF efficiency, and the uniformity of performance of a large number of elements. The application of MMICs to arrays has helped to eliminate many of these concerns [54].

The precision approach radar (PAR) of the AN-TPN-19 system is an example of a hybrid radar system used to guide aircraft during final approach to airports and consists of an offset hyperbolic surface, a phased array, and a feed horn. The system operates at X-band and changing the phase settings of the array's phase-shifters moves the resulting beam. The system requires a limited coverage sector of 15° in elevation and 20° in azimuth [49]. The AN/GPN-22 system has a coverage of 8° in elevation only and is less expensive.

3.3.3. Segmented Reflectors

We have already derived the resolution limit of single aperture systems in section 3.3; the theoretical maximum diameter of an antenna using conventional materials, for instance, is 20 m if operation at 1 millimeter is required. A proposal for an active surface reflector antenna 50 m in diameter capable of operating at 1 millimeter has been put forward recently [55]. The Large Millimeter Telescope will have a HPBW of 5 arc seconds and a gain of 102 dB at 1 millimeter [55]. The proposal calls for a surface that consists of 126 hexagonal segments in a classical Cassegrain configuration and the segments are actively controlled to reduce the effects of gravitation on and/or thermal gradients across the reflector's surface. The design was based on the successful design of the 10 m, M. W. Keck optical telescope the surface of which consists of 36 actively controlled hexagonal segments [56,57]. Each of the 36 mirrors of the M. W. Keck optical telescope is 1.8 m across the corners and is kept in near-perfect alignment by a complex of microcomputers, position sensors, and hundreds of mechanical actuators [56]. At the commissioning stage (14 April 1992) the M. W. Keck was the world's largest optical telescope; it is four times more powerful than the famed Palomar telescope (monolithic, 5.08 m Hale reflector) and the telescope is located atop Mauna Kea, in Hawaii, USA at a height of 4200 m, well above the 'dirty atmosphere'. The new technology telescope (NTT) is 3.58 m in diameter and utilizes 75 mirrors that are actively aligned once every few minutes to maintain perfect alignment [58]. It is perched 2400 m above sea level at La Silla, Chile. Both telescopes are cost effective and astronomically very successful.

In general terms the construction of large segmented reflectors designated for operation at millimeter, infrared, or optical wavelengths can benefit from the evolution of inexpensive and readily available microelectromechanical systems (MEMS). Segmented reflectors represent an inexpensive solution to attaining large collecting areas; clearly there is a point beyond which it is no longer economical to manufacture large monolithic reflectors. Costs and the overall weight define this limit.

3.3.4. Reflectors Utilizing Frequency-Selective Surfaces

We have already noted that frequency-selective surfaces (FSSs) used in conjunction with reflectors allow the user to have access to two or more frequency bands centered at widely separated frequencies. While the communications community undertakes the bulk of the work on FSSs, the same approaches and designs can be used for radar, radar-related, and radioastronomy systems.

Figures 3.4a and b show a symmetrical and an offset Cassegrain system of reflectors, respectively. Both subreflectors are FSSs and one band, centered at f_1, is accommodated at the primary focus, F_1, while the other band, centered at f_2, is accommodated at the secondary focus, F_2, of the systems. Figure 3.4c illustrates how three bands (S-, Ka-, Ku-) can be used in conjunction with only one reflector [59]. For offset systems the receivers/transmitters do not block the incoming/transmitted radiations and the process can be extended so that more bands are accessible to the user when one reflector is used.

A FSS is obtained either with a periodic arrangement of conducting patches, a capacitive grid, or with a periodically perforated conducting screen, an inductive grid. At the resonant frequency the former arrangement is reflecting while the latter is transparent. While the former arrangement has poor cross-polarization characteristics, the design criteria for the realization of inductive grids having very low loss and cross-polarization are formulated in reference [60].

Analyses of systems utilizing one high-gain antenna (HGA) in conjunction with a FSS are given [59,61] where three or four frequency bands are accommodated. Two active FSSs have been studied in reference [62]. In one realization the FSS incorporates switched PIN-diodes and enables the surface to be electronically switched from that of a reflecting structure to a transmitting structure. In the other realization, the FSS is

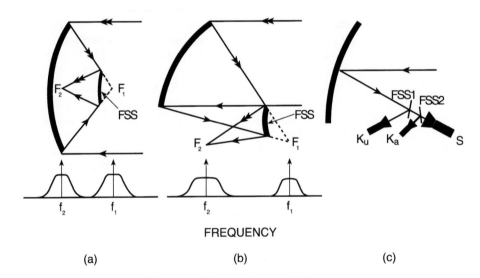

(a)   (b)   (c)

Figure 3.4 Reflectors used in conjunction with frequency-selective surfaces (FSSs). (a) Symmetrical Cassegrain system where the hyperboloid secondary reflector is a FSS. (b) An offset Cassegrain system. The two bands are centered at frequencies f_1 and f_2. (c) A high-gain antenna and two FSSs allow the user to access the S-, Ka-, and Ku-bands, simultaneously (From [59]; ©1992, IEEE.)

printed on a ferrite substrate that is biased with a DC magnetic field. The bias field allows tuning of the resonant frequency by several gigahertz.

The application of high-temperature superconductors (HTSs) as FSSs lower the losses normally associated with conventional FSSs [63]; these losses are significant for systems operating at millimeter wavelengths. Theoretical and experimental studies are outlined in the same reference.

As can be seen, the designer has a plethora of options for accessing four or more bands when only one HGA is used. If one cryodyne houses the HTS-based FSS, and the LNA(s), the front-end losses that often limit a system's performance are minimized. If the system uses an HTS-based multiplexer to further select certain sub bands within a band, the multiplexer can also be included in the cryodyne.

3.3.5. High-Quality Horns

High-quality horns are often used to feed reflectors or lenses. As horns have been studied over a long period [64] we shall assume that the performance of high-quality horns can be taken as ideal. The same horns therefore provide a performance baseline for other antennas that have uniquely attractive characteristics such as lower production cost and conformity to the skin of platforms (aircraft or missiles).

We can easily deduce the requirements for high-quality horns from previous considerations in this book. The feeds should be highly efficient and have equal E- and H-plane radiation patterns, low cross-polarization, and low input VSWR, over as large a bandwidth as possible. Additional requirements are related to real-world considerations: the horns should be compact and easy to manufacture and production costs should be low. If the differences between the E- and H- plane amplitude and phase patterns of circularly symmetric feeds, are approximately zero, the cross-polarization fields are negligible [65].

Both hybrid mode and scalar feeds approximately meet the above requirements [64]. The hybrid mode feed shown in Figure 3.2a has corrugations that are perpendicular to the horn's axis and the cone angle is typically less than 40°. The phase centers for the E- and H-planes coincide and the cross-polarization radiation is negligible over an octave. This is an important property for polarimetric radiometers and radars. The scalar feed shown in Figure 3.2b has a larger cone angle and the corrugations are perpendicular to the flared surface of the cone; the position of the phase center is independent of frequency and the useful frequency band can extend over two octaves. As can be seen, the two types of horns meet different requirements. Most anechoic chambers utilize scalar feeds as transmitting horns [66]. Hybrid mode feeds, on the other hand, are used in multibeam systems because they are smaller.

Hybrid mode feeds having a cross-polarization level of −35 to −40 dB are realizable and design procedures are found in the many papers included in reference [64]. The dimensions of hybrid mode horns as a function of the half illumination angle, θ_c, when the amplitude taper is −10, are illustrated in Figure 3.2c [39]. We have already referred to the latter two figures but it is worth re-iterating here that as θ_c decreases, the radius and slant length of the horns increase exponentially.

Attention is now focused on the derivation of inexpensive manufacturing procedures for hybrid mode feeds with an emphasis on multibeam applications. For research organizations that have access to workshops equipped with numerically controlled lathes, the manufacture of hybrid mode and scalar feeds presents no difficulties. For small research groups in universities, other approaches requiring less sophisticated workshops have been sought. The initial approach was to realize hybrid mode feeds by assembling thin and thick rings of varying internal diameters to form the horn. Such an approach enabled the users to realize high-quality hybrid mode horns in small general-purpose university workshops [39]. Another approach used is to manufacture three segments of the horn separately and then assemble the horn with the aid of conducting glue. The segmentation takes place along the horn's axis and the first segment includes the waveguide, the throat, and the first few corrugations of the horn.

For multibeam systems requiring many horns, operating at millimeter wavelengths, in the focal plane of a high-gain antenna, the manufacture of individual horns by electroforming is prohibitively expensive [67]. To meet the requirements of high performance and low production costs, the platelet horn array was developed [67]. The array of horns is fabricated by diffusion bonding of a large number of platelets that in turn consist of thin sheets of metal in which patterns of holes have been photo-etched [67].

Work toward the design and realization of wideband dielectrically loaded horns is cited in section 3.10.

3.3.6. Polarization: Splitters and Processing

In considering polarization issues we shall restrict attention to the four cardinal polarizations, i.e. the two orthogonal linear or the two opposite circular polarizations, without any loss of generality. The minimum requirement for radioastronomy observations calls for the reception of either the two orthogonal linear or the two opposite circular polarizations. At the other end of the spectrum, the most difficult requirement to meet comes from the radar/communications environment and calls for real-time polarization agility in the receive and transmit modes.

Assuming a receive mode system, the incoming radiation is received either by one antenna from which the two polarizations of interest are derived in ports 1 and 2, or by two separate antennas each accommodating one polarization. If the latter option is selected, the important parameter of interest is the polarization isolation between the two polarizations accommodated in ports 1 and 2. Other parameters of interest are the isolation between ports 1 and 2 and the input VSWR at each port. As the majority of antennas yield the two orthogonal linear polarizations, the two opposite circular polarizations are usually derived from the two linear polarizations by the insertion of 90° phases between either of the antenna ports and the summing point. For the case where the two orthogonal polarizations are derived from one antenna element, high-performance polarization splitters are required and their performance is measured in terms of the bandwidth over which acceptable polarization/interport isolations and input VSWR of the splitter are attained. As the polarization splitters are often inserted before the LNA, low insertion loss is a requirement.

In view of the wide-ranging requirements we have to meet, versatile splitters that can yield a variety of polarization states are also of interest. Some polarization splitters are often referred to as orthomode transducers (OMTs) or polarizers.

3.3.6.1. Polarizers

The bulk of work on polarizers is reported in the early literature, while the performance figures of commercially available polarizers is documented in commercial brochures. In the many references, the input match of the polarizers is given in terms of either their VSWR or their return loss; to facilitate the comparison between the many polarizers we shall report the VSWR and return loss in brackets, i.e. an input VSWR of 2 is referred to as 2 (9.542 dB).

Some of the important pre-1973 work on polarizers is outlined in reference [68]. In reference [69] a circular waveguide dual-mode polarizer yields the two TE_{11} orthogonal modes with an isolation of over 50 dB, over 1GHz centered at 9.1 GHz. Waveguide transitions to the circular guide are used to accommodate the two polarizations and the whole structure is compact but complex; the input VSWR to the two ports over the same frequency range is less than 1.15 (23.127 dB). This polarizer is typical of high-performance polarizers.

Another polarizer employs a square waveguide and the isolation between polarizations is at least 26 dB over a 20% bandwidth centered at 2.9 GHz [70]. Again the performance of this polarizer is typical and the obvious conclusion we can draw is that there is a trade-off between high-performance polarizers and bandwidth over which this performance can be attained.

Next we shall consider a waveguide polarization splitter designed to operate at 18 centimeters for a specialized radioastronomy application [68] because (i) it is a versatile splitter and (ii) it is an example of a narrowband, high-performance polarization

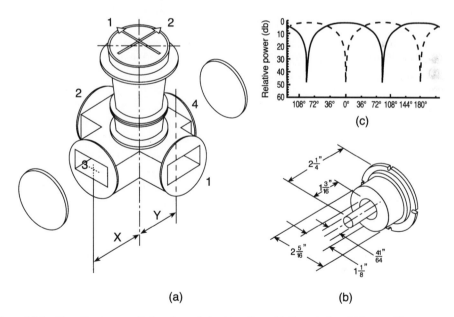

Figure 3.5 An 18 centimeter turnstile junction polarization splitter. (a) General view. (b) Its matching structure at 1666 MHz. (c) The power received in ports 1 (solid line) and 2 (dashed line) when the polarizer is set to receive the two orthogonal linear polarizations and the transmitter is rotated through 360°. ((b) and (c) from [68]; © 1973, IREE (Aust.).)

splitter. The polarization splitter is a six-port turnstile junction [71–73], shown in Figure 3.5a. Ports 3 and 4 are terminated in short circuits, while ports 1 and 2 accommodate the two opposite circular polarizations if the distances X and Y are equal to $7\lambda_g/8$ and $5\lambda_g/8$, respectively, or accommodate orthogonal linear polarizations if the distances X and Y are equal to $3\lambda_g/4$ and $\lambda_g/2$, respectively. The centerline of the matching structure shown in Figure 3.5c coincides with the centerline of the feed horn used to illuminate the reflector.

Table 3.3 Performance figures of the turnstile junction at 1666 MHz [68]

Center frequency	1666 MHz
VSWR	< 1.22 (20.079 dB) from 1653 to 1685 MHz
	Min 1.16 (22.607 dB) at 1666 MHz
Interport isolation	> 30 dB from 1653 to 1677 MHz
(linear polarization)	> 40 dB from 1659 to 1669 MHz
Polarization isolation	> 30 dB from 1663 to 1673 MHz
(linear polarization)	50 dB at 1666 MHz

The input VSWR into ports 1 and 2, the interport isolation and the polarization isolation are listed in Table 3.3 when the turnstile junction is set to receive the two orthogonal polarizations, at 1666 MHz [68]. As can be seen, very high interport and polarization isolation figures are attained over a narrowband of frequencies. In Figure 3.5b the powers received in ports 1 (solid line) and 2 (dashed line) are shown when the turnstile is set to receive the two orthogonal linear polarizations, the linearly polarized transmitter is rotated through 360°, and the transmitted frequency is 1666 MHz. The turnstile junction polarizer exhibits similar performance at 1612 and 1720 MHz when matching structures optimized for these frequencies are used.

A turnstile junction polarizer has been used to compensate for the depolarization caused by rain [74]. The technique was used over a 10 km path and the system operated at 11.7 GHz. One end of the turnstile junction was shorted while the other port used a variable short circuit under computer control. At fades below 15 dB significant isolation improvements between the two channels accommodating the two polarizations were reported, compared to conventional systems where fixed polarizations are used.

A recent reference [75] compares the performance of three OMTs: a finline, a quad-ridged OMT, and a commercial product, the Adams Russel WRD630 OMT. As the OMTs were intended for radioastronomy applications, the following specifications are accepted:

- A minimum input VSWR of 1.432 (15 dB) was stipulated, over a 2.2:1 frequency range, to match the bandwidth exhibited by high-performance corrugated feed horns.
- Minimum interport isolation of 30 dB.
- Maximum cross-polarization level of −20 dB.
- Insertion loss considerably less than 0.5 dB.

All three OMTs met the above specifications but the quad-ridged OMT had superior performance. The commercial OMT matched the performance of the quad-ridged OTM in many respects and can handle higher powers; it is of an ingenious design, but its insertion loss is higher than that of the quad-ridge OMT. Overall, the three OMTs are representative samples of the performance expected from OMTs capable of operation over a 2:1 frequency range.

We have established a baseline for the performance exhibited by narrowband/wideband polarizers. Either requirement can be met if considerable effort is expended and the complexity of the resulting transducers is accepted. Polarizers that meet acceptable performance criteria over a 2:1 frequency range are realizable, and as the bandwidth of operation is narrowed their performance improves. With present technology it is unrealistic to expect high-quality performance from polarizers operating over several octaves.

3.3.6.2. Polarization Processing

The resulting output voltage E_o when the outputs from two linear crossed antennas E1 and E_2 are added is given by the equation

$$E_{oh}^2 = E_1^2 \cos^2 \omega t + E_2^2 \cos^2 (\omega t + \phi) \qquad (3.9)$$

where ω is 2π times the frequency of operation and ϕ is an arbitrary phase inserted before addition. If $E_1 = E_2$ and ϕ takes the values 0°, +90°, +180°, −90°, the resultant E_o has the following polarization states: linear, RHC, linear and LHC. By adjusting ϕ we can therefore attain any polarization state when two crossed linearly polarized antennas are used.

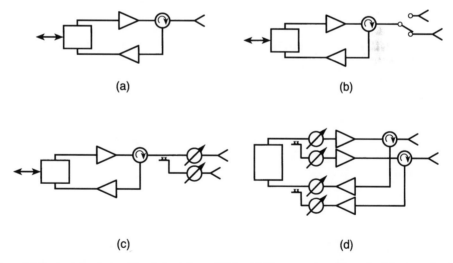

(a)   (b)

(c)   (d)

Figure 3.6 Four radar systems with polarimetric capabilities. (a) The system has either a fixed linear or circular polarization, depending on the polarization of the antenna element. (b) The system can have either of the linear or circular polarizations depending on the polarization of the antenna elements. (c) The two crossed linearly polarized antenna elements are used in a system that can have either linear or circular polarization depending on the settings of the variable phase-shifters. (d) Same arrangement as that of (c) but the phase-shifters are inserted after the LNAs/FPAs.

In Figure 3.6 we illustrate four approaches to deriving various polarization options. Figure 3.6a illustrates an approach to derive a system having fixed linear or circular polarization depending on the polarization of the antenna element used. The system illustrated in Figure 3.6b can be switched between the two linear polarizations derived from two crossed linearly polarized antennas. The system in Figure 3.6c can have either the linear or the circular polarizations depending on the phase settings of the two phase-shifters shown between the two linearly polarized antennas and the summing point. Lastly, the system in Figure 3.6d is similar to that illustrated in Figure 3.6c but the phase-shifters are introduced after the LNAs/FPAs.

Other systems can be derived to attain either full diversity on transmit and receive or full diversity on transmit and adaptive polarization on receive. We have already noted that the latter of the two systems yields the maximum benefit to the user.

3.4. BEAMFORMING LENSES

To the extent that lens apertures do not obstruct the received/transmitted powers in a system, they can be likened to offset reflectors. We have already considered the use of lenses as apertures in the systems illustrated in Figures 2.15d and e. In this section we shall explore beamforming lenses that are often used in conjunction with phased arrays. In some realizations the lens embodies the system's aperture and beamformer. Beamforming lenses are low-cost solutions often used in wideband, wide scan angle applications where a relatively small number of well-isolated staring beams are required. In this section we shall consider the Rotman and Luneberg lenses, two realizations of which are illustrated in Figures 3.7a and b, respectively.

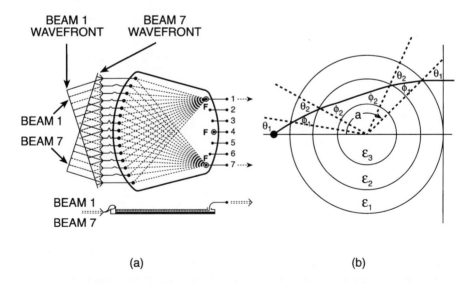

(a)   (b)

Figure 3.7 Beamforming lenses. (a) A linear array used in conjunction with a Rotman lens (From [76]; © 1975, Microwave Journal.) (b) The Luneberg lens combines the aperture and beamforming functions.

3.4.1. Rotman Lenses

Reference [76] is a useful early reference that describes the many types of beamforming lenses most often referred to as Rotman [77] lenses. The same lenses, also referred to as bootlace or Gent [78] lenses, are used mainly in conjunction with linear arrays. Architectures that utilize vertical and horizontal lenses can be used in conjunction with planar arrays [76]. For circular arrays the R-2R or R-KR lenses are more suitable [76].

In Figure 3.7a we illustrate a schematic of a Rotman lens used in conjunction with a linear array. In its rudimentary form, the lens consists of a parallel-plate region fed by a number of coaxial probes placed along its two opposing sides. A number of staring beams, seven in the realization shown, are derived on the right-hand side of the lens because

(i) cables connecting the array antenna elements to the antenna ports of the lens have predetermined lengths; and

(ii) the arcs along which the antenna and beam ports are located have predetermined contours.

Furthermore, the focusing action of the lens is to a first approximation frequency independent and the lens operates over a wide bandwidth.

In the original designs [77,78] the arc along which the beam ports are located was

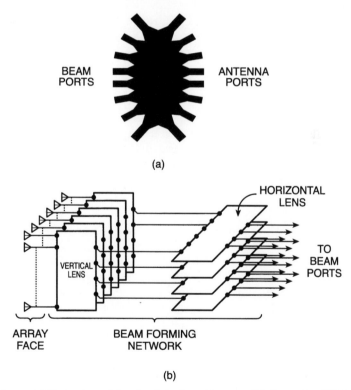

(a)

(b)

Figure 3.8 Rotman lenses. (a) The topology of the printed circuit used in a reversible Rotman lens. When used with an eight-element linear array the lens yields eight staring beams. (From [80]; © 1995, Microwave Journal.) (b) Two-dimensional beamformation used in conjunction with planar arrays. (From [76]; © 1975, Microwave Journal.)

circular and the contour along which the antenna ports are located as well as the cable lengths connecting the antenna ports and the antennas were deduced from the requirement for equal path lengths between antennas and beamports.

A low-cost, high-performance electronically scanned millimeter-wave antenna system utilizing a waveguide Rotman lens has been reported [79]. The lens has 32 active antenna ports and 17 active beam ports, the measured insertion loss ranged between 0.8 and 2.3 dB and maximum sidelobe levels of ≤-30 dB were achieved. Although the system operated in the 33–37 GHz band and the scanning range was ±22.2°, operation at frequencies up to 95 GHz and a scanning range of ±45° are possible with different lens designs.

Other realizations of Rotman lenses utilize microstrip or stripline technology and employ substrate materials with relative permittivity, ε_r, ranging from 2.5 (Teflon–fiber glass) to 233 (cadmium titanate) [76]. All linear dimensions of the RF lens are therefore reduced by a factor equal to $\sqrt{\varepsilon_r}$. Figure 3.8a illustrates the lens topology used for an 8-antenna, 8-beam reversible Rotman lens [80]. In Figure 3.8b the formation of beams from a planar array is shown with the aid of several vertical and horizontal lenses [76]. The MUSTRAC planar array utilized beam-forming lenses in two dimensions [76].

A variant on the original Rotman lens has been proposed [81] in which a non-circular focal arc has been used to reduce the path length errors. Rotman lenses have been used to generate multiple beams in applications ranging from indoor communications [81] to road-toll systems [82].

The applications considered here illustrate the capabilities of several variants of Rotman lenses.

3.4.2. Luneberg Lenses

The Luneberg lens illustrated in Figure 3.7b is a sphere that consists of several concentric shells of dielectric material each having relative permittivity, ε_r, that decreases as the distance from the center increases. If the relative permittivity of the lens varies from a value of 1 at the outer concentric shell to 2 in the central sphere and the exact value of ε_r is given by

$$\varepsilon_r = 2 - \left(\frac{r}{r_0}\right)^2 \tag{3.10}$$

where r is the radius from the center and r_0 is the sphere's radius, the lens will perfectly focus a plane wave from any direction to a point on the opposite side of the lens. The values of 1 and 2 for ε_r are only indicative and other combinations are allowable. As the maximum value of ε_r decreases, the focal radius increases. Luneberg lenses [83] embody the aperture and beamforming functions. Furthermore, there is no foreshortening of the aperture as the lens receives or transmits power toward different elevation/azimuth angles.

References [84] and [85] reflect the recent interest in Luneberg lenses. The number of shells determines the bandwidth of the lens, with a larger number of shells providing a larger range of operating frequencies. There is, however, an upper limit to the number of concentric shells used because there is a small reflection loss at each interface between shells. A 120 centimeter diameter lens has 15 concentric shells and performs adequately over a frequency range from 2 to 20 GHz [85].

The manufacturing process to realize the lenses should be chosen to minimize air gaps between concentric shells, and a shell thickness of one wavelength minimizes losses [86]. The resulting sidelobe level for the Luneberg lenses is approximately at the −20 dB level [79,85], not as low as that derived from Rotman lenses [79], but two approaches have been reported to minimize the resulting sidelobe levels [87].

3.5. PHASED ARRAYS ON STEERABLE MOUNTS

It is often less expensive to mount N antennas, of diameter d, on a mechanically steerable structure to realize a larger-diameter antenna. Depending on the size of d, several systems are realized. In the limiting case when d is very small, a segmented reflector as outlined in section 3.3.3 results. For segmented reflectors the power collected by each reflector is added in space and further processed by one sensor/receiver.

If each antenna mounted on the fully steerable structure is followed by a receiver/sensor, the received powers from all antennas on the mount can by added coherently after some signal processing operations, e.g. beamforming, are performed. In contrast to conventional phased arrays where hundreds or thousands of low-gain antenna elements are used, this class of systems employs either:

(i)   A few high-gain antenna elements (N is between 6 and say 20); we shall refer to this class of systems as annular synthesis antennas, type I, ASAsI.

Figure 3.9 A model annular synthesis antenna. It consists of eight offset antennas mounted on a fully steerable structure and operates at 38 GHz [40].

(ii)   Several moderate-gain antenna elements; this class of systems are referred to as ASAsII.

The synthesized beam or beams derived from an ASA are to a first approximation invariant with elevation/azimuth angles. We shall cover reported work related to ASAsI and briefly outline the options offered by ASAsII.

ASAs that consist of a circular array of N offset antennas mounted on a fully steerable structure have been studied extensively [38–40]. A model ASA where $N = 8$ is shown in Figure 3.9 and the geometric constraints of annular antennas are examined in references [39,40] when N takes different values and hybrid horns are used to feed the offset antennas, or unit collecting areas. If D is the overall array diameter, d is the projected diameter of one offset reflector, and M is the number of the derived independent antenna beams, the following relationships have been deduced [39,40]:

$$d = \frac{D \sin(\pi/N)}{1 + \sin(\pi/N)} \tag{3.11}$$

$$M \approx \frac{N^2}{2} \quad \text{when } N \text{ is even}$$

and

$$M \approx N^2 - N + 1 \quad \text{when } N \text{ is odd}$$

Thus, if N is equal to 8 or 9, the number of derived antenna beams at the system's IF is 32 and 73, respectively. The savings in receiver modules is therefore substantial; furthermore, all LNAs can be accommodated in one cryodyne. The sidelobe levels of this class of antennas can be reduced by using the J_n^2-synthesis method outlined in section 2.4.2.

The model annular synthesis antenna operated at 38 GHz and had the following specifications [40]:

Overall array diameter, D	1.2 m ($D/\lambda = 152$)
Number of quasi-prime-focus antennas used	8
Projected diameter of one quasi-prime focus offset reflector	42λ
The radius of the hybrid mode feeds used	2λ

Annular synthesis antennas have the following attractors:

- One ASA can be used as a standalone imager having a FOV determined by the diameter, d, of the unit collecting areas and a spatial resolution defined by the overall array diameter, D.
- The resulting beams sample the scene perfectly if the Rayleigh criterion between contiguous beams is used. We have already noted in Section 3.3.2 that conventional high-quality focal plane imagers have a problem sampling the scene with contiguous beams.
- The array can be used as a power combiner [88].
- Hybrid mode feeds illuminate the unit collecting areas efficiently.
- Mutual coupling between the feeds of the unit collecting areas can be neglected.
- A conventional paraboloid can be converted into an annular synthesis telescope [39].

- Thermal expansion and contraction affect equally all unit collecting areas and the connections between them and the central processing unit located at the array center.
- A circular array of *N* antenna elements, where *N* is odd, yields the maximum number of spacings between antenna elements.
- When used in conjunction with a high-resolution radioastronomy array, the ASA can provide the range of short spacings (between antenna elements) often missing in conventional radioastronomy arrays.
- A radioastronomy array that consists of ASAs can have an abundance of short spacings [27].

In another realization of ASAs the antenna elements can be inexpensive and light-weight offset reflectors often used for the reception of direct television broadcasts. In this realization the receiving systems are not integrated into one structure. Given that cooling systems are now inexpensive, this is a low-cost realization for the designer to consider. Other systems abound that employ the same design philosophy of utilizing a number of unit collecting areas to realize a diversity of systems, but only a few examples will be considered here.

A system used to track space probes consisted of a steerable structure supporting 2×8 reflectors, each 16 m in diameter [89]. An array designated for satellite mobile communications has been reported [90]; it employed 16, 18, or 19 antenna elements in the form of short back-fire radiators, mounted on a fully steerable mount and operated at L-band under contract from ESA (European Space Agency) [90]. The system was used to generate multiple beams at its IF. The array diameter was 2.1 m and each array element had a gain in excess of 15 dB. The system was deemed suitable for future generations of mobile communication satellites.

Two SPIE (Society of Photo-Optical Instrumentation Engineers) volumes have been dedicated to synthetic aperture systems operating at optical wavelengths [91,92]. The overview paper in [91] points out that the moment of inertia of a synthetic aperture system is lower than that of its monolithic counterpart.

The multiple mirror telescope (MMT) [93] is an important representative of the optical synthetic aperture systems. It uses six 1.8-m mirrors located in a circle and has the light-gathering equivalent of a single 4.5-m mirror. It is located at Mount Hopkins at a height of 2600 m above sea level and was commissioned in May 1979 [93]. The MMT is considered as a precursor of other MMTs having larger overall diameters [93]. Given that a circular array of apertures can yield several radiation patterns, in the form of $J_0(x)$, $J_1(x)$, ... and $J_N(x)$, depending on the phases between each element and the central summing point, the possibility exists to use a $J_1(x)$ pattern for the search of exo-solar planets.

The MMT has been used as a phased array telescope and has the angular resolution of a 6.86 m conventional telescope [94]. Apart from high resolution, the MMT used as a phased array has a high detection sensitivity that results from better discrimination against the sky emission background for IR diffraction-limited images. Recently a resolution of 75 milli-arc second was attained when adaptive optics were used on the MMT [94].

ASAsII can be realized when a number of moderate-gain, frequency-independent spiral antennas having a 10:1 frequency range of operation are used as antenna elements. These types of systems offer the designer:

- A very wide field of view over an extremely wide frequency range.
- A lightweight structure that supports the inexpensive spiral antennas.

- Substantial gain derived for the many antenna elements used.

Further studies of this important class of systems are more than warranted.

3.6. MICROSTRIP/PRINTED BOARD ANTENNAS

We shall now move away from the reflector antennas, the traditional building blocks of radioastronomy arrays, to consider other types of antennas, often used for radar/radar-related and communications applications. When we move from radioastronomy arrays where the antenna elements are widely separated to other arrays where the antenna elements are tightly packed to maximize the power–aperture product of the system, several problems arise; the most serious being mutual coupling between antenna elements and scan blindness toward certain scan angles. For an array operating in the transmit mode, scan blindness is experienced when power directed toward a certain direction is dissipated in surface waves. Heat removal from an array operating in the transmit mode is also a significant problem when the array elements are tightly packed.

There is a plethora of antenna elements used in phased arrays designated for radar or radar-related systems. Apart from the monopoles used in OTHRs, several arrays, e.g. the PAVE PAWS arrays, use crossed dipoles, and the ERIEYE and AWACS systems utilize one or more slotted waveguides that eminently satisfy the requirements for economical, narrowband arrays that scan the surveillance area in one dimension; we shall not consider slotted waveguides further as references [95–97] cover the topic thoroughly. Instead we shall focus attention on generic antennas that can satisfy any of the following three principal requirements:

1. Narrowband operation where the fractional bandwidth is at least a few percent.
2. Wideband operation over 2–3 octaves.
3. Operation over two or more frequency bands centered at widely separated frequencies, e.g. 10:1 in frequency.

Microstrip antennas etched on low-loss dielectric substrates have the following attractors:

- Different realizations of the antennas meet the three principal requirements we have postulated and can be used in arrays having a tile or brick architecture.
- Uniformity of production.
- Low manufacturing costs.
- Lightweight construction.
- Conformity to the skin of the platform, e.g. an aircraft or missile.
- Convenient integration/coupling to the front-ends of MMIC-based modules.
- Eminent suitable to active arrays, where a moderate amount of power is transmitted by each array element.

The antennas were first proposed in the early 1950s [98] and were seriously considered in the 1970s. Theoretical work and design aspects of microstrip antennas are collated in excellent review papers and books [99–103]. However, it is only recently that high-quality microstrip antennas have been realized and in this chapter we shall explore these recent developments.

3.6.1. Simple Patch/Dipole Antennas

A simple microstrip patch and an array of dipole antennas, shown in Figures 3.10 and 3.11, respectively, have a planar structure and their patterns are etched on low-loss dielectric substrates such as Teflon (polytetrafluoroethylene), RT Duroid, silicon, or GaAs. The patch antennas can have several shapes but the circular, rectangular, and square geometries are widely used. Typically, the longer dimension of a rectangular patch is between one-half and one-third of the free-space wavelength of operation, λ; its shorter dimension is about one-half the wavelength in the dielectric substrate material, or $\lambda_g/2$; and the thickness of the dielectric material, h, varies between 0.01λ and 0.05λ. Several methods of feeding these simple realizations of microstrip antennas are illustrated in references [99–103]. We consider these simple microstrip antennas because their performance constitutes a useful baseline from which we can move toward more elaborate realizations of microstrip antennas that meet any of the principal requirements (1) to (3).

Early models treated the microstrip antenna as a cavity or a loaded transmission line. The cavity model lends physical insights into the radiation mechanisms of patch

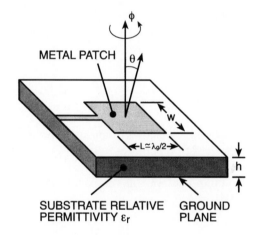

Figure 3.10 A rectangular patch antenna.

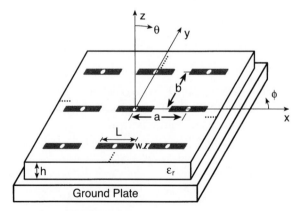

Figure 3.11 The geometry of an infinite array of dipoles.

antennas. The model considers the volume of space under the patch and between the ground plane as a leaky cavity and brings out the type of excitation for the different TM_{nm} modes that can provide the radiation field.

If this model is adopted, the resonant frequency, f_r, of a rectangular patch of length L and width W is given by [104]

$$f_r\big|_{nm} = \frac{c}{2\sqrt{\varepsilon_r}}\left[\left(\frac{m}{L}\right)^2 + \left(\frac{n}{w}\right)^2\right]^{1/2} \tag{3.12}$$

where ε_r is the relative permittivity of the substrate and m and n are integers with values depending on the mode of excitation. The fundamental modes in this case will either be $n = 0$ and $m = 1$ or $n = 1$ and $m = 0$. If $n = 0$ and $m = 1$, equation (3.12) is reduced to

$$L = 0.5\frac{\lambda_0}{\sqrt{\varepsilon_r}} \tag{3.13}$$

a well known result [105].

To account for the fringing fields at the perimeter of the patch, the following empirical formulas have been used for the affective patch dimensions [106]:

$$L_{\text{eff}} = L + \frac{h}{2}, \qquad W_{\text{eff}} = w + \frac{h}{2}$$

where h is the thickness of the dielectric. Empirical expressions for the resonant frequency and the effective relative permittivity, with increased accuracy, have been proposed in reference [102].

Using the same method of analysis, the resonant frequency for a circular patch, of radius r, is given by [104]

$$f_r = \frac{K_{nm}c}{2\pi r_{\text{eff}}\sqrt{\varepsilon_r}} \tag{3.12}$$

where

$$r_{\text{eff}} = r\left[1 + \frac{2h}{\pi r\varepsilon_r}\left(\ln\frac{\pi r}{2h} + 1.7726\right)\right]^{1/2} \tag{3.13}$$

and K_{nm} are the roots of the equation $J_n'(x) = 0$, which is satisfied when x takes the first five values at 1.841, 3.054, 3.832, 4.201, and 5.331.

The radiation patterns of patch and dipole antennas remain approximately hemispherical, even when their input VSWR is unacceptably high; their input VSWR, on the other hand, is strongly dependent on the frequency deviation from resonance. Thus the bandwidth, B, of operation of patch and dipole antennas is defined as the bandwidth over which the input VSWR of the antennas is below a certain value, S, and is given by

$$B = \frac{100(S-1)}{\sqrt{2}\,Q_{\mathrm{T}}}\,\% \quad \text{or} \quad B = \frac{100}{\sqrt{2}\,Q_{\mathrm{T}}}\,\% \quad \text{when } S = 2 \qquad (3.16)$$

where Q_{T} is the total Q factor of the antenna. While most designers accept the above definition of bandwidth, a second definition of the bandwidth is $1/Q_{\mathrm{T}}$ (100%) [107]. To derive Q_{T}, one has to take into account the losses normally associated with microstrip antennas [107]. These losses are related to Q_{T} by [108]

$$\frac{1}{Q_{\mathrm{T}}} = \frac{1}{Q_{\mathrm{r}}} + \frac{1}{Q_{\mathrm{sw}}} + \frac{1}{Q_{\mathrm{d}}} + \frac{1}{Q_{\mathrm{c}}} \qquad (3.17)$$

where Q_{r} is the radiation loss, Q_{sw} is the loss due to surface waves, Q_{d} is the dielectric loss, and Q_{c} is the conductor losses. In reference [107] Q_{sw} and Q_{r} are lumped into a quantity Q_{R}. We shall consider the losses due to surface waves in the next section and expressions for Q_{c}, Q_{d}, and Q_{R} are given in the same reference. The antenna efficiency is defined by [107]

$$\eta = \frac{Q_{\mathrm{T}}}{Q_{\mathrm{R}}} \qquad (3.18)$$

Accepting the second definition of antenna bandwidth, equation (3.18) is reduced to

$$\eta = \frac{1}{Q_{\mathrm{R}}B} \qquad (3.19)$$

For a rectangular patch antenna, the efficiency, η, and bandwidth, $\Delta f(\%)$, as a function of the substrate relative permittivity, ε_{r}, are illustrated in Figure 3.12 as a function of the ratio h/λ_0. Both the efficiency and bandwidth of the antenna improve as ε_{r} decreases. As the substrate becomes thicker, the efficiency and bandwidth of the antenna increase.

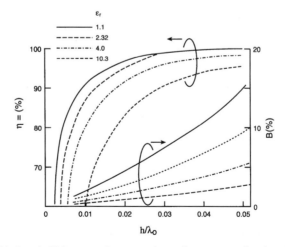

Figure 3.12 The bandwidth, B, and efficiency, η, of a rectangular patch antenna as a function of the substrate's relative permittivity, ε_{r}, and its thickness, h/λ_0. (From [107]; © 1982, Microw Systems News.)

From the foregoing, we can discern a conflict in the integration of the microstrip antennas with the MMIC-based T/R modules that utilize GaAs, a high relative permittivity substrate. This is an issue we shall revisit in section 3.6.2.

3.6.1.1. Array scan blindness

While simple models are used to deduce the characteristics of microstrip antennas, mathematically intensive procedures are required to derive the characteristics of patch or dipole antennas when they are used as elements of phased arrays.

It has been shown that such fundamental scanning characteristics of an array consisting of microstrip antennas as the reflection coefficient, input resistance trends, scan blindness, and grating lobe effects are dictated by the interelement spacings and substrate parameters (height and relative permittivity) and not by the nature of the microstrip antenna element (patches or dipoles) [109]. We shall therefore make no distinction between the two antennas types. Let us begin by presenting a physical insight into the problems at hand before we outline a qualitative account of the performance of arrays having a diverse set of parameters.

Figure 3.13a illustrates a point source of current located in the underside of a metallic patch antenna radiating an EM wave [110]. The waves designated by B radiate out and contribute to the radiation pattern of the antenna. Some of the waves designated by A are diffracted, go back under the patch, and store EM energy. The waves designated by C remain within the dielectric substrate, trapped by the air–dielectric interface. These are the surface waves that propagate along the two-dimensional interface and decay more slowly than the space waves, which spread into space. For an array operating in the transmit mode, some power fed to the antenna is lost, resulting in a decrease of the total antenna efficiency. Additionally, a secondary radiation pattern into space surrounding the antenna results when the surface waves are scattered at the array's physical boundaries. Two methods for the minimization of the secondary radiation pattern have been adopted by designers: (i) to surround the array by dummy antenna elements; and (ii) to place absorbing material along the perimeter of the array as illustrated in Figure 3.13b.

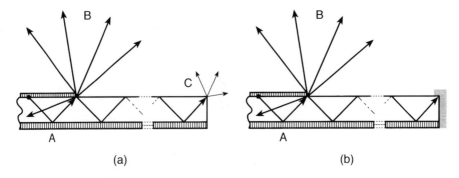

(a)   (b)

Figure 3.13 Radiation mechanisms in an array of patch antennas and one approach used to minimize the secondary array radiation pattern. (a) The radiation mechanisms in arrays of patch antennas from [110]; (c) 1988. Microwave Journal). (b) One approach used to minimize the secondary radiation pattern at the perimeter of the array.

When a second patch antenna is in the vicinity of another, currents are induced in it owing to coupling to both the space and surface waves of the first patch; the second patch therefore becomes a secondary radiator. As the mutual coupling between antenna elements cannot be minimized [111], the designer has to accept that the interelement spacing is another critical array parameter—consider also the discussion related to Figure 3.14b.

Scan blindness of large phased arrays toward a scan angle ψ is caused by surface waves that propagate in synchrony with a Floquet mode of the structure. At ψ, the array impedance is modified to such an extent that the array radiates no power. Consequently, the array FOV is defined by the angle at which the scan blindness is located and not by the position of its grating lobes. In references [109], [112], and [113] the problem of scan blindness has been treated in some detail by considering an infinite array of dipole antenna elements printed on a grounded dielectric slab. The results summarized here are based on the derivation of a Green's function for the array of infinitesimal dipoles scanned to angles θ and ϕ.

With reference to Figure 3.11, the spacing between adjacent dipoles is a and b along the x- and y- directions, respectively, and the coordinates u, v and x_0, y_0 of the m n dipole are given by

$$u = \sin \theta \cos \phi \quad \text{and} \quad v = \sin \theta \sin \phi \tag{3.20}$$

and

$$x_0 = ma \quad \text{and} \quad y_0 = nb \tag{3.21}$$

If we define T_e and T_m by

$$T_e = k_1 \cos (k_1 h) + jk_2 \sin (k_1 h) = 0 \tag{3.22}$$

$$T_m = \varepsilon_r k_2 \cos (k_1 h) + jk_1 \sin (k_1 h) = 0 \tag{3.23}$$

where

$$k_1^2 = \varepsilon_r k_0^2 - \beta^2 \quad \text{Im}(k_1) < 0 \tag{3.24}$$

$$k_2^2 = k_0^2 - \beta^2 \quad \text{Im}(k_2) < 0 \tag{3.25}$$

$$\beta^2 = k_x^2 + k_y^2 \tag{3.26}$$

$$k_0 = 2\pi/\lambda \tag{3.27}$$

The zeros derived from equations (3.22) and (3.23) represent the transverse electric (TE) and transverse magnetic (TM) surface waves of the unloaded grounded dielectric slab, respectively. It is here assumed that the loading of the antenna elements introduces negligible error in the calculation of the scan blindness. Either $T_e = 0$ or $T_m = 0$ is used depending whether E- or H-plane scan angle blindness is required.

The number of surface waves a dielectric material can support is proportional to its thickness. The condition [114]

$$h < \frac{\lambda_0}{4\sqrt{\varepsilon_r - 1}} \tag{3.28}$$

therefore ensures that only the lowest-order surface wave (TM_0) can propagate. The scan blindness angle can be predicted from a comparison of the propagation constants of the surface wave of the dielectric slab and the various Floquet modes.

If β_{sw} is the propagation constant of the first (TM) surface-wave mode of the unloaded dielectric slab and $k_0 < \beta_{sw} < (\varepsilon_r)^{1/2} k_0$, the surface wave resonance occurs when β_{sw} matches a particular Floquet-mode propagation constant. This occurs when

$$\left[\frac{\beta_{sw}}{k_0}\right]^2 = \left[\frac{k_x}{k_0}\right]^2 + \left[\frac{k_y}{k_0}\right]^2 = \left[\frac{m}{a/\lambda} + u\right]^2 + \left[\frac{n}{b/\lambda} + v\right]^2 \tag{3.29}$$

from which the scan blindness angle is derived.

Figure 3.14a illustrates the dependence of the array percentage bandwidth and scan blindness angle of an array of patch antennas when the interelement spacing is $\lambda/2$ and the substrate is GaAs [111]. As can be seen, arrays having thick substrates have increased bandwidths but decreased scan blindness angles that determine the array's FOV. Therefore, trade-offs have to be struck between increased bandwidths on the one hand and extended array FOVs on the other.

Figure 3.14b demonstrates the array scan angle dependence on the substrate relative permittivity when four cases are considered [111]. For cases 1, 2, and 3 the substrate

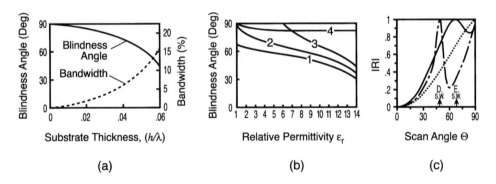

(a)   (b)   (c)

Figure 3.14 (a) The array bandwidth (%) and scan blindness angle as a function of the substrate thickness (h/λ). (From [111]; © 1986, Microwave Journal.) (b) The array scan angle blindness as a function of the substrate's relative permittivity (ε_r), the interelement spacing a, and the substrate thickness (h). (From [111]; © 1986, Microwave Journal.) Case 1, $a = 0.52\lambda$ and $h = 0.06\lambda$. Case 2, $a = 0.5\lambda$ and $h = 0.06\lambda$. Case 3, $a = 0.48\lambda$ and $h = 0.06\lambda$; Case 4, $a = 0.5\lambda$ and $h = 0.02\lambda$. (c) The magnitude of the reflection coefficient of an infinite array of dipoles as a function of scan angle in the E-, H-, and D- (diagonal) planes when the substrate $\varepsilon_r = 2.55$ and the thickness $h = 0.19\lambda$. The arrows indicate the position of the scan blindness angles in the D- and E-planes. (From [113]; © 1984, IEEE.) (For this particular example the scan blindness angle in the H-plane is not calculated.)

height, h, is equal to 0.06λ but the interelement spacing, a, takes the values 0.52λ, 0.5λ, and 0.48λ, respectively; as can be seen, a small variation in a affects the range of the array scan blindness angles. Similarly, cases 2 and 4 have the same interelement spacing, 0.5λ, but the substrate heights take the values of 0.06λ and 0.02λ, respectively; as can be seen, a decrease in substrate height dramatically affects the range of array scan blindness angles. Relatively thin substrates tend to increase the scan blindness angle. When the substrate thickness is set, the designer can substantially vary the array's scan blindness by slightly varying the interelement spacing.

Figure 3.14c illustrates the magnitude of the reflection coefficient $|R|$ of an infinite array of printed dipoles when $a = 0.5774\lambda$, $b = 0.5\lambda$, $\varepsilon_r = 2.35$, and $h = 0.19\lambda$, as a function of scan angle in the E-, H-, and D- (diagonal) planes [113]. The scan blindness angles occur at $\theta = 68.3°$ and $49.3°$ in the E- and D-planes, respectively. Usually the array FOV is restricted to a scan angle equal to the scan blindness angle minus 10°.

Approximate and closed-form expressions for the space-wave power, P_{sp} and the surface-wave power, P_{sw}, have been derived in reference [115] as well as the space-wave efficiency, η, defined as

$$\eta = \frac{P_{sp}}{P_{sp} + P_{sw}} \tag{3.30}$$

When the substrate relative permittivity is 12.8, the space-wave efficiency of an array of infinitesimal dipoles decreases monotonically as h/λ increases; a similar trend is observed when $\varepsilon_r = 2.55$ and h/λ is between zero and 0.1; when $h/\lambda > 0.1$ a plateau in efficiency is reached.

3.6.1.2. Performance of Simple Patch Antennas

Patch antennas can support several modes that generate cross-polarization fields. The use of two or four feed points on one patch contributes to the cancellation of the undesirable cross-polarization fields. As patches having four feed points are too complicated, patches having two feed points are widely used [116–120].

In a linear array of square patch antennas aligned along an axis of symmetry through the centers of all patches, one principal linear polarization is derived by connecting all probes aligned with the axis of symmetry; the other orthogonal polarization is derived by connecting another set of probes perpendicular to the axis of symmetry [116]. The cross-polarization fields of the array of patches in the V- and H-polarization are –27 and –30 dB, at ±0.15° with respect to the array boresight. The array is developed for the advanced SAR (ASAR) of the European Space Agency and has the following additional specifications:

- Frequency of operation: 5.331 GHz.
- Bandwidth ±10 MHz: 3.75%.
- The maximum cross-polarization levels vary between –10 and 15 dB.

Low cost and simplicity are the hallmarks of this approach.

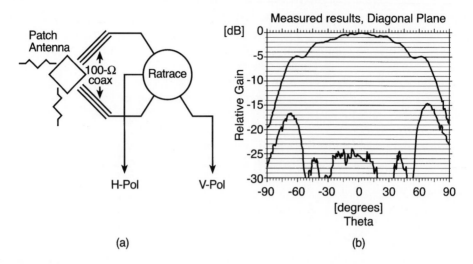

Figure 3.15 The derivation of the two orthogonal linear polarization when a square patch is used. (a) The derivation of the vertical and horizontal polarizations from the square patch. (b) Its co-and cross-polarization radiation
patterns. (Courtesy TNO, Paquay *et al.* [117]).

The patch antenna element developed for the PHased ARray SAR (PHARUS) operating at 5.3 GHz is an important variant of this approach [117]. With reference to
Figure 3.15, the two orthogonal polarizations are connected to a ratrace and the resulting co- and cross-polarization patterns are shown in Figure 3.15b. As can be seen, a
broad minimum of the cross-polarization pattern is attained at the −25 dB level in the
D-plane. The ratrace provides some 30 dB isolation between the two probes, accommodating the two polarizations. The broad cross-polarization pattern and the high
isolation between the two polarizations constitute a benchmark of performance for
simple patch antennas. Comparable results are reported over a 5% (VSWR of 2) bandwidth at 2.4 GHz when the coupling mechanism from the patch consists of slots
[118].

The performance of four dual-polarized square patch antennas has been compared
when the interconnections between patches vary [119]. The optimum geometric
arrangement of interconnections exhibited a cross-polarization level of −24 dB and
an inter-polarization isolation of −33 dB. The bandwidths attained by the four
arrangements ranged between 3% and 8%. Huang [120] implemented a different inter-
connection scheme between the patches and reported an inter-polarization isolation
figure of −40 dB and a cross-polarization level of −28 dB.

Figure 3.16 illustrates three approaches used to derive the circular polarization(s)
from patch antennas. Approaches (a) and (b), where two feed points on the patch are
used, are the most common.

The above examples constitute a useful baseline from which further improvements
can be made toward an increased useful bandwidth of operation. We shall consider several development thrusts in the forthcoming sections.

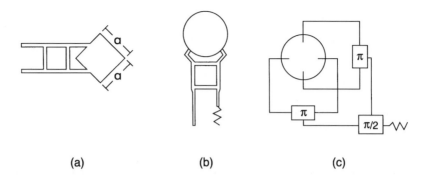

(a)   (b)   (c)

Figure 3.16 Three approaches to derive circular polarization from patch antennas. (a,b) Approaches utilizing two
feed points. (c) Approach utilizing four feed points.

3.6.2. The Evolution of High-Quality Antennas

Before we consider the evolutionary thrusts for high-quality microstrip antennas, it is
useful to outline some design guidelines for arrays consisting of simple patch or dipole
antennas.

Let us begin by defining the regimes of application for the simple patch and dipole.
Here we shall refer to the work of Pozar [121], which has been corroborated by the
Alexopoulos group [122]. The loss due to surface waves for both types of antennas is
comparable. The dielectric losses for both types of antennas have been considered
when the substrates have loss angles tan δ of 0.001 and 0.003. For thin substrates, the
patch antenna has a higher efficiency due to lower dielectric losses. By contrast, the
efficiency of the dipole is comparable and the percentage bandwidth is higher when
dipoles on thick substrates are used. Given that commercially available substrates
come in certain thicknesses, the dipole yields wider bandwidths when the substrate is
thick or when operation at millimeter wavelengths is intended. If we accept this limitation, we can conclude that the dipole/patch antennas are a better choice when
operation at millimeter/centimeter wavelengths is intended. However, the recent trend
toward thin silicon substrates removes the limitation imposed by commercially available substrates and patch antennas operating at millimeter wavelengths are common.
In reference [123], for instance, a rectenna consisting of two patch antennas etched on
a 100 μm-thick, high-resistivity silicon substrate and operating at 76 GHz has been
reported; a similar antenna system operating at the same frequency also employed two
patch antennas [124]. Other antennas used at millimeter wavelengths are the slot
antennas [124,125]. The relative permittivities of silicon and GaAs substrates are 11.7
and 12.9, respectively.

If we ignore all other considerations, the realization of antennas on substrates having
low relative permittivity is preferred. While the array operational bandwidth and efficiency of microstrip antennas improve as the thickness of the substrate increases, the
substrate cannot be too thick, otherwise surface waves will limit the array's FOV. If
transmission lines interconnecting patch antennas are used on the radiating face, radiation from the lines can modify its co- and cross-polarization radiation patterns.

MMIC-based T/R modules are realized on thin, high relative permittivity substrates. If the substrate's relative permittivity and thickness are set, the designer can only vary the spacing between the array elements to avoid the generation of scan blindness. We have already noted that slight variations of the interelement spacing from the half-wavelength criterion result in considerable variations of the scan blindness angle. The integration of simple antenna elements with MMIC-based T/R modules is therefore problematic and involves some performance compromise.

From the foregoing considerations, we can easily conclude that the simple patch antennas have to meet numerous requirements but possess only a few degrees of freedom. The simple patch antennas, however, are eminently suitable for narrowband applications and are easily and inexpensively realized.

Reference [126] is a review paper totally devoted to the design and layout of microstrip structures, including antennas; it also contains a review of analysis techniques for simple patch antennas.

3.6.2.1. Integration Issues

Although we are here interested in the integration of the antennas with the MMIC-based T/R modules, we should consider this issue in the context of other integration issues, namely the integration of the microwave circuits with high-speed digital circuits and microelectromechanical (MEM) subsystems. If one substrate is used for the high-speed digital circuits, MEM subsystems, MMIC-based T/R modules, and antennas, economies result and the array reliability is enhanced. Given that high-speed digital circuits, Si-based MMICs, are emerging [123] and MEM subsystems are realized in silicon substrates, it makes sense to adopt silicon as the substrate of preference for low-cost arrays. GaAs T/R modules are used and will be used whenever the cost constraints are not severe and high performance is of paramount importance. For the antenna–T/R module integration, the two candidate substrates silicon and GaAs have comparable relative permittivities, so integration is problematic if the antennas are realized on low relative permittivity substrates. Before we consider the proposed solutions, let us stipulate that the designer's wish list includes the variation of the substrate's thickness and relative permittivity to meet wide-ranging requirements.

The relative permittivity of the substrates can be artificially modified to lower value either by drilling closely spaced holes underneath and around the antenna or by micromachining the area under and around the antenna [127]. Increased antenna efficiencies have been reported in both approaches.

Low-loss transmission lines and passive components can be fabricated using micromachined circuits ([124] and the references therein). The transmission lines take two forms: the microshield line or the suspended membrane microstrip line. Both lines are fabricated by micromachining techniques on a SiO_2–Si_3N_4–SiO_2 membrane grown on a silicon substrate, and standard lithographic processing techniques are used for the metal pattern definition and cavity etching. The advantages of both these lines are low dielectric losses and low radiation losses at discontinuities. The membrane can be as thin as 1.5 μm [124] and high-performance micromachined resonators and filters operating at millimeter wavelengths have been reported in the same reference [124] and the references therein. The fabrication of micromachined circuits requires a higher technological effort than the fabrication of planar circuits. Both approaches, however, may be cost-competitive because micromachined circuits are self-packaging [124].

Another solution utilizes several layers of ceramic, typically alumina, 250 μm in thickness and several layers of polyimide of 25 μm thickness [128]. The relative permittivities for these substrates are 9.0 and 3.2, respectively, and their loss tangents at 10 GHz are 0.001 and 0.002, respectively. The ceramic/polyimide multilayer arrangement has been developed as a substrate for a multichip module (MCM) for VLSI chips (references included in [128]). Efficient microstrip antennas operating at different frequencies can therefore be realized on the same substrate by selecting the appropriate substrate thickness and effective relative permittivity at the frequency of operation. This is achieved because the antenna can be etched at any polyimide layer and its ground plane can be on any ceramic layer. Integration of the ceramic substrate to MMICs is easy because its thermal expansion coefficient (6.8×10^{-6}/°C) is comparable to those of commonly used semiconductors such as GaAs (6.0×10^{-6}/°C) and Si (3.4×10^{-6}/°C) [128]. Moreover, transmission lines and other passive components can be realized on the same multilayer arrangement. This last solution is a low-cost solution that is eminently suitable to the realization of antennas, transmission lines, and other passive components usually associated with antennas.

3.6.3. Moderate-Bandwidth Antennas

In this section we shall consider novel and important antenna elements that resulted from ingenious approaches taken by many groups of researchers. The common thrust to these approaches is to extend the useful operational bandwidth of microstrip or printed board antennas by increasing the designer's degrees of freedom. The useful bandwidth of operation for these antennas is wider than the bandwidth attained by the simple patch/dipoles with a maximum bandwidth in excess of one octave. Compared to the multi-octave bandwidth of wideband antennas we shall consider in section 3.6.5, the antennas under consideration here have a moderate bandwidth.

3.6.3.1. The Electromagnetically Coupled (EMC) Patch

The EMC patch antenna, shown in Figure 3.17a, consists of a radiating patch and a feeding patch separated by a layer of foam. As can be seen, the designer now has several degrees of freedom, e.g. the diameters of the radiating and feeding patches, D_r and D_f, the relative permittivity of the substrates on which the two patches are realized, and the relative permittivity and height of the foam. Additionally, there are no transmission lines feeding the EMC which is now fed at points A, B, C, or D. With this arrangement the radiating patch is protected from the weather and can be conformal to the skin of the platform. The feed has a wider bandwidth than its conventional counterpart [129] and this attractive feature is directly attributed to its dual-resonant structure. The EMC patch is easy to integrate with an MMIC-based T/R module that uses a high relative permittivity substrate such as GaAs. This important attribute of the EMC patch resolves the difficulty we have allured to, i. e. the integration of the antenna and to the MMIC-based T/R module.

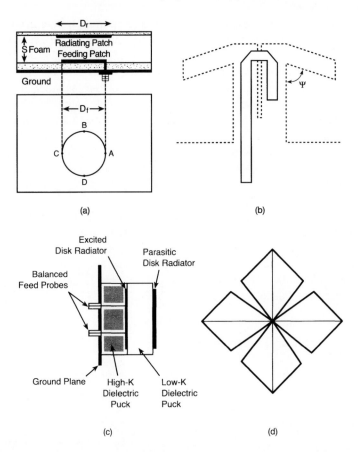

Figure 3.17 Moderate band antennas. (a) The electromagnetically coupled (EMC) patch antenna [129]. (b) The printed board dipole antenna with its coupling structure. (Adapted from [136]; © 1987, Microwave Journal.) (c) Stacked-disk radiator geometry (From [138]; © 1996, IEEE) (d) Wideband crossed dipoles. (Courtesy Dr P. Elliot.)

An EMC patch having a 13% bandwidth and a maximum cross-polarization level of −30 dB at L-band is reported [130] and design guidelines for the EMC patch are given. Similarly, a broadband EMC patch operating at 1.67 GHz has been reported [131]. It has a 1.7 VSWR bandwidth of 14.4% and its cross-polarization level is better than −27 dB over the same frequency range, and comparable performance was obtained by another EMC patch operating at S-band [132]. These are impressive characteristics for low-cost antennas that have all the other desirable properties.

These experimental results are supported by recent theoretical work related to ECM patches [133,134]. Reference [135] is a recent, useful reference for antenna elements normally considered for phased array applications.

3.6.3.2. Printed Board Antennas

The antennas we have considered so far are suitable for the tile architectures, while printed board dipoles are suitable for brick architectures. In Figure 3.17b the printed dipole and its coupling structure are shown [136]. The VSWR 2 bandwidth of the ensemble is 40% and guidelines for its design are presented in the same paper.

The coupling structure used between the antenna and the MMIC-based T/R module often modifies the radiation pattern of the dipole. Studies of the radiation patterns of printed dipoles when ψ takes the values 10°, 30°, 45°, 60°, and 75° have been undertaken. By using the electric field integral equation (EFIE) approach suitably modified to compute the resulting fields of arbitrarily shaped finite conducting/dielectric composite structure embedded in an infinite homogeneous medium such as free space have been calculated [137]. When $\psi = 30°$ the calculated radiation pattern was optimum and measurements corroborated the calculations. It is recalled here that the dipoles of the PAVE PAWS array are also bent. Compared to its microstrip counterparts, the bandwidth of this antenna ensemble is exceptionally wide.

3.6.3.3. Stacked Disk Antennas

Figure 3.17c illustrates the stacked disk radiator [138], which consists of the excited and parasitic disk radiators separated by two pucks having low and high dielectric constants. The balanced feed points and lines from the excited disk yielding the H- or V-polarizations are also shown. The radiator is capable of providing common-phase-center circular or dual-linear polarization and wide scan volume over a wide bandwidth.

The stacked disk radiator can be optimized to have either of the following characteristics:

- A 40% bandwidth with a VSWR of 1.2; or
- A 2:1 bandwidth with a VSWR of 2.

Good E- and H-plane scan performance has been reported up to ±45° and the experimental results were obtained from radiators operating at C-band [138]. Work toward the optimization of the design and the utilization of more than two pucks may lead to increased options for the designer.

3.6.3.4. Novel Dipoles

In references [139] the novel antennas shown in Figure 3.17d are described; the antennas consist of crossed dipoles that are similar to crossed, folded, cage dipoles. Either the two orthogonal linear or two opposite circular polarizations can be derived from these antennas that are one quarter-wavelength or less above the ground plane.

An array consisting of seven crossed, folded, cage dipoles has been realized and operated in the band between 750 and 1750 MHz; the array has the following attractive characteristics [139]:

- Octave bandwidth with active VSWR of less than 2.
- Scan angles of ±30° with respect to its boresight axis.

The foursquare antenna [140] consists of four square patch antennas; each pair of the two diagonal patches yields one of the two linear polarizations. On the basis of the limited information available, the foursquare antenna exhibits similar characteristics to the novel crossed, folded, cage dipole antenna.

These two novel array antenna elements constitute a significant advance over the performance of simple microstrip/printed board dipoles.

3.6.4. Dual-band/Multiband Antennas

We have already made a case for antennas that have acceptable radiation patterns and input VSWRs at two or more frequency bands centered at frequencies $f_1, f_2, \ldots$. In the radar context the requirements stem from dual frequency/multifrequency and dual-polarization SAR systems, or multifunction radars. In the radioastronomy context, high-resolution maps of celestial sources or variable sources are required at different frequencies.

It is recalled that multifrequency SAR systems typically operate at L-, S-, or X-bands. In multimode radars, on the other hand, the two frequencies designated for the surveillance and tracking functions, f_s and f_t are widely separated and can have a ratio of f_s/f_t equal to 10; this latter requirement is here taken as a boundary specification.

If one aperture can accommodate two or more arrays operating at different, widely separated frequency bands, significant economies can be attained. If the bandwidth at each frequency is wide enough, continuous frequency coverage, important for EW applications, is attained.

3.6.4.1. Dual-band Antennas

Some recent work on dual-band microstrip antennas is reported by groups interested in communications, e.g. [141]; while the reported work is interesting, the two frequencies are closely spaced, e.g. f_1 = 1227 MHz and f_2 = 1575 MHz (f_1/f_2 = 1.284).

The authors of reference [142] review recent work in this field with some emphasis on work performed in mainland Europe. A variety of realizations yield f_1/f_2 ratios ranging from 1.6 to 3; one antenna ensemble consisting of a cross resonant at f_2 and a subarray of four patches, shown in Figure 3.18a, designed to operate at f_1, yields an f_1/f_2 ratio ranging from 2.9 to 3.2 [142,143]. By following the design procedure outlined in the latter reference, the ensemble can be used to populate an aperture capable of operation at S- and X-bands simultaneously. Instead of having an arrangement consisting of a cross and four patches, one could have two concentric circular patches designed to operate at widely different frequencies.

The authors of reference [144] propose a multilayer aperture-coupled patch antenna that can be used for dual-polarization operation at C- and X-bands. Two design examples operating at these bands (5.3 and 9.6 GHz) are reported and exhibited a polarization isolation in excess of 28 dB and a port-to-port isolation of 25 dB. For SAR applications, high polarization purity and interport isolation are of considerable importance.

In reference [145] it is demonstrated that the resonant frequency of a circular patch designed to resonate at the high frequency band centered at f_1 can be substantially lowered if a shorting pin is used; the position and dimensions of the shorting pin are critical and have been determined experimentally. The f_1/f_2 ratio can vary between 2.55 and 3.85 when the shorting pin is in place. The approach is experimental and details are given of the antenna realized when the high frequency, f_1, is at L-band.

The attractors of this approach are:

• A small patch designed to operate at f_1 can operate efficiently at a much lower frequency with the aid of a shorting pin.
• An aperture capable of operation at two bands can be realized if it is populated by circular patches of the same size; some of the patches utilize a shorting pin and the distribution of the two sets of patches on the aperture is random or aperiodic.

While considerable work is required in this area, the approaches already reported are important and promising.

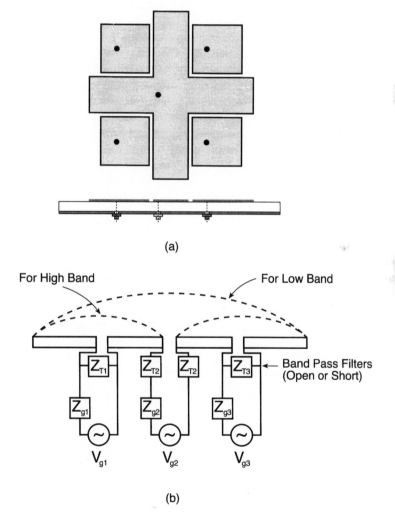

(a)

(b)

Figure 3.18 Dual-band antennas. (a) Cross and subarray of square patches. The low-frequency cross resonates at the low-frequency band, while the patches of the subarray resonate at the high-frequency band. (From [142]; © 1997, IEEE.) (b) Dipole-based dual-band antenna; the concept can be extended to multiband antennas. (Adapted from [146]; © 1995, IEEE.)

3.6.4.2. Multiband Antennas

An array capable of operation at multiple bands is an attractive proposition. We have already reported that efficient antennas operating at different frequencies can be realized on the same multilayer arrangement consisting of several layers of alumina ceramic and polyimide [128]. Two antennas operating at 10 and 20 GHz have been realized on the same multilayer substrate and the realization of other antennas operating at different frequencies is possible.

Without any loss of generality, let us consider a linear array that operates at two frequency bands, proposed in [146] and illustrated in Figure 3.18b. The concept can be extended to planar arrays and to arrays operating at several bands selected on the basis of system requirements. The array consists of two dipoles, each of which resonates at the high-frequency band. The band-pass filters Z_{T1}, Z_{T2}, and Z_{T3} can present either an open-circuit or short-circuit to ports 1, 2, and 3 respectively. The high-frequency band is obtained when: $Z_{T2} = \infty$ and ports 1 and 3 are matched to the high-frequency dipoles. Similarly, operation at the low frequency band is attained when both dipoles are short-circuited to form one dipole resonating at the low frequency. More specifically, the necessary conditions are that the bandpass filters $Z_{T1} = Z_{T3} = 0$, $V_{g1} = V_{g3} = 0$ and Z_{T2} is matched to that of the effective dipole at the low-frequency band. The measurements obtained from an experimental array operating at the UHF, L- and S-bands agree with the analytical work undertaken [146].

3.6.5. Wideband Antennas

In this section we shall consider microstrip/printed board antennas that operate efficiently over one or more octaves. While a variety of antennas are considered here, we have a preference for dual-polarization antennas occupying a minimal aperture real estate. A single dual-polarization antenna occupying an area A, for instance, is preferred to a couple of single-polarization antennas each occupying an area A.

Although there is a significant body of knowledge related to frequency-independent antennas, it is only recently that high-quality microstrip/printed board antennas have become available after a long period of considerable R&D effort. In many instances the resulting antennas are ingenious adaptations of the early frequency-independent antennas such as the log-periodic and log-spiral antennas.

3.6.5.1. Log-Periodic/Spiral-Mode Antennas

A conventional log-periodic antenna consists of a feed line connected to several dipoles resonating at different frequencies. The patch antenna equivalent of the conventional log-periodic antenna consists of a feed line coupled to several patches each resonating at a different frequency. Thus small/medium and large patches resonate at the low/medium and high frequencies, respectively.

The log-periodic antenna reported in reference [147] has the following characteristics:

- Bandwidth 4–16 GHz.

- Gain > 8 dB.
- Beamwidths $40° \times 92°$ at 4 GHz and $30° \times 84°$ at 16 GHz.

The antenna has a wide bandwidth and can be used as element of arrays having limited beamwidth.

The planar spiral mode antennas, e.g. the equiangular and sinuous antennas, have multi-octave bandwidths [148,149], but the antennas radiate to both sides of the spiral plane. The placement of a lossy cavity on one side of the antenna structure absorbs most of the undesired radiation in that direction. However, the lossy cavity is deeper than the radius of the spiral and therefore unsuitable for low-profile applications. When a loss-less cavity or a conducting plane [150] is used, the achievable bandwidth is 40% and 20%, respectively.

Wang and Tripp [151] reported the realization of an experimental spiral-mode antenna operating over the 2–18 GHz frequency range. The antenna had a good input match even when the spiral was 0.1 in above the ground plane. The placement of absorbing material in the periphery of the spiral contributed greatly to the quality of the radiation patterns over the 2–18 GHz frequency range. This important, broadband and low-profile antenna is destined to find many uses in phased arrays having a tile architecture and in EW applications [152].

3.6.5.2. Tapered Slotline Antennas

The planar tapered slotline antenna (TSA) consists of a slotline medium that is gradually opened to a specified width [153,154]. Several realizations of planar TSAs having a linear (LTSA) exponential (ETSA) and tangent of an angle (TTSA) taper are shown in Figure 3.19a. TSAs are also referred to as notch antennas. In some references the ETSA is referred to as a Vivaldi antenna. A tightly bound traveling wave from the excitation end of the structure is gradually decoupled from the guiding line to the radiation field. The main beam of the radiation pattern is in the endfire direction and is largely determined by the axial antenna length and shape of the taper. The antennas are

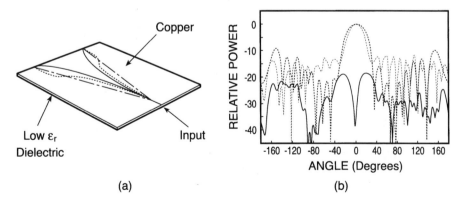

(a)   (b)

Figure 3.19 Planar tapered slotline antennas (TSAs) suitable for arrays having the brick architectures. (a) The TSA having a linear, exponential, and tangent of an angle taper. (b) The co- and cross-polarization of a planar TSA with a linear taper. (From [155]; © 1989, IREE (Aust.).)

members of the class of aperiodic, continuously scaled, slow leaky end-fire traveling wave antennas and the realizations we shall consider are suitable for arrays having either the brick or tile architecture.

TSAs, properly designed, have the following attractive characteristics:

- Acceptable input VSWR and low cross-polarization fields over several octaves.
- Equal E- and H-plane radiation patterns; considering that the antennas are planar, this is a remarkable characteristic.

In Figure 3.19b the co- and cross-polarization radiation patterns of a planar LTSA at 8 GHz are shown; the E- and H-plane radiation patterns are similar and the maximum cross-polarization level is at –21 dB [155].

The other realizations of TSAs suitable for brick architectures are shown in Figure 3.20. The antipodal [156,157], balanced antipodal [158], and 'bunny ears' [159,160]

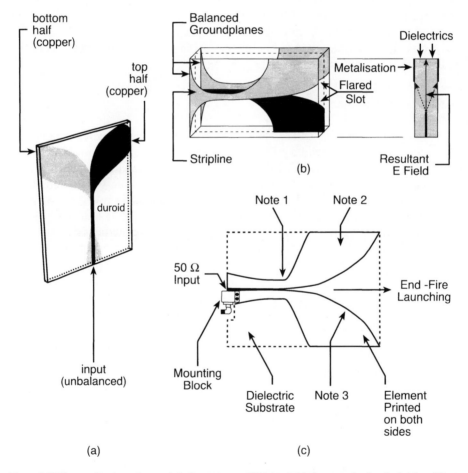

Figure 3.20 Three realizations of tapered slotline antennas (TSAs) suitable for arrays having the brick architecture. (a) The antipodal TSA. (From [164]; © 1993, IEE.) (b) The balanced antipodal antenna. (From [158]; © 1996, IEE.) (c) The 'bunny ears' antenna. (From [159]; © 1995, IEEE.) Notes: (1) Balanced slotline tapered to match radiation resistance; (2) surface current fan-out region; (3) exponential taper for wideband impedance matching.

TSAs are shown in Figures 3.20a, b, and c, respectively. The inputs to the first two antennas are in convenient microstrip and stripline geometries, respectively. The 'bunny ears' antenna is symmetrically printed on both sides of the substrate. Its input transition, the slot line and the launching section are all embodied in a quasi-TEM structure; a Klopfenstein taper is used to transform the 50 ohm antenna input to the radiation resistance, which is 100 ohm [159,161]. The inner contour of the fan-out dipole wings follows an exponential curve.

TSAs have been used in focal plane imagers [162,163]. In reference [164] design guidelines for realizing planar and antipodal ETSAs and TTSAs having equal E- and H-plane HPBWs are outlined. The gain of the antennas considered varied between 8 and 12 dB over the frequency range 3–12 GHz [164]. A comparison of the antipodal and planar TSAs is also undertaken [164]:

- Antipodal TSAs are easy to realize and have superior input VSWR when compared to planar TSAs.
- The planar TSA, illustrated in Figure 3.19a, is difficult to realize because the input slot width has to be accurately controlled before its input VSWR is acceptable.
- Antipodal TSAs exhibit high cross-polarization fields, typically at the –8 to –12 dB level over the same frequency range.
- Planar TSAs have acceptable cross-polarization levels.

Comparisons between planar, antipodal, and balanced antipodal TSAs over the frequency range from 1 to 18 GHz, are reported in reference [158]. The input return loss of the balanced antipodal TSA is lowest, followed by that of the antipodal and planar realizations. Similarly, the cross-polarization level of the balanced antipodal antenna is lowest, followed by that of the planar and antipodal antennas. The cross-polarization level of the balanced antipodal TSA varied between –30 and –15 dB levels over the same frequency range. This is exceptional performance for a low-cost, planar microstrip antenna.

High-gain balanced antipodal TSAs exhibited beam squint in the E-plane radiation pattern of about 15° [158] and only an asymmetric flare reduced the beam squint. For low-gain antennas, having a large FOV, the beam squint might be less significant. Measured radiation patterns of a linear array comprising of seven balanced antipodal TSAs have been reported in the same reference; the array cross-polarization varied between –15 and –25 dB levels.

The average input VSWR of the bunny ear antenna is below 1.5 over the 0.5–18 GHz frequency range [160]; in an array environment, however, the VSWR is below 2 over a 3:1 frequency range [159]. A summary of the array's pertinent characteristics is given in Table 2.21. Theoretical work related to the bunny ear antenna as a stand-alone antenna or in an array has been reported in references [160] and [165], respectively.

A class of E-plane scan blindness in single-polarized arrays of TSAs with a ground plane has been considered in reference [166]. Simple formulas that define the angle of this particular class of scan blindness have been derived that allow the designer to rapidly establish suitable spacings between adjacent TSAs and their lengths so as to avoid this type of scan angle array blindness.

Measurements of the mutual coupling between adjacent planar LTSAs have been reported when the two antennas were co-planar or stacked [167].

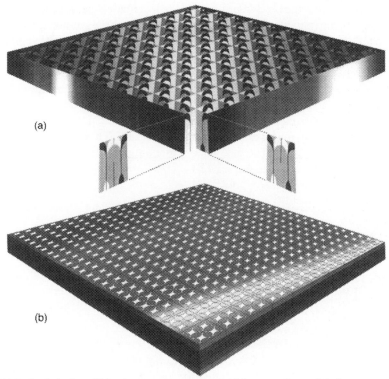

Figure 3.21 Dual-polarization wideband antennas used in arrays. (a) An array of tapered slotline antennas; although the antipodal antennas are shown, other antennas such as the balanced antipodal or bunny ears antennas can be used. The array has the brick architecture. (b) An array of flared-slot antennas. The array has the tile architecture.

The variable parameters were

- Physical separation between the antennas.
- Antenna length, flare angle and substrate thickness.
- Frequency of operation.
- Relative permittivity of the substrates ($\varepsilon_r = 10.5$ or 2.2).

Useful design guidelines have been derived for designers using these antennas as phased arrays or as 'active' arrays for spatial power combining.

A dual-polarization array employing TSAs is shown in Figure 3.21a. While the antipodal TSA is illustrated in the eggcrate arrangement first proposed in reference [168], other types of TSAs can also be used.

3.6.5.3. The Flared-Slot Antennas

The microstrip flared slot antenna [169,170] shares the wideband characteristics of the TSAs we have already considered and is suitable for phased arrays having the tile architecture, as shown in Figure 3.21b. The two orthogonal linear polarizations are available from two orthogonal slots and the two opposite circular polarizations can be derived by using conventional techniques. The additional attractor for the flared slot antenna is that it occupies minimum array aperture real estate. As the slot radiates in both directions, a cavity is used to transform the antenna into a unidirectional radiator; some frequency-dependent squint on the H-plane pattern was observed and the cross-polarization measurements were barely adequate. Further work is required to isolate the effects of cavity design on the antenna return loss, the H-plane radiation patterns and gain [169,170]. Given the popularity of the array tile architecture, this wideband antenna deserves further development work.

Before we consider other types of antennas, it is worth summarizing the developments related to microstrip/printed board antennas. The many realizations of these low-cost, conformal antennas meet many of the requirements we have formulated such as narrowband/wideband operation and acceptable cross-polarization levels. While some problems are still with us, progress toward high-quality microstrip/printed board antennas has exceeded optimistic expectations. It is certain that the future of these antennas operating on a stand-alone basis or as antenna elements of phased arrays is more promising than ever.

3.6.5.4. Circular Disk Monopoles

Circular disk monopoles (CDMs) are found to have wideband impedance characteristics [171]. A 50 mm diameter disk having a height of 1 millimeter exhibited a VSWR of 2 bandwidth that extended from 1.17 to 12 GHz [171]. The H-plane pattern remained nearly omnidirectional with maximum variation in azimuth increasing from 4 to 7 dB. However, as the frequency increased from 2.5 to 9 GHz, the direction of maxima of the conical beam of the E-plane pattern varied from 30° to 60° from elevation. Further studies and developmental work are warranted for this simple CDM that exhibits excellent wideband impedance characteristics.

3.7. SUPERCONDUCTING ANTENNAS/ARRAYS

The advent of high-T_c superconducting materials (HTSs) has initiated interest in superconducting antennas [172,173]. While the applications are many, we shall here confine our considerations to antennas used in phased arrays.

Analytical work [174] has shown that HTSs may be most beneficial when used in the feed and matching networks for microwave and millimeter-wave microstrip arrays. A linear 100-antenna-element HTS array could experience a gain increase of 8–10 dB over an identical copper array [174].

An array of four patches designed and fabricated on lanthanum aluminate (LaAlO$_3$) to operate at 30 GHz was compared with an identical antenna patterned with evaporated gold [175]. The efficiency of the former antenna was 2 dB higher at 70 K and 3.5 dB higher at 40 K [175] than that of the latter. A 20 GHz, circularly polarized microstrip HTS array of four patches exhibited a 1.67 dB gain at 77 K and 3.4 dB at 30 K [176].

An array of dipoles is superdirective (or has a supergain) when its directivity in a reference direction is substantially larger than that of a conventional antenna array, with its elements driven by currents of equal amplitude and phases adjusted for maximum field intensity in the reference direction. These arrays have low radiation efficiency and high Q-factors. In practical applications the low radiation efficiency can be overcome by using HTS in their manufacture. A practical supergain antenna array of closely spaced 4×4 dipoles has been reported [177].

High-T_c superconducting small-sized arrays are practical and the benefits are worth attaining. These arrays can be used at or near the focal planes of reflectors. Larger arrays operating at millimeter wavelengths will also be physically small enough to be cooled to the required cryogenic temperatures. The cooling of physically large arrays is at present problematic.

3.8. SUBARRAYS

The main impetus for forming subarrays is cost reduction; for instance, if four antenna element subarrays are formed, the total number of T/R modules is reduced by a factor of four and the total array cost is dramatically reduced. In the limiting case we can consider an array of antenna elements connected to a summing point via equal lengths of lines. It is arguable whether this arrangement is a phased array because the resulting beam is not steerable, but the array is a low-cost substitute for a reflector aperture. It is useful to determine how many microstrip patch antennas one can interconnect before the losses incurred in the array become prohibitive. The break-even point between the array and a reflector occurs when the number of patch antennas is equal to 256 [178]. The receive-only array consisted of patch antennas operating at 10 GHz and had an interelement spacing of 0.8λ. An exhaustive list of the losses related to the array is also given in the same reference. This work is considered here only because it constitutes a limiting benchmark case.

Let us now explore the important issues related to arrays utilizing subarrays of 2 or 3 antenna elements. The effect the subarrays have on the array scan blindness is reported in reference [179]. Infinite arrays composed of arbitrary subarrays of microstrip antenna elements were considered and the analysis included mutual coupling between elements in the subarray as well as the coupling between subarrays. The magnitude of the reflection coefficient of the radiating elements as a function of scan angle was derived for conventional arrays and for arrays where the number of elements of the subarrays was either 2 or 3. The interelement spacing was $\lambda/2$ and each element was fed in-phase with equal amplitude and a uniform phase progression was applied across the subarrays. The magnitude of the reflection coefficient of the radiating elements, shown in Figure 3.22, decreases significantly when the number of subarray elements is equal to 2 or 3, over scan angles ranging from 0° to 90°. In the same figure, a scan blindness occurs at an angle of 76° when no subarrays are used. As can be seen, subarraying relieves the designer of concerns related to scan angle blindness.

Several authors have explored the possibility of improving the interconnections between the antenna elements of subarrays. Microstrip antenna subarrays consisting of a central aperture-coupled microstrip antenna element feeding additional elements via coplanar microstrip lines have been described [180]; in the same reference, a basic design procedure is outlined. A 2×2 subarray that consists of identical patches electromagnetically driven by a patch etched on a lower substrate has been proposed [181].

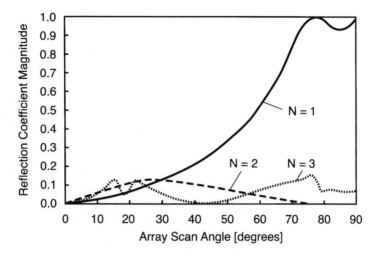

Figure 3.22 The reflection coefficient of the radiating elements as a function of array scan angle for the cases where no subarraying is implemented (*N*=1) and when the number of subarray elements is two (*N*=2) and three (*N*=3) (From [179]; © 1992, IEEE.)

We have already established that subarrays introduce AQ grating lobes, which in turn decrease the array beam efficiency; in reference [179] the array beam efficiency is derived as a function of scan angle for the cases where no subarrays are used and when 2 or 3 antenna elements are used in each subarray. A substantial decrease in the array's beam efficiency is noted and the decrease is proportional to the number of elements used in the subarrays. We shall consider approaches that minimize the resulting grating losses in the next section.

3.8.1. The Marginalization of the AQ Grating Lobes

In the context of subarrays, the manifestation of AQ grating lobes is caused because

- The subarray size and the distances between contiguous subarray centers are constant.
- The required amplitude taper is approximated at the subarray level.

The marginalization of the resulting AQ grating lobes can be achieved by the randomization of both the subarray sizes and the distances between the centers of contiguous subarrays [182]. A 128-element linear array was used to demonstrate the above approach. The array was first divided into 32 subarrays of 4 elements and the resulting three AQ grating lobes, shown in Figure 3.23a, were too high at the –15 to –20 dB levels and too close to the main lobe, at an angular distance of only 25°. The same array was subdivided into subarrays having elements randomly varying from 2 to 6 and the centers of contiguous subarrays also varied randomly. In the resulting array radiation pattern, shown in Figure 3.23b, the amplitudes of the AQ grating lobes have been minimized.

In another approach, the required array amplitude taper is approximated by a two-step process [183]: (i) a taper at the subarray level; and (ii) element amplitude tapers that are identical for every subarray. The proposed approach is inexpensive, practical,

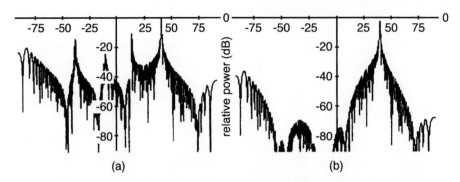

Figure 3.23 The radiation pattern of a 128-element linear array when (a) all subarrays consist of four antenna elements, and (b) the sizes of the subarrays vary from 2 to 6 randomly and the distances between contiguous subarray centers is also random. (From [182]; © 1990, Eur. Micr. Conf. Manag. Committee.)

and efficient in the marginalization of the resulting AQ grating lobes. As the number of identical element tapers used increases from 1 to 2 and 3, the cost and complexity increase but the approximation to the required array amplitude tapers is improved. A 30 dB Taylor taper was implemented on a 70-element array and the absence of any AQ grating lobes was demonstrated [183].

From the foregoing considerations, it is clear that the designer has a few options in meeting the diverse requirements often imposed by operational requirements and financial constraints.

3.8.2. Applications

We have already considered subarrays of four patch antennas in section 3.8. In this section we shall consider other subarrays of similar antenna elements and sequential subarrays.

A subarray of four EMC patches and the associated circuitry comprising of networks in two levels yielded the two circular polarizations, at C-band, and exhibited equivalent performance in both polarization purity and bandwidth to what is achievable with current waveguide approaches [129,184]. Additionally, the unit is estimated to be 2.5 cm thick and to weigh 100 g, while the corresponding figures for the antenna element used in INTELSAT are 30 cm length and 300 g weight. The weight-saving potential of this compact antenna system is obvious and welcome [129]. Given that the antenna system is low-cost, these developments are significant. A reduction of the losses related to similar antenna systems is required before they can be used at the Ku-band.

Sequential subarrays capable of yielding the circular polarizations have been considered in several references e.g. [185,186].

3.9. POLARIZATION AGILITY AND WORK IN PROGRESS

The realization of systems exhibiting full polarization agility (or diversity) involves cost and complexity but can yield maximum benefits to the users. Systems having some polarization agility, on the other hand, can yield some benefits to the user at reasonable costs and complexity. Thus the designer has options to meet the financial and instrumental constraints often imposed on systems.

Given that microstrip/printed board antennas have the many attractive characteristics we have already outlined, our coverage will be confined to the approaches that enable these antennas to exhibit varying degrees of polarization agility. Additionally, our interest is confined to low-RF power circuits and networks related to the attainment of polarization agility.

As wideband, programmable, MMIC-based phase-shifters are readily available, the designer can implement polarization agility at the T/R module level—see section 3.3.6.2. For small arrays, a low-mass meander-line polarizer that mechanically rotates the plane of the linear polarization has been proposed [187]. The merit of this approach is that some polarization agility can be introduced by retrofitting the proposed polarizer to an existing aperture without any modifications to the aperture. Conventional meander-line polarizers have been used to effect linear-to-circular polarization conversion and to cause a 90° rotation of a linearly polarized signal (references cited in [187]).

Several approaches for the implementation of polarization agility at the antenna level have been proposed. The realization of patch antennas that meet all the other requirements normally associated with antenna elements as well as that of polarization agility is a challenging task. Polarization-agile patches in an array application have to meet the additional requirements of being volumetrically attractive and that the phase centers of the two orthogonal polarizations should be coincident.

Two early references explored the realization of polarization-agile patch antennas [188,189]. More recently, two promising approaches utilize patch antennas in conjunction with transistors or varactor diodes to attain polarization agility ([190] and [191] respectively). In both approaches square patches are loaded by two transistors [190] and four diodes [191]. These approaches meet the two additional requirements we have outlined. In both papers the measured co- and cross-polarization radiation patterns are acceptable. Further work on this important aspect is worth undertaking.

Work related to antennas fabricated on ferrite substrates with and without the application of a magnetic field is summarized in reference [192]. References that explore the beam scanning, antenna pattern, and RCS control of antennas by the application of magnetic fields are cited in the same paper. When a uniform magnetic field is applied to a circular patch antenna on a ferrite material, the ferrite microstrip antenna (FMA) yields a circularly polarized radiation pattern in accordance with theoretical predictions. The exploration of the potential of FMAs is a promising research field.

3.10. CONCLUDING REMARKS AND A POSTSCRIPT

Reflector systems and phased arrays perform complementary functions. It is well known that (i) phased arrays remain unchallenged when the system requirement is maximum spatial resolution and (ii) reflectors constitute the building blocks of radioastronomy arrays having maximum spatial resolutions. Similarly, reflectors having a phased array at or near their focal plane constitute affordable and sensitive imagers or radar systems.

Even when reflectors and phased arrays have the same geometric area, the phased array systems offer the user several degrees of freedom, ranging from the interbeam spacing to the resulting beam shapes, but are relatively expensive to realize. Systems based on reflector antennas, on the other hand, offer fewer degrees of freedom but are less expensive. The prime requirements of future radiotelescopes are:

- High sensitivity and adequate spatial resolution.
- The availability of one or more wide fields of view simultaneously.
- Dual polarization capability.
- Operation at many frequency bands simultaneously.

Reflectors yielding one antenna beam are adequately treated in basic antenna treatises and reference books. In this chapter we have therefore considered multiple-beam reflectors that can be used either as stand alone imagers or in conjunction with phased arrays. In a similar vein we have considered approaches that allow simultaneous operation over two or more frequency bands.

As offset systems of reflectors are eminently suited for multiple-beam operation over several frequency bands, they meet most of the above requirements. While the derived high-quality multiple beams are not contiguous, a full map of an extended celestial source can be derived by interleaved observations.

Annular synthesis antennas yield contiguous beams of the scene, an important characteristic for stand alone imaging systems. While conventional phased arrays utilize hundreds or thousands of low-gain antenna elements, ASAsII and ASAsI utilize a number of antenna elements that is significantly lower, but the gain of each antenna element is either moderate or high, respectively. ASAsII deserve further studies. When ASAs are used in conjunction with large synthesis arrays, they constitute low-cost realizations that offer the designer the range of short spacings between adjacent antenna elements, often missing in current radioastronomy arrays. Other antenna elements are also being considered for future radioastronomy arrays.

Regardless of which antenna elements are used, future radiotelescopes will have an instantaneous FOV that is considerably wider than that of conventional radiotelescopes; thus the radio equivalent of the Schmidt optical telescope will be realized. While current radiotelescopes have very fine spatial resolutions, their sensitivity is only adequate. Future radiotelescopes having high sensitivity and moderate spatial resolution will therefore complement the existing radiotelescopes. More explicitly, the minimum sensitivity required is equal to that attained by a phased array radiotelescope fully populating an area of one hectare (10 000 m^2) [193] or 1 km^2 [194]. As discussions progressed within the US community of radioastronomers and members of the SETI Institute, the large-N SKA specification evolved and calls for a phased array radiotelescope extending over N km^2, where N is between 1000 and 2000 [196]. The frequency of operation for the proposed radiotelescope is between 150 MHz and 20 GHz [194].

Another important parameter that emerged from these discussions is flexibility for the radiotelescopes of the new millenium [195]. The proposals are sketchy but one can presume that within that large-N km^2 area, densely populated and thinned arrays operating at different frequency bands are accommodated. If the maximum linear dimension of the array is between 300 and 1000 km [195], the same instrument can offer the observers the options of high sensitivity and high spatial resolution maps, and high sensitivity and low spatial resolution (large-scale) maps.

A major problem facing the designers of current and future radiotelescopes is RFI introduced by the relatively strong signals emanated by terrestrial broadcast, communication, radar and navigation transmitters, as well as by constellations of communications and navigation LEO/MEO satellites. The RFI problem is now generating considerable research interest in the radioastronomy community [196]. With the passage of time the demand for more services will increase and the RFI problem will inevitably become so acute that radioastronomical observations will be compromised. Optical telescopes are affected by light pollution and radiotelescopes by RFI. While the optical telescopes can be located at some distance from city lights, all radiotelescopes, regardless of their position on Earth, are affected by the emissions of the transmitters on board constellations of communications and navigation LEO/MEO satellites. The excision of these unwanted signals in the frequency and spatial domains is therefore mandatory for current and future radiotelescopes.

Considerable effort has been expended toward the definition and mitigation of the effects of RFI on weak signals [197–199]. In particular, real-time adaptive cancellation of the interfering signals as applied in other fields is of particular importance to current and future radiotelescopes. Reference [198] in particular contains over 200 references on this important topic, and reference [199] reports the successful suppression of strong interfering signals from weak signals by the use of this approach.

Apart from the excision of the unwanted signals in the frequency and spatial domains considerable research is directed toward the search for suitable antenna elements for future radiotelescopes [27,200–202]. The essential characteristics of hybrid mode feeds capable of operating over a 2:1 frequency range, respectively, have been outlined in this chapter. The performance of these feeds can be considered as ideal for comparisons with other types of antennas. References [201] and [202] report the design and realization of dielectrically loaded horns that operate over a 5:1 frequency range. The performance of narrowband/wideband polarizers used in single-dish radioastronomical observations and outlined in this chapter can be considered ideal.

Research on adaptive beamforming approaches, robust algorithms for the SKA, and random sparse arrays is reported in reference [203].

Early radar array models were heavy structures that used metal horns as antenna elements followed by metal waveguides. The migration to lightweight structures that consist of dipoles or microstrip antennas and transmission lines is pronounced. Recently, microstrip transmission lines have been substituted with optical fibers and a further decrease in array weight is in the offing. Apart from weight reduction, microstrip antennas are inexpensive to produce and can be conformal to the platform's skin. The migration from an aerodynamically unattractive array housed in an aerodynamically attractive but lossy radome to a conformal aerodynamically attractive array is a significant trend.

The third attractor of microstrip antenna elements is their bandwidth. Recent developments indicate that these antenna elements can meet narrowband, moderate-band and wideband requirements. High-quality performance, for instance, can be expected from wideband aperture-stacked patch microstrip antennas operating over a 2:1 frequency range ([204] and the reference therein). Additionally, microstrip TSAs can comfortably operate over a 3:1 frequency range. Our coverage of microstrip antenna elements is therefore extensive. Overall, the successes achieved in many areas related to microstrip antenna elements have exceeded all optimistic expectations.

Beamforming lenses offer unique solutions to a diverse set of requirements. Luneberg lenses, for instance, perform the aperture and beamforming functions;

additionally, the beamwidth of the derived beams is independent of the angles to which the lens is pointed. Rotman lenses are often used in conjunction with phased arrays and both types of lenses are low-cost solutions that offer large instantaneous bandwidths and wide scanning angles.

Lastly, polarization agility can be attained by the conventional approaches we have considered in this chapter and more direct methods that are being explored.

REFERENCES

[1] Stutzman W. L. and Thiele G. A. *Antenna Theory and Design*. Wiley, New York (1981).
[2] Brueckmann H. Improved wide-band VHF whip antenna. *IEEE Trans. Vehic. Commun.* **VC-15**(2), 25(1966).
[3] Campbell D. V. A low-profile, remote-tuned dipole antenna for the 30 to 80 MHz range. *Research and Development Technical Report, ECOM-4491*, pp. 25–32 (Apr. 1977).
[4] Cooper C. E. Remotely tunable antennas for frequency-hopping. *Electronics & Wireless World* (Jan.) pp. 17–22 (1986).
[5] Fourikis N., Parfitt A. J., Tang T. and Gunn M. W. Microprocessor tuned antennas. *2nd Australian (ATEPB/CSIRO) Symp. Antennas*, Sydney, Australia, paper 21 (1989).
[6] Ruze J. The effect of aperture errors on the antenna radiation pattern. *Nuovo Cimento, Suppl.* **9**(3), 364 (1952).
[7] Ruze J. Antenna tolerance theory—a review. *Proc. IEEE* **54**(4), 633 (1966).
[8] Cogdell J. R. and Davis J. H. On separating aberrant effects from random scattering effects in radio telescopes. *Proc. IEEE* **61**(9), 1344 (1973).
[9] von Hoemer S. Radio-telescopes for millimeter wavelengths. *Astron. Astrophys.* **41**, 301 (1975).
[10] Adatia N. A. Diffraction effects in dual offset Cassegrain antenna. *IEEE Int. Symp. AP-S* p. 235 (1978).
[11] Hannan P. W. Microwave antennas derived from the Cassegrain telescope. *IRE Trans. Antennas Propag.* **AP-9**, 140–153 (May 1961).
[12] Kraus J D. *Antennas*. McGraw-Hill, New York (1988).
[13] Ramsay J. F. Tubular beams from radiating apertures. *Advances in Microwaves*, (L. Young, ed.), vol. 3, p. 127. Academic Press, New York, (1968).
[14] Ramsay J. F. Lambda functions describe antenna/diffraction patterns. *Microwaves, Antenna Des. Suppl.* (June), 69 (1967).
[15] Spencer R. C. *Paraboloid diffraction patterns from the standpoint of physical optics*. Rep. No. T-7. MIT Radiat. Lab., Massachusetts Institute of Technology, Cambridge, MA (1942).
[16] Silver S. *Microwave Antenna Theory and Design*. McGraw-Hill, New York (1949).
[17] Cook J. S. Elam E. M. and Zucker H. The open Cassegrain antenna. Part 1. Electromagnetic design and analysis. *Bell Syst. Tech. J.* **44**, 1255 (1965).
[18] Dragone C. and Hogg D. C. The radiation pattern and impedance of offset and symmetric near field Cassegrain and Gregorian antennas. *IEEE Trans. Antennas Propag.* **AP-22**, 472 (May 1974).
[19] Ohm E. A. A proposed multiple-beam microwave antenna for earth stations and satellites. *Bell Syst. Tech. J.* **53**, 1657 (1974).
[20] Ohm E. A. A proposed multiple-beam microwave antenna for Earth stations and satellites. *Bell Syst. Tech. J.* **53**, 1657–1665 (Oct. 1974).
[21] Terada A. B. and Stutzman W. L. Computer-aided design of reflector antennas: The Green Bank Radio Telescope. *IEEE Trans. Microwave Theory Tech.* **46**(3), 250–253 (Mar. 1998).
[22] Chu T. S. and Turrin, R. H. Depolarization properties of offset reflector antennas. *IEEE Trans. Antennas and Propag.* **AP-21**(3), 1657–1665 (May 1973).
[23] Fourikis N. A parametric study of the constraints related to Gregorian/Cassegrain offset reflectors having negligible cross-polarization. *IEEE Trans. Antennas Propag.* **36**(1), 144 (1988).

[24] Terada M. A. B. and Stutzman W. L. Design of offset-parabolic-reflector antennas for low cross-pol and low sidelobes. *IEEE Antennas Propag. Mag.* **35**, 46–49 (Dec. 1993).

[25] Goldsmith P. F. Focal plane arrays for millimeter-wavelength astronomy. *IEEE MTT-S Dig.* pp. 1255–1258 (1992).

[26] Ruze J. Lateral-feed displacement in a paraboloid. *IEEE Trans. Antennas Propag.* **AP-13**, 660–665 (Sept. 1965).

[27] Fourikis N. Novel radiometric phased array systems. *Microwave Opt. Tech. Lett.* (5 June) 100–108 (1998).

[28] Erickson N. R., Goldsmith P. F., Novak G., Grosslein R. M., Viscuso P. J., Erickson R. B. and Predmore R. A 15 element focal plane array for 100 GHz. *IEEE Trans. Microwave Theory Tech.* **40**(1), 1–11 (Jan. 1992).

[29] Yngvesson K. S., Korzeniowski T. L., Kim Y.-S. Kollberg E. L. and Johansson J. F. The tapered slot antenna—a new integrated element for millimeter-wave applications. *IEEE Trans. Microwave Theory Tech.*, **37**(2), p365–374, (Feb. 1989).

[30] Johansson J. F. *Millimetre wave imaging theory and experiments.* Research Report No. 151. Onsala Space Observatory, Chalmers University of Technology, Gothenburg, Sweden (1986).

[31] Goldsmith P. F., Hsieh C.-T, Huguenin G. R., Kapitzky J. and Moore E. L. Focal plane imaging systems for millimeter wavelengths. *IEEE Trans. Microwave Theory Tech.* **41**(10), 1664–1675 (Oct. 1993).

[32] Swarup G. Alternative array configuration and antenna elements for the square kilometer array interferometer. In *Proc. High Sensitivity Radio Astronomy* (N. Jackson and R. Davis, eds), Jodrell Bank (1996).

[33] Braun R. The square kilometer array interferometer. In *The Westerbork Observatory, Continuing Adventure in Radio Astronomy.* (E. Raimont and R. Gence, eds), pp. 167–183. Kluwer, Dordrecht (1996).

[34] Mansour R. R., Ye S., Dokas V., Jolley B., Thomson G., Tang W.-C. and Kudsia C. M. Design considerations of superconductive input multiplexers for satellite applications. *IEEE Trans. Microwave Theory Tech.* **44**(7), 1213–1228 (July 1996).

[35] Wetherell W. B., Rimmer M. P. General analysis of aplanatic Cassegrain, Gregorian and Scharzchild telescopes. *Appl. Opt.* **11**, 2817 (1972).

[36] Craig W. P., Rappaport C. M. and Mason J. S. A high aperture efficiency, wide-angle scanning offset reflector antenna. *IEEE Trans. Antennas Propag.* **41**(11), 1481–1490 (Nov. 1993).

[37] Patterson D. A. Microprocessors in 2020. *Sci. Am.* (Sep.) 48–51 (1995).

[38] Fourikis N. A new class of millimetre wave telescopes. *Astron. Astrophys.* **65**, 385 (1978).

[39] Fourikis N. Several aspects related to the realisation of annular synthesis telescopes operating at millimetre wavelengths. *J. Electr. Electron. Eng., Aust.* **2**(4), 193 (1982).

[40] Fourikis N. Single structure, steerable synthesis telescopes utilizing offset reflectors. PhD dissertation, University of NSW, School of Electrical Engineering and Computer Sciences (1984).

[41] Hall P. S. and Vetterlein S. J. Review of radio frequency beamforming techniques for scanned and multiple beam antennas. *IEEE Proc., Part H. Microwaves, Optical Antennas* **137**(5), 293 (1990).

[42] Mailloux R. J. Hybrid antennas. In *The Handbook of Antenna Engineering* (A. W. Rudge *et al.* eds.), ch. 5. IEE Peregrinus (1986).

[43] Goldsmith P. F. *et al.* Focal plane imaging systems for millimeter wavelengths. *IEEE Trans. Microwave Theory Tech.* **41**(10), 1664 (1993).

[44] Goldsmith P. F. Focal plane arrays for millimeter wavelength astronomy. *IEEE MTT-S Int. Microwave Symp. Dig.*, p. 1255 (1992).

[45] Erickson N. R. *et al.* A 15 element imaging array for 100 GHz. *IEEE Trans. Microwave Theory Tech.* **40**(1), 1 (1992).

[46] Shoucri M. *et al.* A passive millimeter wave camera for landing under low visibility conditions. *Proc. Natl. Telesystems Conf.*, p. 109 (1993).

[47] Dow G. S. *et al.* W-band MMIC direct detection receiver for passive imaging system. *IEEE M7-T-S Int. Microwave Symp. Dig.*, p. 163 (1993).

[48] Black D. N. and Wiltse J. C. Millimeter-wave characteristics of phase-correcting Fresnel zone plates. *IEEE Trans. Microwave Theory Tech.* **M'IT-35**(12), 1122 (1987).

[49] Brookner E., ed., *Aspects of Modern Radar.* Artech House, Boston, MA (1988).

[50] Brookner E. (ed.) *Radar Technology.* Artech House, Norwood, MA (1977.

[51] Rudge A. W. Current trends in antenna technology and prospects for the next decade. *IEEE Antennas Propag. Soc. Newsl.* (Dec.) 5 (1983).

[52] Sorbello R. M. Advanced satellite antenna development for the 1990s. *AIAA 12th Int. Commun. Satellite Syst. Conf.*, Arlington, VA, *Collec. Tech. Pap.*, p. 322 (1988).

[53] Sorbello R. M. *et al.* MMIC: A key technology for future communications satellite antennas. *Monolithic Microwave Integr. Circuits Sensors, Radar Commun. Syst., Proc SRIE* **1475**, 175 (1991).

[54] Assai F. T., Zaghoul A.I. and Sorbello R. M. Multiple spotbeam systems for satellite communications, *AIAA 12th Int. Commun. Satellite Syst. Conf.*, Arlington, VA, *Collect. Tech. Pap.*, p. 322 (1988).

[55] Cortez-Medellin G. and Goldsmith P. F. Analysis of segmented reflector antennas for a large millimeter wave radio telescope. *IEEE AP-S Int. Symp. Dig.* **4**, 1886 (1992).

[56] Beaty J. K. The new giant of Mauna Kea. *Sky & Telescope* (July) 30 (1992).

[57] Ressmeyer R. H. Keck's giant eye. *Sky & Telescope* (Dec.) 623–625 (1992).

[58] Anon.. Dazzling views from Europe's NTT. *Sky & Telescope* (June) 596–599 (1990).

[59] Zimmerman M. L., Lee S. W. and Fujikawa G. Analysis of reflector antenna system including frequency selective surfaces. *IEEE Trans. Antennas Prop.* **40**(10) pp. 1264–1266 (Oct. 1992).

[60] Besso P., Forigo D. and Gianola P. Frequency selective surfaces with very low loss and cross-polarisation levels. *Eighth Int. Con.f Antennas Propag.* vol. 1, pp. 578–581 (1993).

[61] Wu T.-K. Four-band frequency selective surface with double-square-loop patch elements. *IEEE Trans. Antennas Propag.* **42**(12) 1659–1663 (Dec. 1994).

[62] Chang T. K., Langley R. J. and Parker E. A. Active frequency-selective surfaces. *IEE Proc. Microwaves, Antennas Propag.* **143**(1) 62–66 (1996).

[63] Zhang D. *et al.* Application of high T_c superconductors as frequency selective surfaces: experiment and theory. *IEEE Trans. Antennas Propag.* **41**(6/7), 1032–1036 (June/July 1993).

[64] Love A. W. *Electromagnetic Horn Antennas*. IEEE Press, New York (1976).

[65] Ludwig A. C. Antenna feed efficiency. *JPL Space Programs Summ.* **37–26** (4), 200 (1965).

[66] Caldecott R., Mentzer C. A. and Peters L. Jr. The corrugated horn as an antenna range standard. *IEEE Trans. Antennas Propag.* **AP-21**, 562–564 (July 1973).

[67] Haas R. W., Brest D. H. and Bowers R. J. Multi-beam millimeter-wave focal plane array airborne imaging system. *Proc. SPIE* **2211**, 302–311 (1994).

[68] Fourikis N. An 18 cm turnstile junction as a polarization splitter. *IREE Aust.* **34**(9), 403–405 (Oct. 1973).

[69] Tompkins R. D. A broad-band dual mode circular waveguide transducer. *Trans. IRE* **MTT-4**, 181 (July 1956).

[70] Chen M. H. and Tsandoulas G. N. A wide-band square-waveguide array polarizer. *IEEE Trans. Antennas Propag.*(May), 389–391 (1973).

[71] Allen P. J. and Tompkins R. D. An instantaneous microwave polarizer. *Proc IRE* **47**(7), 1231 (July 1959).

[72] Potter R. S. and Sagar A. A new property of the turnstile waveguide junction. *Proc. Natl. Electron. Conf.* vol. 13, p. 452 (1957).

[73] Meyer M.A. and Goldberg H. B. Applications of the turnstile junction. *Trans. IRE* **MTT-3** (6), 40 (Dec. 1955).

[74] Vogel W. J. Terrestrial rain depolarization compensation experiment at 11.7 GHz. *IEEE Trans. Commun.* **COM-31** (11), 1241 (Nov. 1983).

[75] Skinner S J. and James G. L. Wide-band orthomode transducers. *IEEE Trans. Microwave Theory Tech.* **39**(2), 294–300 (Feb. 1991).

[76] Archer D. Lens-fed multiple-beam arrays. *Microwave J.* **18**, 37–42 (Oct. 1975).

[77] Rotman W. and Turner R. F. Wide-angle lens for line source applications. *IEEE Trans. Antennas Propag.* **AP-11**, 623–632 (1963).

[78] Gent H. The bootlace aerial. *Royal Radar Establishment J.* **20**, 47–57 (Oct. 1957).

[79] Rausch E. O., Peterson A. F. and Wiebach W. A low cost, high performance, electronically scanned mmw antenna. *Microwave J.* **20**, 20–32 (Jan. 1977).

[80] Monser G. J. A reversible Rotman lens useful on short, linear array applications. *Microwave J.* **38** (Jan.) 160–163 (1995).

[81] Tao Y. M. and Delisle G. Y. Lens-fed multiple beam array for millimeter wave indoor communications. *IEEE Antennas Propag. Soc. Symp.* pp. 2206–2209 (1997).

[82] Peik S. F. and Heistadt J. Multiple beam microstrip array fed by Rorman lens. *IEE Ninth Int. Conf. Antennas Propag.* Conf. Pub. No. 407, pp. 348–351 (1995).

[83] Luneberg R. K. *Mathematical Theory of Optics*, pp. 189–212. Brown University Press Providence, CI (1944).

[84] Sanford J. A Luneberg-lens update. *IEEE Antennas Propag. Mag.*, **37**(1), 76–79 (Feb. 1995).

[85] Mitchell M. A. and Sanford J. R. Luneberg lens revival. *Electronics & Wireless World* (May) 456–458 (1989).

[86] Ingerson P. G. Luneberg lenses performance limitations due to fabrication. *IEEE Antennas Propag. Soc. Symp.*, pp. 862–865 (1997).

[87] Sanford J. and Sipus Z. Sidelobe reduction with array fed spherical lenses. *IEEE Antennas Propag. Soc. Symp.*, pp. 670–673 (1995).

[88] Fourikis N. Novel power combining circular arrays operating at mm-wavelengths. *16th Int. Conf. Infrared Millimeter Waves*, Lausanne, Switzerland. *SPIE* **1576**, 503–504 (1991).

[89] Sullivan W. T. III. Radio astronomy's golden anniversary. *Sky & Telescope* (Dec.) 544 (1982).

[90] Molker A. High-efficiency phased array antenna for advanced multibeam, multiservice mobile communication satellite. *77th Int. Conf. Satellite Systems Mobile Commun. Navigat.*, p. 75 (1983).

[91] Fander J. S. Synthetic apertures: an overview. *Proc. SPIE*, **440**, 2–7 (1983).

[92] Johnson R. B., Wolfe W. L. and Fender J. S. *Infrared, Adaptive and Synthetic Aperture Optical Systems*, Arlington Virginia, *Proc. SPIE* **643** (1986).

[93] Cornell J. Six new eyes peer from Mount Hopkins. *Sky & Telescope* (July), 23–24 (1979).

[94] Lloyd-Hart M. *et al.* Direct 75 milliarcsecond images from the multiple mirror telescope with adaptive optics. *Astrophys. J.* **402**, L81–L84 (10 Jan. 1993).

[95] Elliott R. S. The design of waveguide-fed slot arrays. In *Antenna Handbook, Theory, Applications and Design*, ch. 12. Van Nostrand-Reinhold, New York (1968).

[96] Compton, R. J. Jr. and Collin R. E. Slot antennas. In *Antenna Theory* (R. J. Collin and F. J. Zucker, eds.), part 1, ch. 14. McGraw-Hill, New York (1969).

[97] Yee H. Y. Slot antenna arrays. In *Antenna Engineering Handbook* (R. C. Johnson and H. Jasik, eds.), ch. 9. McGraw-Hill, New York (1961).

[98] Deschamps G. A. Microstrip microwave antennas. *USAF Symp. Antennas* (1953).

[99] Carver K. R. and Mink J. W. Microstrip antenna technology. *IEEE Trans. Antennas Propag.* **AP-29**(1), 2 (1981).

[100] Mailloux R. J., McIlvenna J. F. and Kernweis N. P. Microstrip array technology. *IEEE Trans. Antennas Propag.* **AP-29**(1), 25 (1981).

[101] Bahl I. J. and Bhartia P. *Microstrip Antennas*. Artech House, Dedham, MA (1980).

[102] James J. R., Hall P. S. and Wood C. *Microstrip Antenna Theory and Design*. Peter Pelegrinus, London (1981).

[103] James J. R. and Hall P. S. *Handbook of Microstrip Antennas*. Peter Pelegrinus, London (1989).

[104] Lee K. F. and Dahele J. S., Characteristics of microstrip patch antennas and some methods of improving frequency agility and bandwidth. In *Handbook of Microstrip Antennas* (J. R. James and P. S. Hall, eds). Peter Pelegrinus, London (1989).

[105] Munson R. Conformal microstrip antennas. *Microwave J.* **31**, 91 (Mar. 1988).

[106] Lo Y. T., Solomon D. and Richards W. F. Theory and experiment on microstrip antennas. *IEEE Trans. Antennas Propag.* **AP-27**, 137 (1979).

[107] James J. R., Henderson A. and Hall P. S. Microstrip antenna performance is determined by substrate constants. *Microwave Syst. News* (Aug.), 73 (1982).

[108] Griffin J. M. and Lowth J. F. Broadband microstrip patch antennas. *Mil. Microwaves Conf. Proc.* London, p. 237 (1984).

[109] Pozar D. M. General relations for a phased array of printed antennas derived from infinite current sheets. *IEEE Trans. Antennas Propag.* **AP-33**(5), 498 (1985).

[110] Roudot B., Mosig J. and Gardiol F. Surface wave effects on microstrip antenna radiation. *Microwave J.* **31**(3), 201 (1988).

[111] Pozar D. M. and Schaubert D. H. Comparison of architectures for monolithic phased array antennas. *Microwave J.*, **29**, 93 (Mar. 1986).

[112]   Pozar D. M. *Antenna Design Using Personal Computers*. Artech House, Dedham, MA (1985).

[113]   Pozar D. M. and Schaubert D. H. Scan blindness in infinite phased arrays of printed dipoles. *IEEE Trans. Antennas Propag.* **AP-32**, 602 (1984).

[114]   Patel P. D. Approximate location of scan-blindness angle in printed phased arrays. In *Hal Schrank's Antenna Designer's Notebook; IEEE Antennas Propag.* **34**(5), 53 (1992).

[115]   Pozar D. M. Rigorous closed-form expressions for the surface wave of printed antennas. *Electron. Lett.* **26**(13), 954 (1990).

[116]   Johansson F., Rexberg L. and Peterson N. O. Theoretical and experimental investigation of large microstrip array antenna. *IEE Colloq., Recent Developments in Microstrip Antennas*, p. 4/1 (1993).

[117]   Paquay M. H. *et al.* A dual polarised active phased array antenna with low cross polarisation for a polarimetric SAR. *Int. Conf. Radar 92*, p. 114 (1992).

[118]   Zurcher J.-F. *et al.* Dual polarized, single- and double-layer strip-slot-foam inverted patch (SSFIP) antennas. *Microwave Opt. Technol. Lett.* **17**(9), 406 (1994).

[119]   Levine E. and Shtrikman S. Experimental comparison between four dual polarised microstrip antennas. *Microwave Opt. Lett.* **3**(1), 17 (1990).

[120]   Huang J. Dual-polarised microstrip array with high isolation and low crosspolarisation. *Microwave Opt. Technol. Lett.* **4**(3), 99 (1991).

[121]   Pozar D. M. Considerations for millimeter wave printed antennas. *IEEE Trans. Antennas Propag.* **AP-31**, 740 (1983).

[122]   Katehi P. B., Jackson D. R. and Alexopoulos N. G. Microstrip dipoles. In *Handbook of Microstrip Antennas* (J. R. James and P. S. Hall, eds). Peter Pelegrinus, London (1989).

[123]   Rasshofer R. H., Thieme M. O. and Biebl E. M. Circularly polarized millimeter-wave rectenna on silicon substrate. *IEEE Trans. Microwave Theory Tech.* (Special issue on microwave circuits on silicon substrates.) **46**(5), 715–718 (May 1998).

[124]   Russer P. Si and SiGe millimeter-wave integrated circuits. *IEEE Trans. Microwave Theory Tech.* (Special issue on microwave circuits on silicon substrates.) **46**(5), 590–603 (May 1998).

[125]   Strohm K. M., Buechler J. and Kasper E. SIMMWIC rectennas on high-resistivity silicon and CMOS compatibility. *IEEE Trans. Microwave Theory Tech.* (Special issue on microwave circuits on silicon substrates.) **46**(5), 669–676 (May 1998).

[126]   Gardiol F. E. Design and layout of microstrip structures. *IEE Proc., Pt. H* **135**(3), 145–157 (June 1988).

[127]   Papapolymerou I., Drayton R. F. and Katechi L. P. B. Micromachined patch antennas. *IEEE Trans. Antennas Propag.* **46**(2), 275–283 (Feb. 1998).

[128]   Kamogawa K., Tokumitsu T. and Aikawa M. Multifrequency microstrip antennas using alumina-ceramic/polyimide multilayer dielectric substrate. *IEEE Trans. Microwave Theory Tech.* **44**(12), 2431–2447 (Dec. 1996).

[129]   Sorbello R. M. Advanced satellite antenna development for the 1990s. *AIAA 12th Int. Commun. Satellite Sys. Conf.*, Arlington, VA, *Collec. Tech. Pap.*, p. 5 652 (1988).

[130]   Kossiavas G. and Papienik A. A circularly or linearly polarized broadband microstrip antenna operating in L-band. *Microwave J.* **35**, 266 (May 1992).

[131]   Assailly S. *et al.* Low cost stacked circular polarized microstrip antenna. *IEEE Antennas Propag. Soc. Int. Symp.*, p. 628 (1989).

[132]   Terrei C. *et al.* Stacked microstrip antennas, advantages and drawbacks. *PIERS '91*, Cambridge, NM (1991).

[133]   Tulintseff A. N., Ali S. M. and Kong J. A. Input impedance of a probe-fed stacked circular microstrip antenna. *IEEE Trans. Antennas Propag.* **39**(3), 381 (1991).

[134]   Aberle J. T. and Pozar D. M. Phased arrays of probe-fed stacked microstrip patches. *IEEE Trans. Antennas Propag.* **42**(7), 920 (1994).

[135]   Daniel J. P. *et al.* Research on planar antennas and arrays: structures rayonnantes. *IEEE Antennas Propag. Mag.* **35**(1), 14 (1993).

[136]   Edward B. and Rees D. A broadband printed dipole with integrated balun. *Microwave J.*, **130** 339 (May 1987).

[137]   Rao S. M. *et al.* Electromagnetic radiation and scattering from finite conducting and dielectric structures: Surface/surface formulation. *IEEE Trans. Antennas Propag.* **39** (7), 1034 (1991).

[138] Wang A.T. S., Chu R.-S. and Lee K. M. Low profile, broadband, wide-scan, circular-polarized phased array radiator. *IEEE Antennas Propag. Soc. Int. Symp.*, pp. 1150–1153 (1996).

[139] Elliot P. Octave bandwidth scanning array antenna. Patent 5,293,176 (8 Mar. 1994); and Elliot P. Private communication (Nov. 1996).

[140] Buxton C. G., Stutzman W. L. and Nealy J. R. Analysis of a new wideband printed antenna element (the Foursquare) using FDTD techniques. *Digest 1998 USNC/URSI National Radio Science Meeting. Antennas: Gateways to the Global Network*, p 2, (1998).

[141] Pozar D. M. and Duffy S. M. A dual-band circularly polarized aperture-coupled stacked microstrip antenna for global positioning satellite. *IEEE Trans. Antennas Propag.*, **45**(11), 1618–1625 (Nov. 1997).

[142] Maci S. and Biffi Gentili G. Dual-frequency patch antennas. *IEEE Antennas & Propag. Mag.* **39**(6), 13–20 (Dec. 1997).

[143] Salvator C., Borselli L., Falciani A. and Maci S. Dual frequency planar antenna at S and X bands. *Electron. Lett.* **31**(20), 1706–7 (28 Sept. 1995).

[144] Rostan F., Heidrich E. and Wiesbeck W. Dual polarized multilayer aperture-coupled patch antennas for spaceborne application in C- and X-band. *IEEE Antennas Propag. Soc. Int. Symp.*, vol. 3, pp. 476–479 (1994).

[145] Tang C.-L., Chen H.-T. and Wong K.-L. Small circular microstrip antenna with dual-frequency operation. *Electron. Lett.* **33**(13), 1112–3 (19 June 1997).

[146] Chu R. S., Lee K. M. and Wang A. Analysis and design of a multiband phased arrays using multi-feed dipole elements. *IEEE Antennas Propag. Soc. Int. Symp.*, vol. 4, pp. 1814–1817 (1995).

[147] Hall P. S. Microstrip antenna array with multi-octave bandwidth. *Microwave J.*, **29**(3), 133 (Mar. 1986).

[148] Rumsey V. H. *Frequency Independent Antennas.* Academic Press, New York (1966).

[149] DuHamel R. H. Dual polarised sinous antennas. European Patent Appl. 019 8578 (19 Feb. 1986); US Patent 703, 042 (1985).

[150] Nagano H. *et al.* A spiral antenna backed by a conducting plane reflector. *IEEE Trans. Antennas Propag.* **AP-34**, 791 (1986).

[151] Wang J. J. H. and Tripp V. K. Design of multioctave spiral-mode microstrip antennas. *IEEE Trans. Antennas Propag.* **39**(3), 332 (1991).

[152] Tripp V. K., Wang J., Tillery J., Chambers C. and Hirvela G. A versatile, broadband, low-profile antenna. *IEEE Telesystems Conf.*, p. 227 (1993).

[153] Gibson P. J. The Vivaldi aerial. *Proc. 9th Eur. Microwave Conf.*, Brighton, UK, p. 101 (1979).

[154] Prasad S. N. and Mahapatra S. A novel MIC slot-line antenna. *Proc. 9th Eur. Microwave Conf.*, Brighton, UK, p. 120 (1979).

[155] Shuley N. V., Fourikis N. and Lioutas N. Analysis of slot-line endfire travelling wave antennas. *IREE (Aust.) Int. Conf.* Melbourne, Australia, pp. 576–579 (Sept. 1989).

[156] Gazit E. Improved design of the Vivaldi antenna. *IEE Proc., Part H: Microwaves, Optical Antennas* **135**, 89 (1988).

[157] Langley J. D. S., Hall P. S. and Newham P. Novel ultrawide-bandwidth Vivaldi antenna with low cross polarisation. *Electron. Lett.* **29**(23), 2004 (1993).

[158] Langley J. D. S., Hall P. S. and Newham P. Balanced antipodal Vivaldi antenna for wide bandwidth phased arrays. *IEE Proc. Microwave Antennas Propag.* **143**(2), 97–102 (Apr. 1996).

[159] Lee J. J., Loo R. Y., Livingston S., Jones V. I., Lewis J. B., Yen H-W., Tangonan G. L. and Wechsberg M. Photonic wideband array antennas. *IEEE Trans. Antennas Propag.* **43**(9), 966–982 (Sep. 1995).

[160] Lee J. J. and Livingston S. Wide band bunny-ear radiating element. *IEEE Antennas Propag. Soc. Int. Symp.* **3**, 1604–1607 (1993).

[161] Klopfenstein R.W. A transmission line taper of improved design. *Proc IRE* **44**, 31–35 (Jan. 1956).

[162] Yngvesson Y. S., Johansson J. and Kollberg E. L. Millimeter wave imaging system with an endfire receptor array. *Conf. Dig. IEEE Int. Conf. Infrared Millimeter Waves.* p. 189 (1985).

[163] Kim S. Y. and Yngvesson K. S. Characteristics of tapered slot antenna feeds and feed arrays. *IEEE Trans. Antennas Propag.* **38**(10), 1559 (1990).

[164] Fourikis N., Lioutas N. and Shuley N. V. Parametric study of the co- and cross-polarisation of tapered planar and antipodal slotline antennas. *IEE Proc., Pt. H: Microwaves, Opt. Antennas* **140**(1), 17 (1993).

[165] Chu R. S., Wang A. and Lee K. M. Analysis of wideband tapered element phased array antennas. *IEEE Antennas Propag. Soc. Int. Symp.* **1**, 489–501 (1994).

[166] Schaubert D. H. A class of E-plane scan blindness in single-polarized arrays of tapered-slot antennas with a ground plane. *IEEE Trans. Antennas Propag.* **44**(7), 954–959 (July 1996).

[167] Lee R. Q. and Simons R. N. Measured mutual coupling between linearly tapered slot antennas. *IEEE Trans. Antennas Propag.* **45**(8), 1320–1322 (Aug. 1997).

[168] Povinelli M. J. and Johnson J. A. Design and performance of wideband dual-polarized stripline notch arrays. *IEEE Antennas Propag. Soc. Int. Symp.* **1**, 200 (1988).

[169] Povinelli M. J. A planar broad-band flared microstrip slot antenna. *IEEE Trans. Antennas Propag.* **AP-35**(8), 968 (1987).

[170] Povinelli M. J. Further characteristics of wideband dual-polarized microstrip flared slot antenna. *IEEE Antennas Propag. Soc. Int. Symp.* **2**, 712 (1988).

[171] Agrawall N. P., Kumar G. and Ray K. P. Wide-band planar monopole antennas. *IEEE Trans. Antennas Propag.* **46**(2), 294–295 (Feb. 1998).

[172] Hansen R. C. Superconducting antennas. *Seventh ICAP*, London, UK, vol. 2, pp. 555–558 (1991).

[173] Hansen R. C. Superconducting antennas. *IEEE Int. Symp. Dig. Merging Technologies for the 90's*, New York, vol. 2, pp. 720–723 (1990).

[174] Dinger R. J. Some potential antenna applications of high temperature superconductors. *J. Superconduct.* **3**(3), 287–296 (1990).

[175] Richard M. A., Bhasin K. B. Gilbert C., Metzler S., Koepf G. and Claspy P. C. Performance of a four-element Ka-band high-temperature superconducting microstrip antenna. *IEEE Microwave Guided Wave Lett.* **2**(4) (April 1992).

[176] Morrow J. D., Williams J. T. and Davis M. F. A circularly polarized HTS microstrip antenna array. *IEEE Antennas Propag. Soc. Int. Symp.*, pp. 760–763 (1995).

[177] Ivrissimtzis L. P., Lancaster M. J. and Alford N. McN. A high gain YBCO antenna array with integrated feed and balun. *IEEE Trans. Appl. Superconduct.* **5**(2), 3199–3202 (June 1995).

[178] Levine E. *et al.* A study of microstrip array antennas with the feed network. *IEEE Trans. Antennas Propag.* **AP-37**(4), 426 (Apr. 1989).

[179] Pozar D. M. Characteristics of infinite arrays of subarrayed microstrip antennas. *IEEE Int. Symp. Antennas Propag.*, p. 159 (1992).

[180] Duffy S. M. and Pozar M. D. Aperture coupled microstrip subarrays. *Electron. Lett.* **30**(23), 1901 (1994).

[181] Legay H. and Shafai L. New stacked microstrip antenna with large bandwidth and high gain. *IEE Proc. Microwave Antennas Propag.* **141**(3), 199 (1994).

[182] Goffer A. P., Kam M. and Herczfeld P. R. Wide-bandwidth phased arrays using random subarraying. *Proc. 20th European Microwave Conf.*, p. 241 (1990).

[183] Haupt R. L. Reducing grating lobes due to subarray amplitude tapering. *IEEE Trans. Antennas Propag* **AP-33**(8), 846–850 (Aug. 1985).

[184] Chen C. H., Tulintseff A. and Sorbello R. M. Broadband two-layer microstrip antenna. *IEEE Int. Symp. Antennas Propag.*, p. 521 (1984).

[185] Huang J. A technique for an array to generate circular polarisation with linearly polarised elements. *IEEE Trans. Antennas Propag.* **AP-34**, 1113 (1986).

[186] Ito K., Teshirogi T. and Nishimura S. Circularly polarised antenna arrays. In *Handbook of Microstrip Antennas* (J. R. James and P. S. Hall, eds). Peter Pelegrinus, London (1989).

[187] Wu T.-K. Meader-line polarizer for arbitrary rotation of linear polarization. *IEEE Microwave Guided Wave Lett.* **4**(6), 199–201 (June 1994).

[188] Itoh K. Polarimetric integrated antennas composed of microstrip patches. In *Direct and Inverse Method in Radar Polarimetry* (W.-M. Boerner *et al.*, eds) part 2, p. 1335. Kluwer Academic, Dordrecht (1992).

[189] Haskins P. M., Hall P. S. and Dahele J. S. Polarisation-agile active patch antenna. *Electron. Lett.* **30**(2), 98 (1994).

[190] Haskins P. M. and Dahele J. S. Compact active polarisation-agile antenna using square patch. *Electron. Lett.* **31**(16), 1305–1306 (3 Aug. 1995).

[191] Haskins P. M. and Dahele J. S. Four-element varactor diode loaded polarisation-agile microstrip antenna array. *Electron. Lett.* **33**(14), 1186–1187 (3 July 1997).

[192] Roy J. S. *et al.* Circularly polarized far fields of an axially magnetized circular ferrite microstrip antenna. *Microwave Opt. Technol. Lett.* **5**(5), 228 (1992).

[193] The One Hectare Telescope (1HT) Project. http://www.usska.org/papers.html

[194] Concepts for a Large-N SKA. http://www.usska.org/papers2.html

[195] The Square Kilometer Array: Astronomy for the Next Millennium. http://www.usska.org/project.html

[196] Antenna Research. http://www.atnf.csiro.au/people/wbrouw/1kT/antennas.html

[197] Interference Mitigation for Radio Astronomy. http://www.atnf.csiro.au/SKA/intmit/

[198] Partial List of Adaptive Antenna Arrays. http://.ee.vt.edu/rertel/aa/ref.html

[199] Barnbaum C. and Bradley R. F. A new approach to interference excision in radio astronomy: real-time adaptive cancellation. *Astronom. J.* **115**, 2598–2614 (Nov. 1998).

[200] Antenna Research. http://www.atnf.csiro.au/people/wbrouw/1kT/antennas.html

[201] James G. L., Granet C., Forsyth A. R. and Sprey M. A. The new feed system for the SETI institute Woodbory site 30–m diameter Cassegrain antenna. *IEEE APS-Symp.*, pp. 1634–1637 (1997).

[202] K J. Greene and Granet C. Dielectrically-loaded horns used as antenna measurement range illuminators. *6th Aust. Symp Antennas*, Sydney (1999).

[203] Technical Papers. http://www.nfra.nl/skai/archive/technical/index.shtml

[204] Gardiol F. Comments on 'Ku-band 16 × 16 planar array with aperture-coupled microstrip-patch elements.' *IEEE Antennas Propag. Mag.* **41**, 126–127 (Feb. 1999).

4

Transmit/Receive Modules

Science is a way to teach how something gets known, what is not known, to what extent things are known (for nothing is known absolutely), how to handle doubt and uncertainty, what the rules of evidence are, how to think about things so that judgments can be made, how to distinguish truth from fraud, from show. . .

Richard Feynman (1918–1988)

In Chapter 1 we deduced that many important array parameters are defined by the front-end subsystems: the antennas, polarizers, T/R modules, and beamformers. While microstrip/printed board antennas and beamformers can be relatively inexpensive, T/R modules constitute 50% of the overall array cost and 40% and 10% of the costs are apportioned to subarray manufacturing and array integration, respectively [1]. In an era when cost is no longer an independent variable, T/R modules deserve considerable attention. Indeed, one can go as far as stating that we are now witnessing a new era of widespread applications utilizing phased arrays because affordable modules are readily available. As more cost reductions are within reach, detailed considerations of the issues related to T/R modules are warranted.

Given that the transmitter chain of a phased array defines many critical array parameters and the final power amplifiers (FPAs) of the T/R modules are their most expensive constituents, high-power transmitters/amplifiers are treated with the importance they deserve in this chapter. The programmable phase-shifters that allow the array to form one beam in any direction within the array surveillance volume are often included in the T/R modules. We shall therefore consider them in this chapter, but other aspects of beamforming are addressed in the next chapter.

4.1. THE IMPORTANT ISSUES

The array designer is often confronted with many options that seem bewildering. Here we shall attempt to untangle the many complex issues in a heuristic manner by defining the areas of broad agreement before we venture into the areas that need clarifications and/or elaborations. Entering the latter areas is often likened to entering an intellectual minefield.

One of the fundamental decisions an array designer has to take is whether to use solid state devices (SSDs) or tubes. It is widely accepted that:

- SSDs are unchallenged for low-level signal processing functions, e.g. high-speed digital circuits, control circuits, oscillators, phased shifters, buffer amplifiers, and low/medium-power amplifiers. Similarly, SSDs are eminently suitable for low-noise amplifiers (LNAs). While all SSDs require some protection from high microwave powers, the LNAs are especially vulnerable and their protection is imperative.
- Tubes are unchallenged in applications where high power generation/amplification is required at millimeter wavelengths.
- Modern tube-based LNAs exhibit resistance to high microwave powers.
- For high power generation/amplification at centimeter wavelengths, tubes and SSDs offer a diverse set of attractors.

The areas of contention are therefore limited to LNAs and FPAs or high-power oscillators. For radioastronomy arrays operating at centimeter wavelengths, the LNAs often

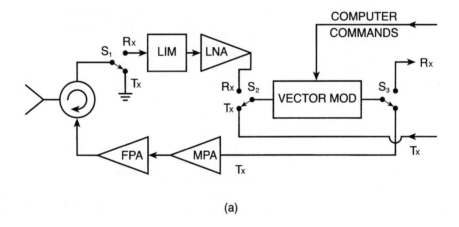

(a)

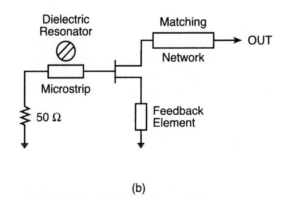

(b)

Figure 4.1 The block diagram of a typical MMIC-based module used in a radar system. The radar generates one antenna beam, which can be scanned anywhere within the array field of view. (a) The constituent parts of the module. (b) The high-stability oscillator used in the transmitter chain.

consist of cryogenically cooled SSDs. When the arrays operate at meter wavelengths, the front-ends have to be protected against lighting. The Culgoora Radioheliograph operated at 80 MHz and used miniature valves as front-end amplifiers [2].

Figure 4.1a is a block diagram of a typical T/R module for an active radar array. The SSD-based LNA is usually protected from high microwave powers by the limiter and the switch S_1 shown in the figure. As can be seen, the transmitter and receiver chains utilize the same vector modulator. A tube-based LNA contender is considered in section 4.4.3.

The additional issues the designer has to consider for the LNA are:

- Volume, weight, and cost.
- Overall insertion loss/noise temperature.
- Integration issues with the remaining constituents of the T/R module.

Figure 4.1b illustrates a dielectric resonator oscillator (DRO) often used to generate the fundamental frequency because it exhibits high spectral purity and stability. The derived signal is modulated, amplified, and buffered before it is sent to all modules. At the module, the vector modulator inserts the appropriate phase to the transmitter signal, which is in turn further amplified by efficient medium and final power amplifiers, designated as MPA and FPA, respectively. If the transmitted frequency is variable, varactor-tuned or yttrium–iron–garnet (YIG) sphere oscillators are used. The active element can be a high-electron-mobility transistor (HEMT), a heterojunction bipolar transistor (HBT), a metal–semiconductor field-effect transistor (MESFET), or a bipolar junction transistor (BJT). Descriptions of these transistors and the related manufacturing processes are found in [3,4]. Gunn and impact avalanche transit time (IMPATT) diode oscillators can be used to derive relatively high powers at microwave- and millimeter-wave bands. Efficient high-power amplifiers can be realized by using either MMICs or tubes, e.g. traveling wave tubes (TWTs). Novel realizations of these tubes in more compact packages are now available for consideration by the array designer—see section 4.4.2.

The drivers for high-power, high-efficiency, and lightweight transmitters operating at centimeter/millimeter wavelengths are military and civilian radars and the satellite-borne phased arrays that are used to usher in the wireless revolution. The designer bases the selection of a technology or a device on the following specific information:

- Knowledge of the platform's specifications: volume, weight, and power constraints.
- Frequency of operation and bandwidth requirements.
- Device output power and power-added efficiency (PAE).
- The uniformity of significant device parameters.
- Measures of transmitter signal purity, stability, and noise emission.
- Reliability expressed as the device MTBF.
- Space operation qualification.
- Overall device volume and weight (including its power supply).
- Costs that include acquisition cost and life-cycle costs (LCCs).

MMICs are the natural evolutionary products of hybrid microwave integrated circuits (MICs) that consist of discrete passive components and solid-state active devices operating at microwave frequencies. Similarly, HEMTs and HBTs are the evolutionary products of discrete FETs and MESFETs. We shall consider how well MMIC-based transmitters or LNAs meet the criteria we have defined in section 4.3.

The microwave power module (MPM), consists of a MMIC-based medium power amplifier and a high-power TWT, combines the attractors usually attributed to tubes, i.e. high output power, and those attributed to MMICs, e.g. high signal purity and stability coupled with low noise. Some designers hailed the MPM as the module for the twenty-first century [5]—see section 4.5. The emergence of high-power hybrid technology holds considerable promise for the generation of respectable power levels at centimeter/millimeter wavelengths [5,6]—see section 4.5 also. In the longer term, the designer can consider some latecomers to the field also—the vacuum microelectronic devices (VMDs) [7] that hold great promise.

The highest current density that can be drawn from thermionic cathodes used in vacuum-tube applications is about 20 A/cm^2 [8]. This compares with contacts between metals and semiconductors whereby current densities greater than 100 A/cm^2 can be injected from the metal into the conduction band of the semiconductor [7]. Although current VMDs cannot compete with available devices now, current densities of 2000 A/cm^2 from Spindt arrays have been reported [8]. Work toward the realization of VMD-based power amplifiers in the 10–100 GHz frequency range seems promising. More specifically, work toward a klystrode capable of delivering 50 W output power at 10 GHz is in progress [9]—see section 4.6 also. The monitoring of progress in VMDs is mandatory for array designers if future surprises are to be minimized.

Cost has a great impact on many aspects of the phased array. For some important military systems and novel high-risk, high-payoff systems, cost can be considered as an independent variable. This is not the case for commercial systems where low cost is used to gain competitiveness. It is not unusual, for instance, for a chief executive officer (CEO) of a high technology organization to declare that next year's product should reach increased levels of quality, cost half as much as this year's product, and enter the market after a development time that is half as short as the development time associated with this year's product. Such multidimensional constraints would seem as limiting the designer's options, but they often drive designers to innovative approaches to reach the desirable goals. Most of us are familiar with the high standards of quality and reliability attained by personal computers, sound systems, microwave ovens, and television sets; these high standards, attained at affordable prices, apply to modern commercial off-the-shelf (COTS) systems and subsystems also.

In a post-cold-war era funds for defense projects are continually shrinking and COTS subsystems are now used by the military. We shall consider the impact of this cost-minimization philosophy on the realization of affordable MMIC modules. If funds are further restricted, low-cost array architectures considered in the next chapter are adopted.

We have now established enough background knowledge to consider separately the issues we have raised.

4.2. SYSTEM ISSUES

Consider a notional 10 000-element passive array powered by one tube, often referred to as the 'bottle'. To meet the required power–aperture product, the bottle and its associated bulky power supplies generate hundreds of kilowatts of power. Bulky, high-power components like power dividers/phase-shifters are used and dangerously high power levels exist in many parts of the array. If the bottle fails, the array is inoperative and the service usually performed by the array, e.g. the defense of a high-value asset or airport management, cannot be performed while the tube is being replaced. The array does not have graceful degradation of performance, it is bulky, and unacceptably high levels of power are present in many of its parts.

In the next sections we shall delineate approaches that overcome these limitations and address other specific system issues.

4.2.1. Graceful Degradation

Instead of relying on one high-power transmitter to provide all the required power, one can have several lower-power transmitters to meet the on-target power density requirement. If N is the number of array elements, one can have either N transmitters, the case of an active array, or m transmitters each feeding n elements, so that $n \times m = N$, the case of a passive array. With these arrangements, arrays having graceful degradation of performance are realized. If some of the array transmitters fail, the array continues to perform its intended function at a reduced efficiency until the required maintenance can be performed.

The COBRA DANE system, for instance, has 96 subarrays, each having 160 active radiating elements, and each subarray is fed by a 160 kW peak power TWT [10].

We have already considered the concept of 'hot maintenance' where the failed transmitters are replaced while the array performs its assigned functions [11]. This approach to attaining high reliability applies to a variant phased array that consists of one aperture working in conjunction with an array of T/R modules. Thus, no interruptions of the array services are experienced. This capability is invaluable when an airport radar has to operate on a 24-hour-a-day and 365-day basis or when a military radar is required to perform its many functions during periods of confrontations lasting for indefinite times.

Let us formulate a measure of graceful degradation for active arrays. If N is the number of array transmitters and F is the number of failed transmitters, the resulting power loss, P, due to failures is given by

$$P = 10 \log\left(\frac{N - F}{N}\right)^2 \tag{4.1}$$

In Table 4.1 are tabulated the power loss, P, when N and F take the listed values. As can be seen, the resulting arrays have considerable tolerance to failures and the tolerance is proportional to the number of its elements.

Table 4.1 Power loss factors in dB when F transmitters out of N fail[a]

N	F									
	100	200	300	400	500	600	700	800	900	1000
1000	0.9	1.9	3	4.4	6	8	10.5	14	20	∞
10 000	–	–	–	–	–	0.5	0.6	0.7	0.8	0.9

[a]Losses lower that 0.5 dB are not listed.

As $N > m$, active arrays have a higher tolerance to failures than their passive counterparts. In Chapter 1 we considered a phased array variant in which one antenna, a
conventional paraboloid, is fed by a transmitter that consists of several MMIC-based
transmitters combined with the aid of a power combiner. The same considerations
related to graceful degradation of performance apply to this special case of phased array
as long as the efficiency of the power combiner, η, is taken into account. The prerequi-
sites for power combiners are:

- High combining efficiency.
- High isolation between transmitters/amplifier.
- Efficient heat sinking.

The first prerequisite implies good impedance matching at the output of all transmit-
ters/amplifiers and no differential phase and/or amplitude variations between the
transmitters/amplifiers used. With proper design, η can be rendered negligible [12].
Power combining has been used to combine the powers of a multitude of SSDs; 22 kW
of peak power was, for instance, generated in the 2.7–2.9 GHz frequency band by using
state-of-the-art power combiners [11]. The powers from several mini-TWTs have been
combined and the trend is destined to continue. Combining efficiencies of up to 90–95%
have been reported when four mini—TWTAs operating in the 2–8 GHz band were combined [13] and the same combiner can be used to combine the powers of SSDs. As we
have seen, some power combining can be implemented at the array level and some at the
target (e.g. [14]). Quasi-optical power combining is now a firmly established approach to
increasing the power of SSDs operating at centimeter and millimeter wavelengths [15].

The migration to active arrays resulted in the employment of low power rating components. As can be seen in Figure 4.1a, the programmable vector modulator shared
between the receiver and transmitter chains is located before the FPA and among low
power rating subsystems.

The significant advantage of active arrays when compared to their passive counterparts is their capability of generating multiple beams. The other attractors of active
phased arrays are:

- A higher tolerance to transmitters that fail.
- Lower powers are maintained throughout the array face.
- Minimal losses between the antennas and the LNAs/FPAs.
- Low power rated programmable phase-shifters and their insertion loss need not
  affect the array noise temperature.

By contrast, the phase-shifters of passive arrays are high power rated and have to exhibit
minimum insertion losses if the array noise temperature is not to be significantly degraded.

4.2.2. Self-Healing Arrays

Despite the many advantages that active phased arrays possess, there are a couple of
inherent drawbacks the designer has to address. When a large number of T/R modules
is used, the designer has to ascertain that:

(i)   the performance of all modules is monitored at all times; and

(ii)   the array performance is optimized at all times, even when some modules develop faults.

In the context of multifunction arrays, these issues assume a profound importance because one multifunction array substitutes several arrays that afford some redundancy to the high-value platform. If the multifunction array does not perform its assigned functions efficiently, in harsh weather and electronic environments, the high-value platform has a diminished defense capability.

Fault monitoring and correction schemes for operational phased arrays have been proposed in references [16] and [17]. The most common faults are due to phase-shifter settings and the proposed schemes isolate the faulty modules and readjust the complex weights of the remaining operating elements to compensate for the faulty module(s). At the appropriate time the faulty modules are substituted with fully tested modules.

4.2.3. The Implementation of Amplitude Tapers

Array sidelobes can be minimized by tightly controlling the amplitude/phase errors occurring in the different parts of the array and by amplitude/spatial tapers across the array. When an amplitude taper is required, the issue of importance is to maintain a high PAE for the module FPA while its output power is varied according to the required taper function. We have already considered one cost-effective approach to meeting the requirement by the manufacture of amplifiers capable of delivering 5–6 different power levels. With this approach the required amplitude taper is approximated and optimum PAEs are attained for all FPAs.

The following two approaches have been suggested by I. Bahl [18]. The first approach requires good power devices designed for class-B operation and design skills to maintain high PAE as the device delivers decreased powers from its rated output power. The other approach has been demonstrated but not used (early 1998 status). Most transistors, e.g. FETs, HEMTs, and HBTs, behave linearly when the supply voltage is decreased from the designed level. In an array environment, several power supply voltages will be required. A 5–6 dB variation in power can result when power supply voltages of 5, 6, 7, 8, 9, and 10 V are available. This approach meets the requirement without sacrificing the device PAE. Both approaches increase costs, however. Given the importance of phased arrays, low-cost solutions to this requirement are expected in the not too distant future [18]. Until then, the approaches suggested in Chapter 2 will be adopted.

In Chapter 2 we noted that an array might require amplitude tapers that vary in a dynamic manner. If the FPA consists of several amplifiers connected to a power combiner, different array amplitude tapers can be attained [14]. With this arrangement, some amplifiers in power combiners associated with antenna elements located some distance away from the array center are disabled for the implementation of the required taper.

4.2.4. Performance at a Cost

If three or four apertures are needed for a radar phased array to have a hemispherical coverage, the costs attributed to T/R modules increase proportionally. Keizer [19] proposed a phased array radar in which the transmitter chain is shared by the four apertures of the system. Given that each aperture of the system transmits for a fraction of the PRF, this proposal has considerable merit because it substantially reduces the overall array costs.

While MMIC-based modules have several attractors, there are several cost penalties associated with their manufacture that we shall consider in the next section. We have already noted that cost considerations are of pivotal importance for a variety of phased array applications and approaches to minimizing these costs are explored in section 4.3.1.

Photonics technology offers several attractions to array designers and references [20] to [23], aimed at closing the knowledge gap between array designers and photonics experts, accelerated the adoption of photonic technology into phased arrays. The most obvious application of photonics is in the area of signal distribution to and from the array face. While the radiating face of conventional arrays looks elegant, a bulky and heavy maze of coaxial cables and power lines dominates the back view of phased arrays. Optical fibers can substitute the coaxial cables to carry RF/digital signals to and from the T/R modules. Apart from the obvious weight and volume advantage, the fibers offer immunity to EM pulse (EMP) and EM interference (EMI). Significant R&D related to the distribution of RF/digital signals between communication nodes carried by the communications community is readily available to the phased array community; this being the case, we shall not consider this aspect of R&D further.

4.3. MMIC-BASED MODULES

We have already considered the typical T/R module, shown in Figure 4.1a, often used in radar arrays to generate one inertialess beam that can be scanned with the aid of programmable vector modulators. The module may include a microprocessor, not shown in the figure, that is part of the array's distributed logic. The microprocessor at each module is set according to the module position in the array and the signals sent by the computer are simply related to the required array scan angles.

Referring to Figure 4.1a, the SPDT switches S_1, S_2, and S_3 are synchronized to allow the module to either transmit or receive radiation. The same vector modulator and settings are used when the module operates in the transmit or receive mode. Other more sophisticated realizations of the T/R module include polarization processing and system-level integrated circuits (SLICs) that facilitate reliable operation, improved performance, and ready application [24]. If the beamforming is implemented at the IF of the system or at optical wavelengths, suitable frequency translations have to be implemented at the module level.

Hybrid microwave integrated circuits (MICs) have been used since the mid 1970s while monolithic MICs (MMICs) have been used since the early 1980s. The relative merits of MMICs over their hybrid counterparts are mainly in the areas of volume, weight, high-volume production capability, reproducibility, and reliability. Conservatively, MMIC-based subsystems have 5–0 times the advantage in size and weight over their hybrid counterparts [3]. The total chip area for typical MMIC-based modules operating at C-band when the power output is 15 or 8 W is 100 or 75 mm² [3]. These are indicative chip areas that serve the purpose of establishing approximate baseline cases.

The attraction to MMIC-based T/R modules is almost irresistible because they possess the following characteristics:

- Essential module parameters such as phase and amplitude are tightly controlled by the manufacturing processes. This is a fundamental requirement for active phased arrays having thousands of antenna elements since it often reduces the array sidelobes. The same characteristic is vital for the monopulse tracking of targets. As the manufacturing knowledge base increases and deepens, the control of these module parameters becomes tighter.
- Minimal space occupation, an attribute that is particularly important if operation at millimeter wavelengths is required.
- Superior reliability, long life, and light weight. Here the comparisons are made not only with hybrid MICs but also with tubes.
- Production costs decrease as
  (a) the overall yield of modules increases, which in turn is closely dependent on the breadth and depth of knowledge related to the manufacturing processes;
  (b) the integration of the many module functions increases; and
  (c) the number of T/R modules required for a production run increases.

For some time the conventional package for MMICs was the small outline integrated circuit (SOIC) shown in Figure 4.2a. The chip faces up, i.e. away from the substrate, while its connections to the outside world are mounted onto the substrate. In the flip chip package, developed by IBM, shown in Figure 4.2b, the chip faces the substrate and connections to the outside world are implemented in a variety of ways with the aid of flawless solder bumps, tape automated bonding, compliant bumps, or pressure contacts. Compared to the conventional packages, flip chips are 25% smaller and 22% lighter. More importantly, their electrical performance is enhanced because their leads to the

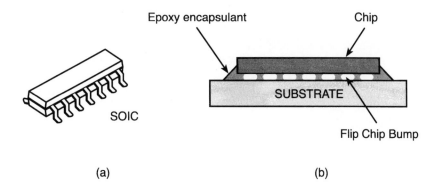

(a)   (b)

Figure 4.2 Two popular packages for MMICs. (a) The small outline integrated circuit (SOIC). (b) The flip chip.

outside world are shorter while assembly yields are excellent. The flip chip package usually requires the use of coplanar waveguide transmission lines on the mounting surface. Unfortunately, model libraries for coplanar waveguide and flip chip mounted transistors are not as well developed as the corresponding libraries for microstrip circuit elements and the more conventional packages. As long as this deficiency exists, it will inhibit the widespread use of flip chip packages for MMICs [25].

Table 4.2 lists the essential physical and electrical parameters of silicon and GaAs, the two substrates materials commonly used to realize MMICs. As can be seen, GaAs exhibits very good radiation hardness when compared to silicon. Radiation hardness against the hazards encountered in space is gaining importance in satellite-borne phased arrays designated to support the wireless revolution. Recent investigations on the radiation hardness of SiGe HBTs seem promising but more work is required to explore the many aspects of this issue [26]—see section 4.3.1.5 for other aspects related to this material.

Although the relative permittivities and dielecric loss factors of the two substrates are comparable, their thermal conductivities differ substantially, so the silicon substrate is more suitable for dissipating thermal power generated by power devices.

Table 4.2 The essential electrical and physical parameters of the commonly used substrates silicon and GaAs

Parameters	Silicon	GaAs
Semi-insulating	No	Yes
Ease of handling	Very good	Good $\Rightarrow$ very good[a]
Radiation hardness	Poor	Very good
Relative permittivity	11.7	12.9
Dielectric loss factor (at 90 GHz)	1.3×10^{-3}	0.7×10^{-3}
Thermal conductivity (W cm^{-1} K^{-1})	1.45	0.45
Electron mobility (cm^2/V^s)	700	4300
High-field drift velocity (cm/s)	10^7	6×10^6
Resistivity (Ω cm)	10^3–10^5	10^7–10^9
Density (g/cm^3)	2.33	5.32

[a] Post Microwave and Millimeter Integrated Circuit (MIMIC) Program environment — see section 4.3.1.1.

High-frequency operation of SSDs depends on their electron mobility and high-field drift velocity. Although the latter parameter is comparable for the two substrates, the electron mobility of GaAs is six times that of silicon GaAs-based MMICs are now considered eminently suitable for LNAs and FPAs at centimeter and millimeter wavelengths. There is, however, room for new SSDs to challenge these established devices.

The SiC MESFET and GaN MODFET are the newcomers to the field of high-power SSDs. The former demonstrated outstanding performance at X-band and below, while the latter demonstrated K-band performance comparable to the GaAs pseudomorphic PHEMT ([27 and references therein). Though this technology seems promising, a great deal of fundamental work needs to be done before it can be considered for satellite applications.

The design, testing, and fabrication costs of MMICs is high for the following reasons:

- High up-front costs for facilities that use GaAs as the chip material.
- High recurring costs associated with the materials used, testing at several production stages, module assembly, and packaging [25].
- High non-recurring engineering (NRE) costs that result because the designers cannot fine-tune the MMICs to optimize their performance. As a consequence, it is generally difficult to achieve spec-compliance with one fabrication run.
- Long lead times that include the design of a new module, fabrication runs, and testing at many stages of the production run.

In what follows we shall delineate various approaches taken to minimize module costs.

4.3.1. Cost Minimization Approaches

Module cost minimization is a multidimensional problem that can be best appreciated by considering the many aspects of module manufacture, e.g. the quality of the materials used, methods of maximizing their effective use, processing yield, tests undertaken at different production phases, packaging, and the implementation of quality assurance for all processes and practices.

Reference [25] outlines trends in the development of MMICs and packages for active electronically scanned arrays (AESAs). Naturally, cost minimization approaches occupy a central focus area. In the same reference it is accepted that:

- The cost of GaAs MMICs has been reduced by more than an order of magnitude in the period between 1988 and 1996.
- Modules for the Ground Based Radar (GBR) system, the largest MMIC based array produced today, are being manufactured at the end of the module production cycle for significantly less than US$1000 each.
- Assuming continued progress in the areas outlined in the paper, it is anticipated that the total cost of a typical T/R module operating over the 7–11 GHz frequency range with 10 W of power output and 25–30% PAE could be reduced to approximately US$200 in production volumes greater than 100 000.

As can be seen, significant progress in module cost reduction has been made and further cost reductions are attainable in the near future. One of the major thrusts for affordable T/R modules is the Microwave and Millimeter Integrated Circuit (MIMIC) Program.

4.3.1.1. The MIMIC Program

In production, silicon is preferred to GaAs owing to low costs, established production lines, and the high standard of technological experience. However, the funding of the MIMIC program to the tune of half a billion US dollars by the Advanced Research Program Agency (ARPA)/Tri-Service in the United States resulted in a significant increase and widening of the knowledge base of processes related to the manufacture of GaAs MMICs [28,29]. More specifically, the aim of the program, which lasted from 1987 to 1995, provided resources and manufacturing infrastructure in order to transition GaAs IC technology from a research and development status to a mature technology status.

European countries and Japan also supported similar programs and goals. As a result of these programs, the manufacture of GaAs MMICs is now well-established in many countries. Other programs we shall consider in the following sections are aimed toward specific problem areas in the manufacture of MMIC-based modules. It is important to appreciate here that solutions of the many associated problems depends critically on continued research thrusts over a relatively long period.

4.3.1.2. End-to-End Chip and Assembly Yields

The evolution of the high-yield affordable module presents a raft of challenging inter-disciplinary problems. For a start, the module has a large number of active devices and passive components, and EM phenomena that can be neglected when the ICs operate at lower frequencies have to be considered. Novel fabrication methodologies are required to meet the dual challenges of attaining high manufacturing yields and the tight control of the important module parameters.

Historically, the quest for high-yield MMICs attracted a high priority because (i) the costs attributed to MMICs constitute a high proportion of module costs; and (ii) typical MMIC yields were 10% in 1982 [30]. In what follows we shall outline the various research thrusts that resulted in high yield MMICs.

The recessed gate process is most commonly used to fabricate MESFET MMICs, while the self-aligned gate (SAG) process is used to process efficiently devices optimized for different circuits, e.g. digital, microwave small-signal, and power on the same wafer at the same time. One particular embodiment of the SAG process, the multifunction SAG (MSAG) process, results in high-yield MMICs [3]. This fabrication process has been applied to MESFETs used in low-noise and power amplifiers with great success [31]. Reference [32] reported excellent results using the MSAG processes to attain high-yield, low-cost GaAs MMICs.

Design guidelines for MMICs have been assembled and are based on the premises that there are considerable variations in the many transistor parameters and the sheet resistance within a wafer and from wafer to wafer. These variations determine the process windows for the derived MMICs. In reference [33] the assembled design guidelines are outlined and several designers have their sets of guidelines. The assembled guidelines are often incorporated into computer-aided design/manufacturing, (CAD/CAM) packages.

Several approaches to maximizing the yield and minimizing the cost of MMICs have been reported [31,34–40]. The improvements in yields resulted from

- the use of design-technological process optimization [31];
- better materials and processes [34 and 5.35]; and
- comprehensive test procedures at critical manufacturing stages.

As MMICs contain a multitude of passive components, the work related to high-yield passive components [34] is significant. The yields attained range from 80% to 100% for passive components and from 30% to 90% for MMICs. Yields are by and large inversely proportional to the chip area occupied by the MMIC and to increased complexity. The yields of a LNA and a vector modulator are, for instance, 70% and 30%, while their chip areas are 2.73 and 13.7 mm^2, respectively [41].

Recently, low-noise and ultra-high-speed digital ICs have been fabricated using the InAlAs/InGaAs/InP HEMT process [42]. The process assures uniformity by taking advantage of a 0.1 μm T-shaped gate and an InP recess-etch stopper. For the low-noise amplifiers operating at 62 GHz, and the ultra-high-speed digital circuits having a toggle frequency of 36.7 GHz, the corresponding fabrication yields are 75% and 63%, respectively. Apart from high yield, the process is ideal for applications where uniformity and stable operation are also important. The latter fabrication process as well as the MSAG process can be used for analog and digital ICs; given the ever-increasing integration of analog and digital circuits, this last characteristic is important.

There are basically two design philosophies for the realization of affordable T/R modules:

(i)   The integration of as many module functions as possible onto one chip. From a manufacturing engineering point of view, this is a valid approach that is adopted in many systems.
(ii) The multichip module (MCM) consisting of several chips that are appropriately interconnected.

In 1987 a single-chip, X-band T/R module was heralded as a major cost saver for phased arrays [43]. Its total area was 58.5 mm^2, half of which was occupied by the phase-shifter, and the average chip yield was only 14%. Since that time, affordable, efficient, high-yield modules having one or two chips have been mass produced [44].

Compared to the MCM, the high integration modules have significantly fewer parts, die attach operations, and wire/ribbon bonds, and require fewer RF tests [44]. This set of advantages contributed to the affordability of T/R modules. There is a widespread consensus that high-level integration is the key to lower module costs, and programs such as the High-density Microwave Packaging (HDMP) Program contribute toward the lowering of the T/R module costs [45].

4.3.1.3. The HDMP Program

The objective of the HDMP Program is to develop and establish the reproducibility of complex shape, lightweight, and high-density T/R microwave modules [45]. The focus of the program is for systems designated to operate in the 7–11 GHz frequency band on spaceborne/airborne platforms such as the next generation of F-22 and comparable Navy aircraft and Army helicopters. The program can be seen as the natural progression to the quest for ever-increasing integration.

The more specific aims of the HDMP program are [45]:

• Cost reduction of the high-density and lightweight T/R module by at least an order of magnitude without sacrifice of the required performance levels. A target cost estimate less than US$200 per channel is set.
• The extensive use of computer-aided design, manufacturing, and testing capabilities to attain higher yields that often result in lower costs.
• The incorporation of advanced technologies and packaging techniques in order to meet system requirements and module reliability and hermiticity.

While the three contractor teams, [46–48] adopted different approaches to meet the requirements of the HDMP Program, all use multichip assemblies (MCAs) that consist of subarrays of modules fabricated simultaneously and housed within a single package. All contractor teams use the tile architecture and are developing 2×2 MCAs, but it is expected that the MCA assemblies will be extended to 4×4 and 6×6 configurations in the future. Interestingly, GaAs and silicon chips are used in their packaging technology, the weight of each element is about 20% of a conventional element, and one MCA has a volume of ~5 cm^3 (0.3 cubic inches: 1 in $\times$ 1 in $\times$ 0.3 in) [46].

The Hughes Aircraft Company estimated that the 1995 cost of a module, between US$350 and US$400, can be reduced to just over US$100 on the assumptions that 7000 modules are produced and their 3-D packaging technology is used [46]. These are extremely encouraging cost projections for the widespread use of phased arrays.

Historically, a key ingredient to understanding the issues in hand is continuous learning and improvement. While the quality of wafers was barely adequate and production engineers could not tightly control the many processes, it was economical to realize T/R modules consisting of several small chips that had a reasonable yield. As the quality of wafers increased and methods of tightly controlling the many processes involved were derived it became economical to produce larger chips.

At a more fundamental level Cohen [49] suggests the realization of a computer-integrated manufacturing facility to perform array architecture and design trade-off studies. From the foregoing considerations, it is not hard to appreciate that work on important and challenging phased array problems has just begun.

4.3.1.4. The Multichip Module

The other design philosophy advocates that the module assembly contains the antenna and a small number of chips that perform the many module functions. We have already seen that several microstrip antennas afford the designer the freedom to integrate the antenna to the MMIC-based module. Before the recent availability of reliable interconnects between chips [50], this approach lowered the module reliability; now multichip modules (MCMs), reliably interconnected, are common.

This design philosophy allows designers:

- to use a finite set of fully tested, 'standard' chips to realize a variety of systems [51], a very attractive proposition; and
- to design, realize, and test T/R modules over short intervals of time.

Typically, a prototype dual-band multichip receive module was designed, fabricated, and tested within a period of six months at a cost of US$950 [51]. This design philosophy is particularly attractive for small research groups residing in universities or research institutes since it allows them to explore important aspects of research in phased arrays without any access to IC design and fabrication facilities. It is commonly acknowledged that diversity in research is a basic ingredient of success.

While the average module for radar phased arrays consists of 5–8 chips, an EW module was reported to have 72 MMICs [52].

Both design philosophies are valid and allow designers coming from diverse backgrounds to realize tailor-made modules.

4.3.1.5. Other Important Approaches, Initiatives and Trends

Now that chip yields are high, considerable effort is focused in the following areas that also significantly affect costs [3]:

- Accurate models and integrated 'first-pass design' computer-aided design tools.
- On-wafer integrated automatic testing and prescreening of power ICs.
- Single-chip and multichip packages for high-performance, low-cost applications.
- Automatic assembly of chips into packages and modules.

and

- The development and extensive use of computer-aided manufacturing and testing tools.
- The evolution of novel packaging technologies, e.g. flip chips; we have already noted that this packaging technology results in high assembly yields.
- The development of improved array components such as circulators, power supplies, and capacitors that are required to meet overall array requirements undertaken under the aegis of the Microwave and Analog Front End Technology (MAFET) Program [49].
- A possible alternative material to GaAs is Si–Ge. If adequate performance margins can be achieved, Si–Ge offers the potential for producing MMICs that can be an order of magnitude lower in cost than those made from GaAs [25].
- Additional approaches to reducing the recurring cost and NRE costs associated with module production are delineated in reference [25].

All these programs and approaches ensure that affordable modules will accelerate the adoption of phased arrays in many systems.

For very large quantities, the manufacturing cost of chips depends on their geometric area. For small chips between 1 and 5 mm^2, the cost is less than 1 \$US/mm^2. For larger chips between 5 and 50 mm^2, the cost is several US\$/mm^2 [3]. The complexity of the chip often defines the exact costs. The typical module cost can be broken down into the following parts: 30% for an MMIC consisting of eight chips; 25% parts; 15% testing; 15% assembly; and 15% controller [3].

4.3.2. High-Stability and High-Purity Oscillators

For fixed-frequency operation a transistor is often used in conjunction with a frequency-determining element such as a dielectric resonator oscillator (DRO) or a surface acoustic wave (SAW) cavity resonator. DROs provide stable operation from 1 to 100 GHz as fixed-frequency oscillators. Additionally, they are simple to design, have high efficiency, and are compatible with MMIC technology. For variable-frequency operation, a varactor or a YIG sphere is used as the frequency-determining element.

Transistor oscillators provide reliable, low-noise performance up to millimeter wavelengths, while Gunn and IMPATT diode oscillators provide higher power levels at millimeter wavelengths. We have already noted that a variety of transistors can be used. Compared to a GaAs MESFET oscillator, a BJT- or HBT-based oscillator has 6–10 dB lower phase noise very close to the carrier frequency [3]. Higher powers are simply attained by connecting a buffer amplifier and medium- and high-power amplifiers to the oscillator. The buffer amplifier minimizes frequency pulling as it isolates the oscillator from its load.

In Table 4.3 the typical levels of the single-sideband (SSB) phase noise are listed for DRO stabilized oscillators operating at C-, X- and Ku-bands. The spectral performance of oscillators operating at X-band is shown in Figure 4.3 [53]. The phase noise sideband level as a function of frequency offset from the carrier frequency is shown for two different oscillators. Oscillator (1) is a DRO oscillator operating at X-band with a resonator loaded Q of 5000. Oscillator (2) consists of a crystal oscillator operating at a fundamental frequency of 100 MHz and a multiplier; the resonator loaded Q is 75 000. As can be seen, the phase noise of oscillator (2) is lower than that attributed to oscillator

Table 4.3 Essential parameters of DRO stabilized oscillators [3]

Parameters	C-band	X-band	Ku-band
Frequency (GHz)	5.027	10.740	13.120
Output power (dBm)	+12	+16	+10
SSB phase noise (dBc/Hz)			
100 kHz	−115	−110	−100
10 kHz	−88	−80	−70
Frequency stability[a] (ppm/°C)	±2	±4	±5
Power stability[a] (dB)	±0.75	±1.0	±0.5

[a] Measured from −54 to +85°C.

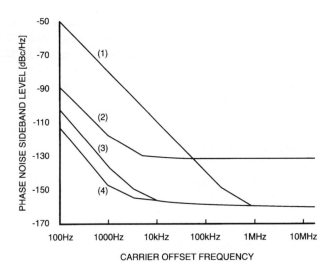

Figure 4.3 The spectral performance of oscillators operating at X-band. (1) DRO-based fundamental oscillator exhibiting a loaded Q of 5000. (2) Crystal oscillator at 100 MHz has a loaded Q of 75 000 is followed by a frequency multiplier chain. (3) Predicted phase-noise performance of an oscillator attainable through the use of an ultra-high-Q microwave resonator; unloaded Q = 300 000. (4) Same as in (3) but unloaded Q = 1.5×10^6. (From [53]; © 1992, IEEE.)

(1) at frequencies close to the carrier frequency and the reverse is true at frequencies above 80 kHz. The noise floor of the crystal-based oscillator is attributed to the multiplication process. The predicted phase noise performance attainable through the use of ultra-high-Q microwave oscillators is illustrated by the curves labeled (3) and (4), where the loaded Qs are equal to 300 000 and 1.5×10^6, respectively.

Considerable improvements in the spectral performance of DROs are therefore attainable when the loaded Q increases significantly. In reference [53] it is shown that with high-quality, large-area $YBa_2Cu_3O_7$ (YBCO) films and simple styrofoam mounts, unloaded X-band Q values of over 250 000 are attainable when a sapphire dielectric resonator operated at 77 K [53]. These resonators can be used to realize high-purity and high-stability oscillators required in future phased array systems.

4.3.3. Typical Low-Noise Front-Ends and Power Amplifiers

When compared to parametric amplifiers, tube amplifiers, and masers, solid-state front-ends are preferred because they exhibit excellent electrical characteristics at room and at cryogenic temperatures, have small volume and low-complexity, and are lightweight. Maser amplifiers, for instance, require elaborate 4 K cryogenic systems; by contrast, SSDs yield ultra-low-noise figures when cooled to 20 K. While tubes and parametric amplifiers exhibit unacceptable complexity and bulk, most tubes can withstand high microwave power without damage—see section 4.4.3.

Hybrid LNAs exhibit the lowest noise figure while MMIC-based LNAs, are compact, and can be inexpensively mass-produced. Cryogenically cooled hybrid LNAs and mixers followed by MMIC-based LNAs are therefore used in applications such as radioastronomy where the lowest possible noise figure is of paramount importance. The noise figure versus frequency characteristics of typical MMIC LNAs available from different manufacturers are shown in Figure 4.4a [27]. Figure 4.4b shows the noise temperature attained by subsystems utilizing InP HEMT-based hybrid amplifiers and SiS mixers followed by LNAs operating at an IF as a function of frequency and when the devices are cooled to 20 K [54]. These are the typical LNAs used in radioastronomy applications. As can be seen, SSDs operating at room temperature or at cryogenic temperatures have excellent low-noise characteristics at centimeter/millimeter wavelengths in addition to the other desirable characteristics we have already delineated.

A limiter or protector is used either to attenuate the incoming signals or to blank the receiver when the radar transmits. Given that the limiter is located before the LNA, its insertion loss should be minimal. PIN-diodes are commonly used in conjunction with a quarter-wave transmission line to perform the limiting function [55]. Minimum losses of the order of 0.7 dB [55] are incurred with the single-ended realization over a relatively narrow band. PIN-diode limiters fabricated using molecular beam epitaxy (MBE)-grown layers can provide less than 0.2 dB small-signal insertion loss and greater than 15 dB limiting at 10 GHz [56,57].

A variable attenuator-limiter utilizing planar PIN-diode fabrication in GaAs has been developed at X- and Ka-bands; the limiter exhibited 20 dB of variable attenuation and its insertion loss at Ka-band was typically 1.4 dB [56,58]. As can be seen, the protection is not absolute for all power levels transmitted by a jammer and the limiter's insertion loss has to be taken into account in the calculations of the system's noise figure.

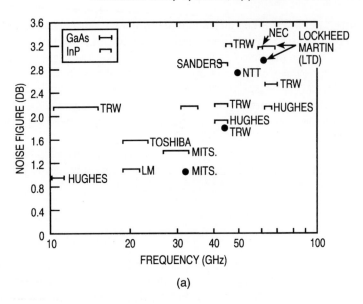

(a)

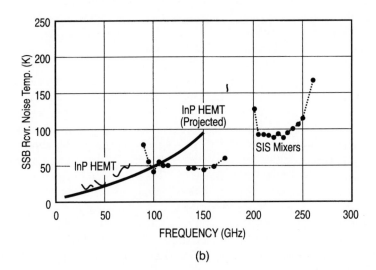

(b)

Figure 4.4 Noise figure/single-sideband (SSB) receiver noise temperature versus frequency attained from low-noise subsystems. (a) Typical MMIC LNAs available from different manufacturers. (From [27]; © 1998, IEEE). (b) Cryogenically cooled low-noise amplifiers and mixers. (From [54]; © 1995, IEEE.)

The electrical parameters of importance for solid-state radar transmitters are:

- High pulsed power output and PAE.
- Phase, frequency, and amplitude stability.
- Linearity of transmission phase with input power up to the 1 dB power compression point [3].

- The maintenance of the channel temperature below 150°C for good stability and long life by the provision of adequate cooling arrangements.

We can discern the following approaches to attaining high powers using SSDs:

- Three-terminal power amplifiers operating at centimeter and millimeter wavelengths.
- Two-terminal oscillators operating at millimeter wavelengths.
- Conventional and spatial power combining of the output of several amplifiers.

MMIC amplifiers operate in Class A, AB, and B operation depending on the active devices used and the frequency of operation.

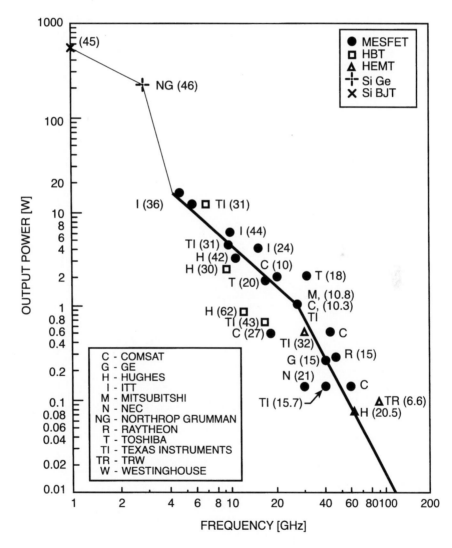

Figure 4.5 Power output versus frequency of typical MMICs [3,59,60]. The corresponding device PAE as a percentage is given in brackets.

In Figure 4.5 [3,59,60] the power output of single-chip MMICs as a function of frequency and device PAE is shown. Each entry is defined by the manufacturer and the PAE is given in brackets. Different types of SSDs are used at different bands and new SSDs often displace the traditional contenders with the passage of time. With reference to Figure 4.5, the silicon bipolar junction transistors (BJTs) yielded the highest power at frequencies below 5 GHz in the recent past; now SiGe HBTs [60] and silicon carbide transistors [61] contest the same frequency range. MESFETs, HBTs, and HEMTs dominate the frequency range above 5 GHz. As the frequency of operation increases,

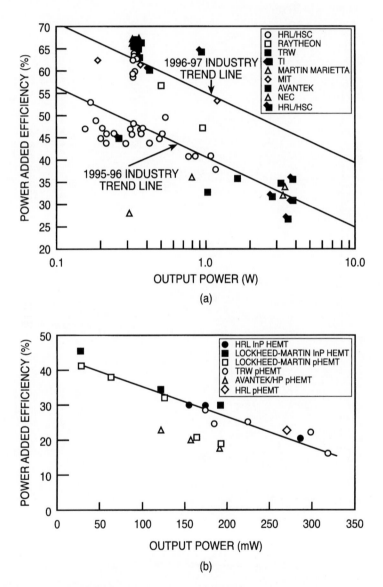

Figure 4.6 Power output versus device PAE of several MMICs operating at (a) K-band and (b) at V-band. (From [27]; © 1998, IEEE.)

however, the device power output and PAE fall. For a given device, the maximum power output is not attained when the PAE is maximized. The newer and less mature GaN MODFET has demonstrated K-band performance similar to that attained by the other more established transistors [27].

The output power versus the PAE of SSDs at the K- and V-bands is shown in Figures 4.6a and b, respectively [27]. Of considerable importance is the industry trend line for the 1996–97 period shown in Figure 4.6a; the improvements in device power output and PAE attained over one year are significant. The requirements imposed by the phased arrays on board the many constellations of satellites ushering in the wireless revolution will act as a catalyst for the realization of SSDs having higher powers and PAEs at K-band and at higher frequency bands. In the same reference the typical performance figures for Q-band power amplifiers are given; the boundary coordinates of device power output and PAE are 1 W and 38%, respectively.

As can be deduced, the power output of single-chip power amplifiers at millimeter wavelengths is minute. Two terminal devices such as Gunn and IMPATT diodes operating as oscillators at millimeter wavelengths yield power outputs below 3 W and have maximum PAEs of 22% [62].

Table 4.4 High-power narrowband systems/MMICs and wide-bandwidth MMIC modules

Entry no.	Frequency [GHz]/ Frequency band	Power output [W]	PAE [%]	Reference
1	0.42–0.45	850 000[a,b] BMEWS		[63]
2	1.2–1.4	2000[b]	45–55	[59]
3	1.2–1.35	28 000[a,b] RAMP		[63]
4	2.7–2.9	22 000[b]		[11]
5	L- or S-band	40	64	[64]
6	2.8 (SiGe transistor)	200	46	[60]
7	C-band	55	26	[65]
8	C-band	12	36	[3]
9	C-band	1.7	70	[3]
10	X-band	6	44	[3]
11	X-band[c]	8	>45	[66]
12	X-band[d]	5	32–36	[67]
13	9.5–9.9[e]	70	>30	[68]
14	10.8–11.6	16		[69]
15	30	2		[3]
16	34.5	6	24	[70]
17	7–11	5	44.4 average	[71]
18	6–18	5.1 (pulsed)	10–30[f] 12–27[g]	[72]
19	58–65	1–50 W	22 and 16[h]	[73]

[a] Peak power.
[b] Power addition at the module level.
[c] Hybrid amplifier.
[d] Large volume production amplifier; bandwidth of 28%.
[e] Power addition of six MMIC amplifiers.
[f] The PAE of the single-ended stage.
[g] The PAE of the balanced stage.
[h] Correspond to the elements of the building block.

Several approaches are used to either increase the bandwidth of operation of a module or boost the output power of a system with the aid of conventional power combining techniques on or off the chip. Table 4.4 lists a representative sample of high-power narrowband systems/MMICs and wide-bandwidth MMIC modules. RAMP, the acronym used in the table, stands for the radar modernization program. The first 16 entries correspond to narrowband systems/modules, while the last three entries correspond to typical wideband modules. The physical size of the HBT MMIC power amplifier, in entry 17, is 3.0 mm × 3.4 mm, a remarkable achievement for the power it delivers.

The last two entries require some elaboration. Entry 18 corresponds to an EW amplifier and consists of two 2.5 W channels, realized on the same chip, and an external combiner that combines the output powers from the two channels. Each channel consists of one traveling wave amplifier, two single-ended amplifiers, two balanced amplifiers, on-chip power combiners and matching circuits. The PAE efficiencies listed for entry 18 (10–30% and 12–27%) correspond to the single-ended and balanced amplifiers, respectively. These PAEs should be considered as preliminary and more work is under way to improve them [72]. The amplifier represents the state of the art for EW solid-state power amplifiers.

The final entry corresponds to a chip set developed as a building block for amplifiers that can deliver 1–50 W at V-band; again the chip set represents state of the art for power amplifiers operating at V-band.

As can be seen, several SSDs can be used to realize either narrowband high-power amplifiers or wideband medium-power amplifiers. At the low frequency end, the number of amplifiers that can be combined before the combining losses become significant is very high. The number of modules used for the BMEWS radar, for instance, is 2500 [63], while the number of wideband, high-power amplifiers that can be combined is at best 16 at the high frequency end [72]. Currently this limit is imposed by transmission line losses and the stringent manufacturing tolerance requirements associated with the building blocks used. The limit should be seen as a baseline figure that can be increased if these issues are addressed.

Spatial or quasi-optical power combining is often used to increase the power available from one SSD, and reference [15] outlines the work undertaken in this area.

4.3.4. Programmable Phase-Shifters/Vector Modulators

In this section we shall consider a narrowband phased array designated to yield one antenna beam that can be scanned with the aid of programmable phase-shifters/vector modulators. For each T/R module a vector modulator provides the complex weights to each antenna element, so that all antenna elements are pointed to the required scan angles. Given that the array meets the requirements outlined in section 2.2.1.2, phase shifts are used that range from 0° to 360° in increments that depend on the required phase tracking accuracy.

In an analog phase-shifter a set of voltage settings is used to obtain the required phase shifts. Specific phase shifts, e.g. 0°, 22.5°, 45°, etc., are obtained when a digital phase-shifter is used. Both types of phase-shifter realizations are usually monolithic. Often the phase and amplitude errors introduced by the module are largely determined by the specifications of the programmable vector modulators. Given the importance of active arrays, the phase-shifters used in passive arrays are treated only briefly here; as active arrays utilize digital phase-shifters or vector modulators, our coverage of analog phase-shifters is also brief.

The requirements for phase-shifters/vector modulators are:

- Phase/amplitude accuracy over the system bandwidth and the range of phase shifts.
- Minimum amplitude variations between the many phase-states.
- Minimal volume, weight, insertion loss, and power consumption.

Given that the phase-shifters of passive phased arrays are located behind the antenna elements, the minimum insertion loss requirement is of paramount importance, because it decidedly affects the system noise figure. The same requirement is not as important for active arrays because amplifiers placed prior to the phase-shifters ensure that the system noise figure remains unaffected. It is, however, a requirement to minimize the insertion loss of phase-shifters so that the gain requirements of the amplifiers used to offset these losses are not too demanding or costly.

The power rating of phase-shifters used by passive phased arrays should be appropriate for the high powers transmitted. Additionally, a requirement exists for the phase-shifter settings to be repeatable and resettable. Often temperature rises can affect these two important phase-shifter characteristics. The temperature rises can be due to the ambient temperature and/or to the temperature rise caused by the transmission of high average RF power. If the same phase-shifter is used on both the transmit and receive paths, it ought to be reciprocal.

Ferrite phase-shifters are often used in passive phased arrays. Their defining characteristics are considerable bulk, low insertion loss, and substantial power-handling capability. Useful descriptions of ferrite phase-shifters are included in reference [74] and the critical parameters of currently available ferrite phase-shifters are given in reference [75]. Compared to analog and digital phase-shifters, ferrite phase-shifters have the lowest insertion loss and can handle the highest power; analog solid-state-based phase-shifters have comparable or slightly higher insertion loss and can handle significantly lower powers, but are smaller in volume. Digital phase-shifters used in conjunction with active arrays have the highest insertion loss.

Analog phase-shifters utilize capacitors, lumped capacitors, and Lange couplers in conjunction with MMIC compatible Schottky barrier or varactor diodes. The same phase-shifters can have wide bandwidths and their insertion loss ranges from 1.5 to 3 dB depending on the frequency of operation ([e.g. [76]).

4.3.4.1. Digital Phase-Shifters

Digital phase-shifters/vector modulators are usually used in conjunction with active phased arrays. We can distinguish four main types of phase-shifters, which are illustrated in Figures 4.7a to d: the switched line, reflection, loaded-line, and low-pass/high-pass configurations. Two of them are compact wideband realizations: the switched-line and lowpass-/high-pass phase-shifters. The other two, the reflection (b) and loaded line (c), are narrowband realizations, although the former can be used to introduce small phase changes over an octave.

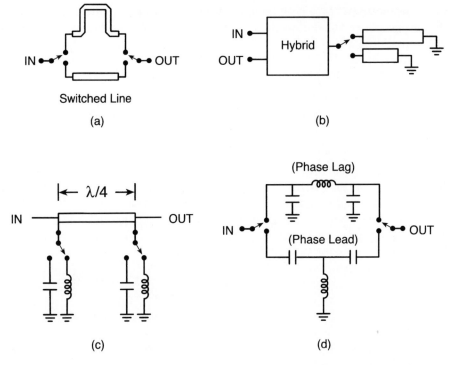

Figure 4.7 The four basic types of phase-shifters: (a) switched line; (b) reflection; (c) loaded line; and (d) low-pass/high-pass realizations.

Table 4.5 Typical digital phase-shifters and vector modulators

Frequency (GHz)	Bits	Amplitude error (dB)	Phase error (deg)	Insertion loss (dB)	Chip size (mm × mm)	Reference
IF	12	0.02	0.2			[77]
L-band	5	0.5 rms	2.5 rms	8.5		[78]
2.2–2.3	5	±0.5	7 rms	4.9	5.9 × 2.9	[79]
3–6	6	±1	< 1	10	3.8 × 3.3	[80]
X-band	4	0.3 rms	2 rms	7		[78]
35–37	4	±1.8	1.5 rms	12.8	3 × 7.8	[81]
4–18[a]	5	1.5	< 9		10.2 × 12.7[b]	[82]
5.5–8.5[a]	5	±1	< 5	13	2.4 × 1.22	[83]
X-band[a]	6	0.3 rms	3 rms	–	5 × 2.7	[41]

[a]Vector modulator.
[b] Dimensions of the carrier.

The essential characteristics of six typical digital phase-shifters are listed in Table 4.5 [77–83,41]. It is clear that the accuracy obtained when the phase-shifter is at IF (first entry) is much higher than that obtained when the same function is performed at RF. We have already established that phase-shifters at the IF can meet the requirement for low sidelobe level corresponding to small phased arrays. However, a couple of issues have to be addressed. The IF has to be chosen on the basis that the instantaneous array bandwidth, B, can be easily accommodated at the selected IF. If B/IF is approximately equal to 10%, this requirement is easily met.

If the downconversion takes place in the module, local oscillator (LO) signals can be brought to the module on the transmission line normally dedicated to connect the reference transmitter frequency to the module on transmit. If optical fibers are used for the distribution of signals to and from the modules, a dedicated 'line' can be used to bring the LO signal to the module. The use of optical fibers in phased arrays to distribute RF, IF, and digital signals to the many array elements is an attractive proposition.

4.3.4.2. Vector Modulators

The last three entries of Table 4.5 correspond to vector modulators. The first vector modulator is particularly wideband. There are essentially two preferred realizations for vector modulators. In the first realization, the vector modulator implements the required phase-shift by the addition of two vectors aligned with the X–Y coordinates which represent 0°, 90°, 180°, and 270° phases. To derive a given phase, the appropriate quadrant is first chosen, then the two vectors of the quadrant are attenuated by the required amount before they are added [83]—see section 5.3 and also Figure 5.1. With the exception of the baluns that are used to provide 180° phase changes, all constituent components of the vector modulator are easily fabricated by the use of standard MMIC technology. As standard monolithic baluns are physically too big [83], conventional off-chip baluns have been used [84]. Recently a compact monolithic balun consisting of two crossed interdigital couplers has been realized [84]. Its bandwidth extends from 7 to 19 GHz, and occupies an area of 1 mm $\times$ 2 mm.

In the other vector modulator realizations the required phase change is attained by simply routing the incoming signal through two cascaded phase-shifters; the first phase-shifter introduces the 90°, 180°, and 270° shifts and the second one introduces the finer phase shifts [82].

4.3.5. Other T/R Module Components

The focus of the MAFET program is the development of improved array components such as circulators, power supplies and capacitors needed to meet overall array requirements. Volumetrically attractive circulators are required at all wavelengths but especially at millimeter wavelengths where the half-wavelength distance between antenna elements is short.

Circulators are often used to perform the duplex function of radar phased arrays. The requirements for duplexers are:

- High isolation between the transmitter and the receiver over the bandwidth of operation.
- Minimum bulk consistent with the power handling capability requirement.
- Low insertion loss and VSWR (<1.1).

With present technology, a circulator that has to handle high powers is bulky. If the power is not high, the designers would prefer small-size inexpensive circulators, compatible with the MMIC technology.

Conventional ferrite circulators for active phased arrays generally consist of microstrip transmission lines and ferrite 'plugs' placed at suitable locations; in other realizations ferrite substrates are used. The circulators have typical insertion and isolation figures of 0.3–0.5 dB and 20–30 dB, respectively [85], depending on the bandwidth required and the frequency of operation. The critical parameters of conventional and miniature circulators are presented in the same reference. Several approaches have been taken to decrease the size of the circulators. Lumped-element and distributed-element circulators have been realized at UHF/VHF bands and at frequencies above 20 GHz, respectively [85].

Interest in producing a ferrite circulator that is comparable in size and cost with monolithic microwave circuits has been announced [86]. The intrinsically high Q of YIG resonators and the compatibility of film technology with monolithic circuits [86] are key attractors for this technology. A circulator operating over the 6–18 GHz band has been developed [87] and its insertion loss and isolation over the band were 1 dB and 14 dB respectively. The two factors that contributed to the wideband operation were: (i) The introduction of a secondary lower saturation magnetization ferrite to reduce the spatial non-uniformity within the ferrite, and (ii) A careful selection of materials and geometry to minimize low-field losses over wide excursions of frequency and temperature. The dimensions of the circulator were $0.14'' \times 0.275'' \times 0.6''$.

The realization of MMIC-based circulators is the natural progression to miniaturization. The approach described in reference [88] is based on the use of active out-of-phase divider and in-phase combiner to realize a quasi-circulator. In other realizations of MMIC-based circulators, distributed amplifiers are used [89,90]. Although the bandwidths of the above devices extends over many octaves, their insertion loss is higher than 5 dB and their power-handling capability is of the order of a few dBm. These limitations, however, cannot diminish the importance of these approaches.

As isolators and circulators enter the commercial market [91], their volume, cost, and weight are expected to decrease.

4.4. TUBES

Although klystron tubes were invented in the late 1930s and traveling-wave tubes toward the end of the Second World War, updated versions of these tubes are still used in some passive phased arrays. The main attractor of tubes is their capability to generate high RF power at meter, centimeter, and millimeter wavelengths. One or more tubes can generate enough transmitted power to meet the transmitter power requirements for a passive phased array radar system.

After a long period of 'benign neglect' of R&D related to tubes, in the West, there is now a renaissance of interest in vacuum technology and in subsystems that utilize tubes and SSDs.

Here we list catalysts that renewed and maintained interest in several aspects of vacuum technology:

- The Microwave & Millimeter-wave Advanced Computational Environment, (MMACE) Program, initiated in 1992, is a tri-Service /ARPA Vacuum Electronics Initiative directed at the development of a comprehensive computational design environment for microwave power tubes [92]. The interrelated problems posed by microwave power tube design necessitate a design and analysis environment incorporating shared common data, a 'master' geometry source, usable standards, and a common user interface for all design tools.
- R&D thrusts that resulted in mini-TWTs [93].
- Russian R&D efforts that resulted in the availability of a variety of novel tubes including the multibeam tubes [94], high-power gyro-klystrons [95], super-miniaturized low-power vacuum microwave devices [96] and low-noise self-protecting electrostatic amplifiers (ESAs) [97].
- The evolution of the microwave power module (MPM) that utilizes MMIC-based medium-power amplifiers and a high power TWT amplifier [5]. Other permutations of TWTs, MMIC-based amplifiers, solid-state multipliers and a quasi-optical power combining technique seem promising [6]; the former approach yields considerable powers at centimeter and millimeter wavelengths, while the latter approach holds the promise of generating respectable power levels at millimeter wavelengths—see section 4.5.

As detailed and comprehensive descriptions of conventional vacuum tubes are given in reference [98], here we shall consider them only briefly as a basis for comparisons with their more recent counterparts. For some time slow-wave, linear beam devices such as klystrons, extended-interaction klystrons, TWTs and backward-wave oscillators enjoyed considerable popularity mainly because the phase of the transmitted power was easily controlled within a pulse and from pulse to pulse with the aid of appropriate phase-lock loops. Phase-locked radar transmitters therefore increased a system's SNR by integrating the returns of a number of transmitted pulses.

Slow-wave, cross-field devices such as magnetrons and fast-wave devices like gyrotrons did not enjoy as much popularity because the phase and frequency of pulses emanating from these devices are time-variant within one pulse. The recent availability of signal processors allows the designer to overcame these shortcomings [99–101]. While gyrotrons can provide very high powers at centimeter/millimeter wavelengths, magnetrons can be used for low-cost radars [101–102]. These important advances are based on the utilization of SSD-based signal processors in conjunction with vacuum tubes.

4.4.1. Klystrons and Traveling-Wave Tubes

A linear electron beam between the cathode and collector is established within a klystron/TWT amplifier (TWTA). In a klystron the interaction between the input RF field and the electron beam causes velocity modulation and bunching of the electrons. The bunched beam, in turn, causes induced currents to flow in the output circuit of the klystron and energy is thus extracted from the device. While a conventional klystron has one interaction gap per cavity; the extended-interaction klystron (EIK) has several distributed interaction gaps per cavity, which in turn provide more efficient bunching of the electrons that leads to higher interaction efficiency and gain-bandwidth products. Klystrons can either operate as narrowband amplifiers or oscillators.

A helix TWT amplifier consists of a cathode, a collector, and a helix that allows the RF to interact with the electron beam. The modulation or electron bunching again induces higher-amplitude RF currents on the slow-wave structure. Helix TWTs offer the designer the option of moderate power outputs over a relatively wide bandwidth, while coupled cavity TWTs (CCTWTs) offer maximum power outputs over a narrow bandwidth.

Table 4.6 Essential characteristics of klystrons and TWTAs operating at centimeter/millimeter wavelengths

Entry no.	Characteristics	Klystron	TWTA	Reference
1	Average power at L-band	1 MW (amplifier)		
		5 MW max peak	12 kW CCTWT	[98]
2	Average power at X-band	>10 kW(amplifier)	10 kW CCTWT	[98]
3	Bandwidth (%)	1–10	5–15[a]	[98]
4	Gain (dB)	30–65	30–65	[98]
5	Efficiency (%)	up to 65	up to 60	[98]
6	Cathode voltage	up to 125 kV	up to 42 kV	[98]
	for peak power	5 MW, L-band	0.2 MW, L-band	
7	Thermal noise, typically	–90 dBc in a 1	–90 dBc in a 1	[98]
		MHz bandwidth	MHz bandwidth	
8	X-band klystron amplifier[b]	250 kW		[103]
9	35 GHz EIK	3 kW (peak)		[104]
10	100 GHz EIK	2.5 kW (peak)		[104]
11	220 GHz EIK	80 W (peak)		[104]
12	Space communications TWTAs[c]		600 W at 1 GHz	[105]
13			300 W at 10 GHz	[105]
14			100 W at 20 GHz	[105]
15			20 W at 40 GHz	[105]
16			5 W at 70 GHz	[105]
17	Terrestial communications, TWTAs[d]		8k W at 4 GHz	[105]
18			1.3 kW at 10 GHz	[105]
19			100 W at 30 GHz	[105]
20			25 W at 60 GHz	[105]

[a] Up to 100% for helix TWTs.
[b] Subsystem of a solar system radar. Frequency 8.51 GHz, 20 MHz, 1 dB bandwidth, and 45% efficiency.
[c] Helix slow-wave circuits. High linearity or low phase and gain distortion. Bandwidth 1–5%.
[d] Helix or coupled-cavity slow-wave circuits. Bandwidth 1–5%. Low distortion (high linearity).

Table 4.6 lists the essential parameters of several medium- to high-power klystrons and TWTAs that are used in radar and communication systems. The first seven entries list critical parameters such as cathode voltages and efficiencies of klystrons and TWTAs. Entry 8 corresponds to a state-of-the-art klystron amplifier used on a unique radar designated to study the solar system. Entries 9–11 and 12–20 correspond to high-power EIKs and TWTAs used in radar and communication systems, respectively. As can be seen, substantial powers can be generated at reasonable efficiencies. Bulk is always associated with the klystrons/TWTs, their associated power supplies, and cooling systems. If the voltages required are too high, lead screens that further contribute to additional weight are needed to protect the users from X-ray radiation.

4.4.2. Miniaturization of Vacuum Microwave Devices

A summary of Russian research undertaken in the field of miniaturization and super-miniaturization of vacuum tubes is outlined in reference [96]. The essential attractive vacuum tube parameters, e.g. insensitivity to temperature, radiation, and high amplitude EM fields, are maintained at any level of miniaturization. Low-power super-miniaturized klystrons or minitrons, BWOs and TWTs having accelerating voltages as low as a few tens of volts are theoretically viable. The development of miniature TWTAs was accelerated by the overall volume constraints imposed by the MPM [5], which we shall consider in detail in section 4.5.

The other thrust for miniaturized TWTAs originated from the requirements for Ku-band earth stations [93]. The resulting mini-TWTA based on a novel collector exhibits super-low dissipation under low-duty pulse operation and reduced volume. (Typical data rate is 64 kb/s and a 1% duty ratio.) TWTAs meet the requirements for small-capacity time domain multiple access (TDMA) satellite Earth stations. The average power consumption of the new collector is 1/7 of that corresponding to a conventional collector. The authors claim that the mini-TWTAs when compared to conventional TWTAs offer lower running costs and a volume reduction of 40%.

These are significant advances in the art of realizing TWTAs designated to meet specific requirements.

4.4.3. Electrostatic Amplifiers

Work on the electrostatic amplifiers (ESAs), also known as cyclotron wave electrostatic amplifiers (CWESAs), undertaken in the West and in the former Soviet Union/Russia is outlined in reference [106]. Initial work originated with R. Adler [107,108] and C. L. Cuccia [109] in the West and with S. P. Kantyuk in the Soviet Union [97].

In contrast to the conventional longitudinal grouping of electrons into dense bunches, the operation of the ESA is based on the principle of transverse grouping of the electron beam in a longitudinal magnetic field. The essential constituent parts of the ESA are shown in Figure 4.8a: the electron gun, electron beam, input and output resonator cavities, the amplifying structure, and the collector. The input energy is coupled into the electron beam in such a manner as to induce cyclotron motion of the electrons. This transfer takes place when the input signal, electron cyclotron, and input resonant frequencies are equal. Similarly, an exponential growth of transverse oscillations of electrons takes place in the periodic electrostatic structure when the cyclotron oscillation of the electron beam equals the periodic electrostatic field generated by the amplifying structure. The output cavity couples the energy from the beam to the output load. Often a solid-state amplifier follows the ESA to increase the overall gain [97]. Figure 4.8b shows the typical dimensions of an ESA operating at 4 GHz.

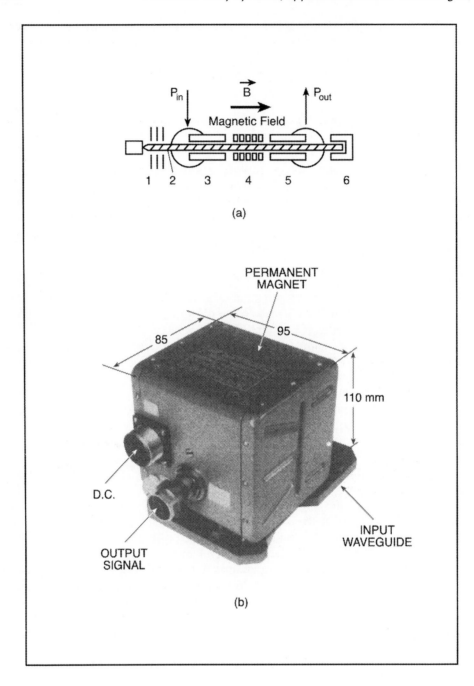

Figure 4.8 The electrostatic amplifier (ESA). (a) A simplified diagram of the ESA: (1) electron gun; (2) electron beam; (3) input resonator; (4) amplifying structure; (5) output resonator; (6) collector (From [97]; © 1993, IEEE.) (b) The dimensions of a typical ESA operating at 4 GHz. (Courtesy Professor V. A. Vanke.)

ESAs are commercially available in the 0.7–14 GHz range and their attractive characteristics are low noise, wide dynamic range, resistance to high EM radiation, and linear amplitude and phase characteristics [97].

Table 4.7 Essential characteristics of three ISTOK ESAs [110]

Parameter	IES-2GC-0A	IES-6GC-0A	IES-9GC-0A
Frequency range (GHz)	1.9–2.1	5.9–6.2	9.2–9.6
Maximum noise figure (dB)	1.5	4	4
Gain (dB)	25–30	22–25	22–26
Gain ripple (dB)	3	2	2
Linear gain input power (W)	10^{-5}	10^{-5}	10^{-5}
Permitted input microwave power (kW)	10 pulse 200 special order	10 pulse	10 pulse
Recovery time after microwave power action (ns)	>90		>90
Amplitude – phase noise at ± 1 kHz (dB/Hz)	–120	–120	–120
Heater voltage (V)	9 ± 5%	2.5 ± 1%	2.5 ± 1%
Heater current (mA)	350	350	350
Collector voltage (V)	400 ± 5%	400 ± 5%	400 ± 5%
Maximum collector current (mA)	1	1	1
Weight (kg)	5	2.5	3.5
Operating temperature range (°C)	–60 to 70	–60 to 70	–60 to 70

The essential characteristics of three ESAs are listed in Table 4.7 and additional specifications of typical ESAs are given in reference [106]. Excluding the waveguide input, output, and DC ports, typical dimensions of an ESA operating at 4 GHz are 85 mm×95 mm×110 mm [111]. As can be seen the ESAs have the following characteristics:

- They can withstand significant amounts of EM radiation and recover in relatively short times.
- They have a relatively narrow bandwidth.
- They require relatively high collector voltages when compared to SSDs but draw low current.
- They have substantial weight.

In summary, ESAs constitute an attractive and readily available alternative for array designers to consider.

4.4.4. Multiple-Beam Tubes and Gyro-Klystrons

Reference [112] cites important work on multiple-beam klystrons (MBKs) prior to 1993. The impetus for the development of multiple-beam tubes is based on increased powers and efficiencies, and lower voltages and weight. Compared to one-beam klystrons (OBKs), MBKs can have typically half the beam supply voltage and 1/10 of the weight. The transmitter of the US AWACS system, for example, uses a klystron amplifier with beam voltage of 70 kV. The Russian version of the US system uses a multiple-beam device having a beam voltage of 35 kV [113].

Table 4.8 ISTOK 6-Cavity, 36-Beam Klystron [114]

Frequency (GHz)	3.3
3 dB bandwidth (MHz)	200
Peak power (kW)	500–800
Beam voltage (kV)	28–32
Weight of klystron (kg)	25
Weight of solenoid (kg)	95
Klystron dimensions (mm)	220×700

The pertinent parameters of the ISTOK 6-cavity, 36-beam klystron are listed in Table 4.8. Russian work on MBKs, multiple-beam TWTs, and BWOs is delineated in reference [115]. R&D in the area of high-power gyro-amplifiers at millimeter wavelengths in Russia has been ongoing for many years and was considered to be more extensive than anywhere in the world in 1996 [95]. The following gyro-klystrons have been reported [95]:

- Three-cavity, 250 kW peak power, half-power bandwidth of 1.5% at 35 GHz, exhibited a gain of 40 dB and 32% efficiency.
- Two-cavity, 750 kW peak power and 24% efficiency at 35 GHz (to be extended to 1 MW).
- Four-cavity, 2.5 kW CW operating at 92 GHz with instantaneous half-power bandwidth 0.3%; weighs only 70 kg.
- Peak power of 65 kW, efficiency of 34%, and instantaneous half-power bandwidth of 0.3% at 93 GHz.

These gyro-klystrons were developed for radar applications. The derived powers from gyro-klystrons at high efficiencies are considerable at millimeter wavelengths.

Comprehensive information about devices developed in Russia was hard to obtain before the early 1990s. With the passage of time, some information about ESAs and multiple-beam tubes such as klystrons, TWTs, and BWOs became available to the West. If the same pattern is replicated, comprehensive information about gyro-klystrons will become available in the West in the not too distant future. In the meantime it is useful to know the basic capabilities of gyro-klystrons.

Interest in Russian high-power devices is steadily growing and a collaborative project between the UK Ministry of Defence, MoD, GEC-Marconi, and the Russian Academy of Sciences has been reported [116]. The joint experiment was aimed at the detection of small, fast-moving targets traveling just above the sea surface. The core radar transmitter, a Russian-designed relativistic BWO, used Cherenkov radiation to supply the electromagnetic field and operated at 10 GHz [117]; it can produce a peak power of 500 MW with a pulse duration of 5 nS.

Other applications of high-power microwave transmitters are outlined in reference [118].

4.5. THE MARRIAGE OF VACUUM AND SOLID-STATE TECHNOLOGIES

We have already considered some useful subsystems derived from the marriage of solid-state and vacuum technologies in section 4.4, and in this section we shall consider other subsystems derived from the marriage at the technological level. In the next section we shall consider the merging of these two technologies at a more fundamental level.

The MPM combines the attractive characteristics of vacuum tubes and those attributed to SSDs. More explicitly, vacuum tubes can generate respectable powers over wide bandwidths and SSDs are eminently suitable to perform the functions of low-noise and medium-power amplification over wide bandwidths, the equalization of the resulting power over the required bandwidth, the supply of suitable power for the many subsystems, and the control circuitry required to protect and optimize the system under a diverse set of operational conditions.

The R&D toward the realization of the first MPM began in 1991 [119–122] and the specifications on the first MPM designated for radar/ECM applications were

Frequency range	6–18 GHz
Power output	100 W
Overall efficiency	>33%; and
Overall envelope size	$6 \times 4 \times 0.32$ in

Low cost and high reliability were implicitly assumed. Since then, the realized MPMs have exceeded reasonable expectations. The MPM consists of one miniature TWTA, also known as the vacuum power booster (VPB), and the following solid-state subsystems:

- MMIC-based solid-state low-noise, medium-power amplifier and equalizer.
- Switched-mode power supply, also known as the power conditioner.
- Various circuits that protect and optimize the performance of the MPM under a diverse set of operational requirements.

The inclusion of the power supply in the MPM improves the graceful degradation of a phased array system. The major achievement was to realize miniaturized TWTs 5.5 in (140 mm) long and 0.288 in (7.3 mm) high [5]; the second most important achievement was the manufacture of very compact power conditioners that have unprecedented packing density of the order of 90 W/in^3 (5.5 W/cm^3) [5]. Lastly, the use of MMICs and application-specific integrated circuits (ASICs) enables designers to meet the required reliability, acquisition and life-cycle costs, and area/volume constraints [5].

The pertinent characteristics of the Northrop Grumman MPMs operating at different bands are summarized in Table 4.9. The last two entries in the table correspond to the Decoy MPM (DMPM), and the Millimeter Wave MPM (MMPM). As can be seen, substantial power outputs can be generated at moderate efficiencies and MPMs are ideal candidates for radar and EW applications. Considering the relatively short development times, MPMs constitute a significant quantum step toward higher performance in the many diverse areas we have already outlined.

MPMs have been used in linear phased arrays [5]. If MPMs are used in planar arrays where the half-wavelength criterion between adjacent array elements is observed, the

Table 4.9 Pertinent characteristics of the Northrop Grumman MPMs [5]

MPM	Frequency range (GHz)	Nominal power (W)	Saturation gain (dB)	Noise figure (dB)	Efficiency (%)	Length (inches)	Width (inches)	Height (inches)	Weight (pounds)
Low-band	2–6	100	55	10	30	9.0	4.5	1.2	2.5
C-band	4–6	125	55	10	50	7.0	6.0	1.0	2.2
Radar	7–11	125	55	10	50	6.0	2.0	1.0	0.6
High-band	6–18	100	65	13	30	6.9	5.5	0.8	1.5
Radar	6–18	100	65	13	30	6.0	4.0	0.32	1.2
DMPM	5–17.5	100	55	12	25	6.5	1.5	0.75	0.45
MMPM	18–40	100	50	13	30	8.0	3.5	0.8	1.2

overall dimensions of the MPMs have to be altered. Furthermore, components such as a transfer switch or circulator, an LNA chain, and a set of phased shifters have to be added to the MPM. And information related to the long-term reliability of the miniature TWT is required—see section 4.7 also. At Northrop Grumman work continues toward the optimization of TWT performance by utilizing tools developed under the MMACE framework [123]. Typical costs for a 9–16 GHz MPM produced by Varian, the VZM-6192PI are (1996 values) US$45 000, US$25 000, and US$15 000, for quantities of 1, 100, and 1000, respectively [6]. The critical parameters of narrow-band MPMs are considered in section 4.7.

In summary, the designer has a viable option to consider, especially if operation at centimeter/millimeter wavelengths is contemplated and substantial powers are required.

Although MPMs operating at millimeter wavelengths are a reality, another approach of generating substantial powers at millimeter wavelengths has been explored. The output of a MPM operating at centimeter wavelengths is multiplied using a large number of non-linear solid-state devices and quasi-optical power combining to derive respectable powers at millimeter wavelengths ([6] and the references therein); the initial aim is to generate about 5W at 99 GHz.

4.6. VACUUM MICROELECTRONIC DEVICES

RF vacuum microelectronics combines the advantages of electron transport in vacuum with gated electron emission structures derived from solid-state microfabrication [124]. Reference [7] is an excellent review paper on vacuum microelectronic devices (VMDs) and contains references to R&D efforts undertaken in this field up to 1994. This area holds considerable promise because current densities of 2000 A/cm^2 from Spindt arrays have been reported [8].

Microfabricated field emitter arrays (FEAs) are being used in an ongoing DARPA/NRL (Naval Research Laboratory) program as a means for gating or prebunching electrons in a microwave amplifier tube. The goals of the program are to demonstrate 10 dB gain at 10 GHz and 50 W output in a gated klystrode amplifier tube with 50% efficiency [9]. Conventional klystrodes combine the attractors of gridded tubes and klystrons, deliver 500 kW peak power, and exhibit efficiencies up to 70% at 425 MHz [125].

4.7. COMPARISONS

Designers of phased arrays consider many array aspects before they accept a SSD/vacuum tube into a phased array. The overriding consideration is to meet the many diverse requirements and we have already considered some of the obvious criteria used to discriminate one technology from another.

The MTBF of a solid-state T/R module is typically in excess of 77 000 hours [63], but only some special TWTs operated for over 7 years [126]. The acquisition cost of an active array radar (AAR), at US$1000 per radiating element, is comparable to that of a tube system, but the LCC of the AAR is half that of a tube type [44]. Further reductions in the cost of MMIC-based modules will pave the way for the widespread application of active phased arrays operating over narrow/widebands at meter/centimeter wavelengths.

Typical power output and PAEs (or efficiencies) for SSDs and tubes have been considered and it is evident that solid-state transmitters produce minuscule powers at reduced PAEs at millimeter wavelengths.

The system MTI stability is calculated from measurements of the transmitter's residual SSB phase noise and AM noise. High MTI stability is required for enhanced target detection in general and for the detection of low-RCS returns from gust fronts and dry microbursts in particular. The MTI stability of a solid-state transmitter generating 22 kW at S-band is reported to be 73.8 dB, when the same figure for its synthesizer is 75.4 dB [11]. This level of stability, it is claimed, exceeds that of any available tube or solid-state transmitter. The controlled uniformity of MMIC parameters is a significant attractor because it impacts on the array's performance. Moreover, the tolerances on the essential MMIC parameters become tighter as the manufacturing processes are better understood.

Other more subtle criteria will be introduced as we consider the following examples that serve the purposes of illustrating the complexity involved in selecting one technology over another.

We begin our comparisons by considering a relatively simple communications system. The authors of reference [69] used a 16 W solid-state power amplifier system operating at X-band for terrestrial point-to-point digital communication links, instead of a TWTA, for the following reasons:

- The amplifier's mean time before failure (MTBF) is 50 years.
- The total selling price of the solid-state amplifier including its power supply is less than the purchase price of the TWT alone.
- The solid-state amplifier met the intermodulation distortion levels of –42 dBc even when operated 1 dB above the required 38 dBm output power.
- The solid-state amplifier performed better than its TWT counterpart with respect to bit error rate (BER), AM and PM noise, noise figure, AM–PM conversion, group delay, and intermodulation distortion.

From this foregoing 1996 application, however, one cannot assume that high-power solid-state amplifiers are superior to TWTAs for all applications and at all frequency bands. The example simply illustrates the complexity involved in selecting one device over another for a specific application.

Although reference [55] is a 1991 reference, it deals comprehensively with the many considerations related to X-band radar phased arrays designated to operate on board fighter aircrafts. An active array utilizing SSD-based T/R modules is preferred when compared to a passive TWT-based array. The comparison is based on the assumptions that the radar range, the transmit duty cycle, and the peak two-way sidelobe level are equal for both systems. Furthermore, the authors consider a raft of criteria before reaching their conclusion. Some of the important criteria that served as discriminators between the two systems are:

- Their relative weights and prime powers used.
- Their noise figures.
- The losses incurred in the passive array.

While MMIC- and TWT-based systems have evolved in many significant ways since 1991, the authors outlined an important approach of comparing two competing systems. Other designers can emulate their approach, taking into account the most recent parameters that define MMIC-based and tube-based systems at any epoch.

Reference [127] provides a 1998 comparison of power amplifier technologies for narrowband systems operating at the center frequency of 13.4 GHz designated for spaceborne applications. More specifically, the applications envisaged include large constellations of satellites such as Teledesic, Orbicomm, Globalstar, and Iridium, deep space probes such as Pluto Fast Flyby, and remote sensing satellites such as Cassini, Windsat, and Mars Global Surveyor.

The contender technologies are solid-state amplifiers based on GaAs and AlGaAs MMICs, the MPM amplifier, and the TWTA. The many criteria used for the comparison are listed in Table 4.10. The efficiencies, volumes, and weights cited include the power converters required to power the four amplifiers. The space qualification

Table 4.10 Comparison of projected Ku-band power amplifier technologies at 13.4 GHz [127]

Parameters	GaAs MESFET	AlGaAs PHEMT	MPM	TWTA
Bandwidth (MHz)	± 50	± 50	± 50	± 50
Power in/out, (mW/W)	1/40	1/>40	1/>65	1/65
Efficiency[a] (%)	10–20	20–25	>50	>50
Harmonics (dBc)	60	60	10	20
Spurious (dBc)	80	80	80	50
Amplitude ripple (dB)	±0.2	±0.2	±0.2	±0.2
Volume[a] (cubic inches)	54	50	42	166
Weight[a] (pounds)	8	7.5	6	8
Prime power (VDC)	+30	+30	+270	+30
Duty cycle (%)	0–33	0–33	0–100	0–33
Reliability	Established	Established	In progress[b]	Established
Space heritage	Yes	Limited	None	Yes
Construction technique	Integral DC–DC converter	Integral DC–DC converter	Integral DC–DC converter	Separate HVPS
Thermal control	Radiation/conduction	Radiation/conduction	Radiation/conduction	Radiation/conduction

[a] Includes power converter.
[b] Space qualification in progress.

reliability of the MPM amplifier is in progress. The criterion 'space heritage' or the failure-free on orbit performance has emerged as an important discriminant for the MPMs. Interestingly, no cost estimates are given by the authors.

The authors of the same reference conclude that there is no clear winner, or 'silver bullet' emerging from the comparisons. While some might think that this is an undesirably ambivalent situation in which to find ourselves, the authors hold that all three technologies—SSD, MPM, and TWTA—will be used for many years to come.

Given the rapid changes we are witnessing it is more important to delineate the criteria for selection rather than to select one technology over another at an arbitrary reference epoch.

4.8. CONCLUDING REMARKS

One of the important decisions an array designer faces is whether to use SSDs or tubes to perform the many array functions. It is widely accepted that:

- SSDs are unchallenged for low-level signal processing functions, e.g. high-speed digital circuits, control circuits, oscillators, phased shifters, buffer amplifiers, and low-power/medium-power amplifiers. Similarly, SSDs are eminently suitable for LNAs. While all SSDs require some protection from high microwave powers, the LNAs are especially vulnerable and their protection is imperative in an electronic environment where intentional/unintentional jammers are encountered.
- Tubes are unchallenged in applications where high power generation or amplification is required at millimeter wavelengths.
- The ESAs offer low-noise amplification and high resistance to high microwave powers; however, their weights and volumes have to be taken into consideration for phased array applications.
- For high power generation/amplification, tubes and SSDs offer a diverse set of attractors.
- For systems operating below 3 GHz, high powers can be generated when conventional power addition of many SSDs is implemented. Currently the upper frequency limit is imposed by transmission line losses and the stringent manufacturing tolerance requirements associated with the building blocks used. If solutions to these design issues are found, the upper frequency limit will increase.

Newcomers to the field such as mini-TWTs, multiple beam tubes, gyrotrons, and MPMs also have to be considered. Furthermore, progress in developments related to vacuum microelectronic devices has to be monitored if future surprises are to be minimized.

In an era when progress is rapid it is more important to delineate the criteria for selection rather than to select one technology over another at an arbitrary reference epoch. Many of the important criteria for selection have therefore been delineated in this chapter.

A decision to use SSDs or tubes is usually based on a raft of considerations and not on just one device parameter. For space applications, the space qualification of a device is one of the important criteria the designer has to consider.

In this chapter we have delineated the many thrusts that will decrease the cost of MMIC-based modules. Similarly, the controlled uniformity of MMIC parameters is a significant attractor since it impacts on the array's performance. As the manufacturing processes are better understood with the passage of time, the tolerances on the essential MMIC parameters become tighter. While affordable MMIC-based active arrays are a reality now, the same type of arrays will become irresistible as the cost of MMIC-based modules is lowered over time. These developments will pave the way for the widespread application of active phased arrays operating over narrow/widebands at meter/centimeter wavelengths.

REFERENCES

[1] Cohen E. D. Active electronically scanned arrays. *IEEE Natl. Telesyst. Conf.* p. 3 (1994).

[2] Sheridan K. V. and Sparks J. B. The Culgoora Radioheliograph. 4—The radio-frequency system. *Proc. IREE (Aust.)*, **28**(9), 302–307 (Sept. 1967).

[3] Fisher D. and Bahl I. *Gallium Arsenide IC Applications Handbook.* Academic Press, London, San Diego, Orlando, FL (1995).

[4] Kirkpatric C. G. Making GaAs integrated circuits. *Proc. IEEE* **76**(7), 792–815 (July 1988).

[5] Smith C. R. Power module technology for the 21st century. *Proc. NAECON*, vol. 1, pp. 106–113 (1995).

[6] High power hybrid technology millimeter-wave sources. http://www.ece.ucdavis.edu/~rosenau/research.html.

[7] Brodie I. and Schwoebel P. R. Vacuum microelectronics devices. *Proc IEEE*, **82**(7), 1006–1034 (July 1994).

[8] Brodie I. Vacuum microelectronics—the next ten years. *Tech. Dig.* IVMC '97, Kyongju, Korea, pp. 1–6 (1997).

[9] Spindt C. A., Holland C. E., Schwoebel P. R. and Brodie I. Field emitter array development for microwave applications. II. *J. Vac. Sci. Technol.* B **16**(2), 758–761 (Mar./Apr. 1998).

[10] Brookner E. (ed.). *Aspects of Modern Radar.* Artech House. Norwood, MA (1988).

[11] Rivera D. J. An S-band solid-state transmitter for airport surveillance radars. *IEEE 1993 Natl. Radar Conf. Rec.*, p. 197 (1993).

[12] Gupta M. S. Degradation of power combining efficiency due to variability among signal sources. *IEEE Trans. Microwave Theory Tech.* **40**(5), 1031 (1992).

[13] Mallavarpu R. and Puri M. P. High power CW with multi-octave bandwidth from power-combined mini-TWTs. *IEEE M*TT-S Int. Microwave Symp. Dig.*, p. 1333 (1990).

[14] Fourikis N. Novel shared-aperture phased arrays. *Microwave Opt. Techn Lett.*, (20 Feb.) 189–192 (1998).

[15] Mink J. W. and Rutledge D. G. (Guest eds). Special Issue on Quasi-Optical Techniques. *IEEE Trans. Microwave Theory Tech.*, **41**(10) (Oct. 1993).

[16] Liu S. C. A fault correction technique for phased array antennas. *IEEE* Antennas Propag. Soc. *Int. Symp.* p. 1612 (1992).

[17] Lee K.M., Chu R. S. and Liu S. C. A performance monitoring/fault isolation and correction system of a phased array antenna using transmission line signal injection with phase toggling method. *IEEE Antennas Propag. Soc. Int. Symp.*, p. 429 (1992).

[18] Bahl I. Personal communication (Jan. 1998).

[19] Keizer W. P. M. N. New active phased array configurations. *Conf. Proc. Mil. Microwaves*, p. 564 (1990).

[20] VanBlaricum M. L. Photonic systems for antenna applications. *IEEE Antennas Propag. Mag.* **36**(5), 30 (1994).

[21] Seeds A. Microwave optoelectronics. Tutorial review. *Opt. Quantum Electron.* **25**, 219 (1993).

[22] Seeds A. Optical technologies for phased array antennas. Invited paper. *IEICE Trans. Electron.* **E76-C**(2) 198 (1993).

[23] Glista A. S. Airborne photonics, a technology whose time has come. *AIAA/IEEE 12th Digital Avion. Syst. Conf., DASC*, p. 336 (1993).

[24] Shalkhauser K. A., Windyka J. A., Dening D. C. and Fithian M. J. System-level integrated circuits for phased array antenna applications. *IEEE MTT-S Dig.* pp. 1593–1596 (1996).

[25] Cohen E. D. Trends in the development of MMICs and packages for active electronically scanned arrays (AESAs). *1996 IEEE Int. Symp. Phased Array Systems and Technology*, Boston, MA, pp. 1–4 (1996).

[26] Cressler J. D. SiGe HBT technology: A new contender for Si-based RF and microwave circuit applications. *IEEE Trans. Microwave Theory Tech.* **46**(5), 571–589 (May 1998).

[27] Greiling P. and Ho N. Commercial satellite applications for heterojunction microelectronics technology. *IEEE Trans. Microwave Theory Tech.* **46**(6), 734–737 (June 1998).

[28] Peterson J. MIMIC Program spawns GaAs infrastructure. *Microwaves RF*, (Jan.), 55 (1995).

[29] E. D. Cohen, MIMIC from the Department of Defense perspective. *IEEE Trans. Microwave Theory Tech.* **38**(9), 1171 (1990).

[30] Pengelly R. S. Broadband monolithic microwave circuits for military applications. *Military Microwaves* (Oct.), p. 244 (1982).

[31] Fisher D. G. GaAs IC applications in electronic warfare, radar and communications systems. *Microwave J.* **31**, 275 (May 1988).

[32] Bahl I. J. *et al.* Multifunction SAG process for high-yield, low-cost GaAs microwave integrated circuits. *IEEE Trans. Microwave Theory Tech.* **38**(9), 1175 (1990).

[33] Podell A., Lockie D. and Moghe S. Practical GaAs ICs designed for microwave subsystems. In *The Microwave System Designer's Handbook*, 4th ed., vol. 16, no. 7, p. 327. EW Communications Inc., Palo Alto, CA (1986).

[34] Cetronio A. and Graffitti R. A reproducible high yield technology for GaAs MMIC production. *Alta Freq.* **55**(3), 173 (1986).

[35] Mayousse C. *et al.* 'Design and Technology' optimisation for high yield monolithic GaAs X-band low noise amplifiers. *Proc. 17th European Microwave Conf.*, p. 267 (1987).

[36] Landbrook P. H. *MMIC Design: GaAs FETs and HEMTs,* ch. 1. Artech House, Norwood, MA (1989).

[37] Shiga N. *et al.* X-band MMIC amplifier with pulse-doped GaAs MESFETs. *IEEE Trans. Microwave Theory Tech.* **39**(12), 1987 (1991).

[38] Bar S. X. *et al.* Manufacturing technology development for high yield pseudomorphic HEMT. *IEEE GaAs IC Symp.*, p. 173 (1993).

[39] Mondal J. High performance and high-yield Ka-band low-noise MMIC using 0.25-μm ion-implanted MESFETs. *IEEE Microwave Guided Wave Lett.* **1**(7), 167 (1991).

[40] Wang H. *et al.* High-yield W-band monolithic HEMT low-noise amplifier and image rejection downconverter chips. *IEEE Microwave Guided Wave Lett.* **3**(3), 281 (1993).

[41] van den Bogaart F. L. M. and Bij de Vaate J. G. Production results of a transmit/receive-MMIC chip set for a wideband active phased array radar at X-band. *Conf. Proc., Military Microwaves 92*, p. 138 (1992).

[42] Umeda Y., Enoki T., Osafune K., Ito H. and Ishii Y. High-yield design technologies for InAlAs/InGaAs/InP-HEMT analog-digital ICs. *IEEE Trans. Microwave Theory Tech.* **44**(12), 2361–2368 (Dec. 1996).

[43] Wisseman W. R. *et al.* X-band GaAs single-chip T/R radar module. *Microwave J.* **30**(9), 167 (1987).

[44] Brukiewa T. F. Active array radar systems applied to air traffic control. *Proc. IEEE Natl. Telesyst. Conf.*, San Diego, CA, p. 27 (1994).

[45] Cohen E. D. High density microwave packaging program. *IEEE MTT-S Dig.*, pp. 169–172 (1995).

[46] Wooldridge J. High density microwave packaging for T/R modules. *IEEE MTT-S Dig.*, p. 181–184 (1995).

[47] Reddick J. A. *et al.* High density microwave packaging program. Phase 1—Texas Instruments/Martin Marietta team. *MTT-S Dig.* pp. 173–176 (1995).

[48] Costello J. A. The Westinghouse high density microwave packaging program. *IEEE MTT-S Dig.*, pp. 177–180 (1995).

[49] Cohen E. D. Active electronically scanned arrays. *Proc. IEEE Natl. Telesyst. Conf.*, San Diego CA, p. 3 (1994).

[50] Tang R. and Burns R. W. Array technology. *Proc. IEEE* **80**(1), 173 (1992).

[51] Axness T. A., Coffman R. V., Kopp B. A. and O'Haver K. W. Shared aperture technology development. *Johns Hopkinks Tech. Dig.* **17**(3), 285–294 (1996).

[52] Garbe S. *et al.* A 6.8–10.7 GHz EW module using 72 MMICs. *IEEE MTT-S Dig.*, p. 1329 (1993).

[53] Driscoll M. M. *et al.* Cooled, ultrahigh Q, sapphire dielectric resonators for low-noise, microwave signal generation. *IEEE Trans. Ultrasonics, Ferroelectrics and Frequency Control* **39**(3), 405–411 (May 1992).

[54] Pospieszalski M. W., Lakatosh W. J., Nguyen L. D., Lui M., Liu T., Le M., Thomson M. and Delaney M. J. Q- and E-band cryogenically-coolable amplifiers using AlInAs/GaInAs/InP HEMTs. *IEEE MTT-S Dig.*, p. 112 (1995).

[55] McQuiddy Jr. D. N. *et al.* Transmit/receive module technology for X-band active array radar. *Proc. IEEE* **79**(3), 308 (1991).

[56] Sharma A. K. Solid-state control devices: State of the art reference. *Microwave J.*, Sup. 7.72; *GHz* p. 95 (1989).

[57] Seymore D. J., Heston D. D. and Lehmann R. E. Monolithic MBE-grown GaAs PIN diode limiters. *IEEE Microwave Millimeter Wave Monolithic Circuits Symp. Dig.* p. 35 (1987).

[58] Seymore D. J., Heston D. D. and Lehmann R. E. X-band and Ka-band monolithic GaAs PIN diode variable attenuator limiters. *IEEE Microwave Millimeter Wave Monolithic Circuits Symp. Dig.*, p. 147 (1988).

[59] Anon. Microwave Modules and Devices, A 2kW, L-band radar module using 550 W "quad" building blocks. *Microwave J.* **28**(6), 157 (1985).

[60] Potyraj P. A., Petrosky K. J., Hobart K D., Kub F. J. and Thompson P. E. A 230-Watt S-band SiGe heterojunction bipolar transistor. *IEEE Trans. Microwave Theory Tech.* **44**(12), 2392–2397 (Dec. 1996).

[61] Morse A. W. *et al.* Recent application of silicon carbide to high power microwave. *IEEE MTT-S Dig.*, pp. 53–56 (1997).

[62] Eisele H. and Haddad I. Two-terminal millimeter-wave sources. *IEEE Trans. Microwave Theory Tech.* **46**(6), 739–746 (June 1998).

[63] Considine B. C. Solid-state transmitters take advantage of continuing advances in device technology. *Electron. Prog.* **29**(2), 24 (1989).

[64] Pusl J. A., Widman R. D., Brown J. J., Hu M., Kaur N., BeZaire M. and Nguyen L. D. High efficiency L and S-band power amplifiers with high breakdown GaAs-based pHEMTs. *MTT-S Digest*, vol. 2, pp. 711–714 (1998).

[65] Pearce W. L. C-band transmitter module delivers 55–W output. *Microwaves RF* (Mar.) 101–104 (1997).

[66] Zoyo M., Galy C., Darbandi A., Lapierre L. and Sautereau J-F Design of a very compact space borne X-band high power high efficiency hybrid amplifier. *25th European Microwave Conf.*, Bolognia, Italy, vol. 1, pp. 27–30 (1995).

[67] Raicu D., Basset C. R., Baughman C. R., Chung Y., Day D. S. and Hua C. High-efficiency, flat gain 5-Watt X-band MMIC, for large volume production. *24th European Microwave Conf.*, Cannes, France, vol. 2, pp. 1053–1058 (1994).

[68] Sweeder J., Truitt A., Nelson R. and Mason J. S. Compact, reliable 70-Watt X-band power module with greater than 30-percent PAE. *IEEE MTT-S Dig.*, pp. 57–60 (1996).

[69] Vincent J. B. and van der Merwe A. 16W solid state MMIC X-band amplifier for TWT replacement. *1996 IEEE AFRICON* vol. 2, pp. 749–752 (1996).

[70] Ingram D. L., Stones D. I., Elliott J. H., Wang H., Lai R. and Biedenbender M. A 6-W Ka-band power module using MMIC power amplifiers. *IEEE Trans. Microwave Theory Tech.* **45**(12), 2424–2430 (Dec. 1997).

[71] Komiak J. J. and Yang L. W. 5-Watt high efficiency wideband 7 to 11 GHz HBT MMIC power amplifier. *IEEE 1995 Microwave and Millimeter Wave Monolithic Circuits Symp.*, pp. 17–20 (1995).

[72] Barnes A. R., Moore M. T. and Allenson M. B. A 6–18 GHz broadband high power MMIC for EW applications. *IEEE MTT-S Dig.*, pp. 1429–1432 (1997).

[73] Hwang Y. *et al.* 60 GHz high-efficiency HEMT MMIC chip set development for high-power solid state power amplifier. *IEEE MTT-S Dig.*, pp. 1179–1182 (1997).

[74] Stark L. Microwave theory of phased array antennas: a review. *Proc. IEEE* **62**(12),1661 (1974).

[75] Hord W. E. Microwave and millimeter-wave ferrite phase-shifters: state of the art reference. *Microwave J.*, sup. 81 (1989).

[76] Sharma A K. Solid-state control devices: state of the art reference. *Microwave J.*, 95 (1989).

[77] Aumann H. M. and Willwerth F. G. Intermediate frequency transmit/receive modules for low-sidelobe phased array application. *Proc. 1988 IEEE Natl. Radar Conf.*, p. 33 (1988).

[78] Borkowski M. T. and Leighton D. G. Decreasing cost of GaAs MMIC modules is opening up now new areas of application. *Electron. Prog.* **29**(2), 32 (1989).

[79] Rhodes M. and Lane A. A. Monolithic five-bit phase shifter for Artemis space application. *IEE Colloq. Active and Passive Components Phased Array Systems*, p. 10/1 (Apr. 1992).

[80] Komiak J. J. and Agrawal A. K. Design and performance of octave S/C-band MMIC T/R module for multifunction phased arrays. *IEEE Trans. Microwave Theory Tech.* **39**(12), 1955 (1991).

[81] Slobodnik, A. J., Jr. Webster R. T. and Roberts G. A. A monolithic GaAs 36 GHz four-bit phase shifter. *Microwave J.*, **36**, 106 (1993).

[82] Norris G. B. A fully monolithic 4–18 GHz digital vector modulator. *IEEE MTT-S Int. Microwave Symp. Dig.*, p. 789 (1990).

[83] Ali F. *et al.* A single chip C-band GAaS monolithic five bit phase shifter with on chip digital decoder. *IEEE MTT-S Int. Microwave Symp. Dig.*, p. 1235 (1990).

[84] Tsai M. C. A new compact wideband balun. *IEEE MTT-S Dig.*, p. 141 (1993).

[85] Rotrigue G. P. Circulators from 1 to 100 GHz: State of the art reference. *Microwave J*, 115 (1989).

[86] Webb D. C. Status of ferrite technology in the United States. *IEEE MTT-S Int. Microwave Symp. Dig.*, p. 206 (1993).

[87] Blight R. E. and Schloemann E. A compact broadband circulator for phased array antenna modules. *IEEE MTT-S Int. Microwave Symp. Dig.*, p. 1389 (1992).

[88] Hara S., Tokumitsu T. and Aikawa M. Novel unilateral circuits for MMIC circulators. *IEEE Trans. Microwave Theory Tech.* **38**(10), 1399 (1990).

[89] Katzin P. *et al.* 6 to 18 GHz MMIC circulators. *Microwave J.* **35**(65), 248 (1992).

[90] Robertson I. D. and Aghvami A. H. Novel monolithic ultra-wideband unilateral port junction using distributed amplification techniques. *IEEE MTT-S Int. Microwave Symp. Dig.*, p. 1051 (1992).

[91] Murakami Y. Microwave ferrite technology in Japan: current status and future expectations. *IEEE MTT-S Dig.*, p. 207 (1993).

[92] The Tri-service MMACE Program: Overview of the program and prototype design system. http://mmace.nrl.navy.mil.8080/mmace/mmace_docs/mm-prog-overview.html

[93] Mita N. A mini-TWT amplifier with super low power dissipation for Ku-band. *24th European Microwave Conf.*, Cannes, France vol. 2, pp. 1854–1859 (1994).

[94] Pobedonostsev A. S., Gelvich E. A., Lopin M. I., Alexeyenko A. M., Negirev A. A. and Sazonov B. V. Multiple-beam microwave tubes. *IEEE MTT-S Dig.*, pp. 1131–1134 (1993).

[95] Office of Naval Research Europe S&T Reports, London, UK, ONREUR (96-1-R), March 6 1996. http://www.ehis.navy.mil/nbnews61.htm

[96] Gulyaev Y. V. and Sinitsyn N. I. Super-miniaturization of low-power vacuum microwave devices. *IEEE Trans. Electron Devices* **36**(11), pp. 2742–2743 (1989).

[97] Budsinsky Yu. A. and Kantyuk S. P. A new class of self-protecting low-noise microwave amplifiers. *IEEE MTT-S Dig.*, pp. 1123–1125 (1993).

[98] Sivan L. *Microwave Tube Transmitters*. Chapman and Hall, London (1994).

[99] Manheimer W. Application of gyrotrons to radar and atmospheric sensing. *Proc. 7th Int. Conf. Infrared Millimeter Waves*, p. 142 (1993).

[100] Lhermitte R. M. Cloud and precipitation remote sensing at 94 GHz. *IEEE Trans. Geoscience Remote Sensing* **29**(3), 207 (May 1988).

[101] Li H., Illingworth A. J. and Eastment J. A simple method of Dopplerizing a pulsed magnetron radar. *Microwave J.* **37**(4), 226 (1994).

[102] Christie D. J. Advances in solid-state magnetron modulation. *Microwave J.* **36**(1), 111 (1993).

[103] Freiley A. *et al.* The 500 kW CW X-band Goldstone solar system radar. *IEEE MTT-S Int. Microwave Symp. Dig.*, p. 125 (1992).

[104] *Millimeter wave extended interaction klystrons (amplifiers and oscillators) and transmitter subsystems*. Varian Booklet (Nov. 1990).

[105] Hansen J. W. US TWTs from 1 to 100 GHz: state of the art reference. *Microwave J.*, **32**, 179 (1989).

[106] Vanke V. A., Matsumoto H. and Shinohara N. A new microwave input amplifier with high self-protection and rapid recovery. *IECE Trans. Electron.*, **E81–C**(5) (May 1998).

[107] Adler R. Parametric amplification of the fast electron wave. *Proc IRE*, **46**(10), 1756 (1956).

[108] Hrbek G. and Adler R. Low-noise DC-pumped cyclotron-wave amplifier. *Proc 5th Int. Congr. Microwave Tubes*, Paris, p. 17. Academic Press, New York (1965).

[109] Cuccia C. L. Parametric amplification, power control and frequency multiplication at microwave frequencies using cyclotron-frequency devices. *RCA Rev.* **21**(2), 228 (1960).

[110] ISTOK ESAs. Essential characteristics. http://www.istok.com/ESA_chart.html

[111] Personal communication: Professor Vladimir A. Vanke, Faculty of Physics, Moscow State University (June 1998).

[112] Gelvich E. A., Borisov L. M., Zhary V., Zakurdayev A. D., Pobedonostsev A. S. and Poognin V. I. The new generation of high-power multiple-beam klystrons. *IEEE Trans. Microwave Theory Tech.* **42**(1), 15–19 (Jan. 1993).

[113] ISTOK multiple-beam technology. http://www.istok.com/multibeam.html

[114] ISTOK 6 cavity, 36 beam klystron. http://www.istok.com/IKS-9007.html

[115] ISTOK Microwave Company. Products. http://istok.com/company.html

[116] Wardrop B. A Russian experimental high-power, short-pulse radar. *GEC J. Technol.* **14**(3), 1–12 (1977).

[117] Nusinovich G. S., Antosen T. M. Jr., Bratman V. L. and Ginzburg N. S. Pinciples and capabilities of high power microwave generators. In *Applications of High-Power Microwaves*, ch. 2. Goponov-Greknov A. V. and Granatstein V. L. (eds.). Artech House, Norwood, MA (1994).

[118] *Applications of High-Power Microwaves*. Artech House, Norwood, MA (1994).

[119] Abrams, R. H. Jr. and Parker R. K. Introduction to the MPM: what it is and where it might fit. *IEEE MTT-S Int. Microwave Symp. Dig.*, p. 107 (1993).

[120] Cosby L. MPM applications: A forecast of near- to long-term applications. *IEEE MTT-S Int. Microwave Symp. Dig.*, vol. 1, p. 111 (1993).

[121] Brees A. *et al.* Microwave power module (MPM) development and results. *Proc. EEE Int. Electron Devices Meet.*, Washington, DC, p. 145 (1993).

[122] Christensen J. A. *et al.* MPM Technology developments: An industry perspective. *IEEE MTT-S Int. Microwave Symp. Dig.*, p. 115 (1993).

[123] Whaley D. R., Armstrong C. M., Groshart G. and Stotz R. High performance TWT development for the microwave power module. *IEEE Int. Conf. Plasma Science, Abstracts*, p. 104 (1996).

[124] Jensen K. L., Abrams R. N. and Parker R. K. Field emitter array development for high frequency applications. *J. Vacuum Sci. Technol. B* **16**(2), 749–753 (1998).

[125] Preist D. H. and Shrader M. B. A high-power klystrode with potential for space application. *IEEE Trans. Electron Devices* **38**(10), 2205–2211 (Oct. 1991).

[126] Woods R. L. Microwave power source overview. *Proc. Military Microwaves Conf.* London, p. 349 (1986).

[127] Smith M. C. and Dunleavy L. P. Comparison of solid state, MPM and TWT based transmitters for spaceborne applications. *Proc. Southeastcon '98. Engineering for a New Era*, Orlando, FL, pp. 256–259 (1998).

5

Beamformers

One learns by doing the thing; for though you think you know it, you have no certainty until you try.

Sophocles (496?-406BC)

As the term implies, the beamformer forms one or more antenna beams anywhere within the array's surveillance volume. A beamformer therefore also scans the beam(s) to the required azimuth and elevation scan angles. If the radiation pattern of the antenna elements is hemispherical, the array surveillance volume is defined by the required scan angles; additionally, the beamformer adjusts the beam shape of the resulting beam(s) with the aid of the amplitude/phase weights associated with each antenna element. The beamformer of a typical passive array can be bulky and its insertion loss usually affects the overall array noise temperature. By contrast, the beamformer of an active array can be transparent to the system and volumetrically attractive. These two beamformer characteristics lend the designer several degrees of freedom, which we shall explore in this chapter.

We have already considered MMIC-based phase-shifters and vector modulators, the building blocks of beamformers, in Chapter 4. As their volume is small, the designer can easily form one or several scanning/staring beams. Similarly, the designer has the freedom to select the beamformer's frequency of operation. The beamformer's frequency of operation can be equal to the RF of the array or to a convenient IF. For the implementation of digital beamforming, the RF band is often translated into a band centered at an IF where the IF<<RF; conversely, if beams are formed with the aid of a photonic subsystem, the beamformer operates at a band centered at an IF where the IF>>RF. Between these extremes of frequency, beamformers can be formed at any convenient IF.

It is useful to recall here that the array beam shape is defined by many array characteristics including the physical dimensions of the array and the amplitude/phase weights associated with each antenna element. While the sharpest sum and difference beams are required to perform the tracking function of an airborne radar system, a broader beam is required for the terrain following operation. Similarly, the array sidelobe level is defined primarily by the phase and amplitude errors occurring in the many parts of the array, including the beamformer. If the calibration of these errors is implemented, the resulting array sidelobe level decreases further. We have already seen that the effects of mutual coupling between adjacent antenna elements can be minimized if a suitable calibration procedure is adopted. As adaptive nulling of jammers is treated in other books in some detail, we shall not consider this topic further.

Some of the important reasons why a designer abandons the conventional beam-former architectures, we have considered in Chapter 4 are:

- The cost constraints are extreme. While affordable phased arrays are a reality now, lower-cost arrays are required for some applications.
- The angle of the vector between an airborne platform equipped with a phased array and a communication satellite is not always known and is changing as a function of time.
- The requirements for:
    (i)   lightweight and volumetrically attractive beamformers;
    (ii) arrays to have many simultaneous and independent beams/nulls; and
    (iii) two or more simultaneous and independent beams centered over two or more
          widely separated bands.

Ultimately the decision to use one type of beamformer over another depends on a set of requirements. We have already considered beamforming lenses that meet an important set of requirements, e.g. low cost, a moderate number of independent and simultaneous beams, wide-angle scanning, and wide instantaneous system bandwidth, in Chapter 3.

The selection of the beamformer usually depends on the following bits of information:

- Cost, weight, and volume constraints.
- Whether the array's scan angles as a function of time can be known in real time.
- The array's instantaneous and overall bandwidths.
- Operation at one or more frequency bands centered at frequencies $f_1, f_2, \ldots$, and f_N.
- Number of scanning/staring beam(s)/nulls required.
- The number of array antenna elements and the required sidelobe level.

In this chapter the basic beamformer issues and some elaborations are first considered before the simplest and progressively more complex beamformer architectures are explored. Applications drawn from radioastronomy, ESM/ECM, satellite communications, and radar arrays utilizing the different beamforming architectures are also delineated.

5.1. THE BASIC ISSUES AND SOME ELABORATIONS

Staring beams are independent and simultaneous antenna beams that are often contiguous in the space domain. The term independent needs some elaboration because, while one can generate a multitude of antenna beams, for some applications there is no point in generating antenna beams that have redundant information. In the radioastronomy context, for instance, no redundancy is required for the derivation of radiometric maps of celestial sources. The separation between contiguous beams is therefore equal to the half-Rayleigh limit of resolution ($\lambda/2D$, where D is the array diameter) [1]. For other systems, the separation between contiguous antenna beams is application dependent.

An additional loss related to the beam spacing that reduces to zero for orthogonal beam sets has been considered [2–6]. Mutually orthogonal beams satisfy the condition that the average value over all angles of the product of one beam response with the conjugate of the other must be zero. The crossover between contiguous beams satisfying this condition is at about –3.9 dB. Most practical phased arrays will not have exactly orthogonal beams and will therefore incur an additional loss. Beams generated by a uniform amplitude distribution call for a –4 dB crossover point. In practice, beams with distributions yielding sidelobe levels lower than –13 dB call for much lower crossover levels [5].

Although the beamformer's insertion loss can be transparent in an active phased array, the maintenance of the system's noise temperature and dynamic range, which is intrinsically related to the array SNR, are issues worthy of the designer's consideration. As beamformers operating at optical wavelengths exhibit significant insertion losses, the designer should ascertain that the system's noise figure and dynamic range are maintained through the many subsystems required to perform the beamforming/scanning functions. Naturally the minimization of the insertion loss of any beamformer is a good design goal.

In Chapter 3 we explored several array illumination functions that result in narrow beamwidths. A phase gradient across a linear array results in a wider array beamwidth; a phase front shaped like a gable can also be used to broaden the array's radiation pattern, and more complicated phase distributions are required when the array is planar. In airborne/spaceborne platforms the following beamformer characteristics volume/weight and DC power consumption assume an added importance.

References [6–8] are excellent contributions to the art of beamforming. In what follows we shall provide a roadmap for the many beamforming architectures and delineate their regimes of applicability.

5.2. THE FORMATION OF STARING BEAMS

Staring beams are required for many radioastronomy, radiometric, and ESM applications. The generation of many staring beams, for instance, enhances the POI of an ESM system in the spatial domain. For many applications the generation of as many staring beams as possible is required.

The beamformer architectures we shall explore in this section have narrow instantaneous system bandwidths.

5.2.1. Resistive Network Realizations

Beamformers utilizing resistive networks are relatively easy to realize. Figure 5.1 illustrates a typical beamformer architecture where N antennas yield M staring beams and the required phases, ϕ_{MN} between each antenna N and beam M are implemented with the aid of resistive networks; the network used is also known as a crossbar beamformer network (BFN). The system operates in the receive mode and after suitable amplification the incoming signal from each antenna is split into four lines that enter the beamforming network. The four lines introduce 0°, 90°, 180°, and 270° phase-shifts to the received signals. With this arrangement any phase ϕ_{NM} can be introduced between

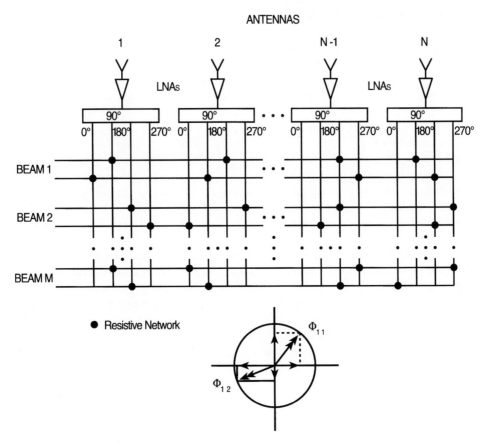

Figure 5.1 Beamformer architecture for the realization of *M* staring beams from an array having *N* antenna elements when resistive networks are used.

an antenna *N* and *M* beam port. First the two vectors that define the appropriate quadrant are selected and are in turn attenuated with the aid of resistive networks to derive the required phase. The symbol • in Figure 5.1 denotes the resistive network used to introduce the appropriate phase between an antenna and a beam. In the same figure we illustrate how Φ_{11} and Φ_{12} are introduced between antenna 1 and beam 1 and between antenna 1 and beam 2.

At the commissioning stage, the one mile long Sydney University cross-type radio-telescope [9] had 11 staring pencil beams [10,11] formed by using resistive networks between the appropriate antenna and beam ports.

The architecture of the crossbar BFN is important because MMIC-based vector modulators can substitute the resistive networks. With this arrangement a significant number of agile antenna beams can be formed—see section 5.3. The simplicity, versatility and elegance of the above architecture have been recognized by the radar [12] and satellite communications [13] communities.

5.2.2. Transmission Line Realizations

With reference to Figure 5.2, M staring beams can be derived from a phased array having N elements by the introduction of $M \times N$ suitably calibrated transmission lines between the antennas and beams. After suitable amplification, the signals from each antenna are divided in M ways and the M beams are formed at the summing points. The $M \times N$ lengths of transmission lines are accurately calculated from knowledge of the array geometry and the required interbeam spacing. For wideband operation, the transmission lines are equal to the delays required between the antenna elements and the summing point of each beam. For narrowband operation, the transmission lines introduce the required phases between the antenna elements and the summing points. The beamformer is therefore a real-time analog Fourier transform in the spatial domain [14].

Using this BFN architecture, 15 [15] and 48 [14] stationary antenna beams have been formed for radioastronomical applications. In the latter application, the 48 stationary pencil beams were derived from the Culgoora Radioheliograph, a phased array that consisted of 96 antenna elements forming a 3 km circle, along a N–S line. The derived beams were in turn swept along an E–W direction in 64 discrete steps with the aid of programmable phase-shifters introduced in the LO lines used to downconvert the array's 1 MHz RF band centered at 80 MHz into the IF band centered at 7 MHz. With this arrangement two radio pictures of the Sun, each consisting of 48 × 64 picture points, corresponding to the two opposite circular polarization, were produced every second. The beamformer consisted of 48 × 48 = 2304 accurately cut coaxial cables and occupied a volume of 1 m × 3.2 m × 2 m [14].

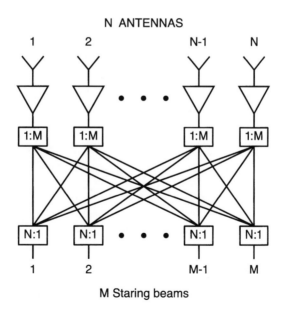

Figure 5.2 Beamformer architecture for the realization of M staring beams when $M \times N$ transmission lines are used between the N antenna elements of a phased array and the M beam ports.

The bulk and weight of these beamformers prevented designers from realizing a higher number of antenna beams from phased array-based radiotelescopes. The beamformer architecture, however, is important because the coaxial cables can be substituted by optical fibers that are lighter and less bulky – see section 5.5.2. Additionally, this architecture allows the designer to adopt approaches that decrease the required number of optical fibers significantly.

5.2.3. Blass and Butler Matrices

The Blass [16–21] and Butler [22–28] matrices have been used in many applications to generate a number of staring beams. The Blass matrix [16], shown in Figure 5.3, consists of a set of traveling-wave feed lines connected to the antenna elements of a linear array crossing another set of lines connected to the beam ports. At the crossover points directional couplers, denoted by the symbol •, are used. Assuming a receive mode, a signal applied at an antenna progresses along the feed line to the end termination. At each crossover point a small signal is coupled into each beam line. The topology of the matrix is such that beams are formed in different directions and the aperture illumination is defined by the coupling coefficients of the couplers.

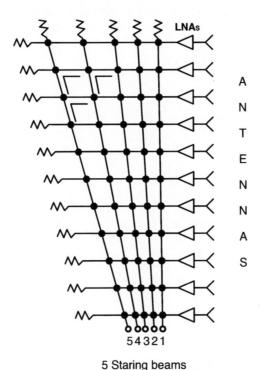

5 Staring beams

Figure 5.3 The Blass beamformer used to generate five staring beams from an array that has 11 antenna elements. • denotes a directional coupler.

Although a design aim is to minimize the powers dissipated in the terminations, the range of coupling values available to the designer is limited [17]. A synthesis procedure that estimates the efficiency of arbitrary beam crossover levels has been reported [18]. Shaped beams have been synthesized with the aid of Blass matrices [19]; moreover, the Blass matrix concept has been adopted to form a planar, two-dimensional multiple-beam microstrip patch array [20,21].

The Butler matrix [22], shown in Figure 5.4, consists of fixed phase shifts interconnected to hybrids and yields orthogonal beams. The lengths of the lines used are in units of $\pi/8$ radians and 180° hybrids are used. The crossovers of the interconnecting lines, shown in the figure, can be eliminated by reconfiguring the matrix into a checkerboard network [26] and the utilization of 180° hybrids has several advantages over the 90° hybrids [23]. A reduction in the hybrid count can be achieved using reflective matrices [24,26], and wide-bandwidth matrices with up to an octave frequency range have been reported [27,28]. Butler matrices have been realized in different media including waveguide for high-power use [29], microstrip [30], and integrated optic form [31]. With the available technology, 64×64 Butler matrices seem to be the largest possible when low relative permittivity microstrip technology is used [7].

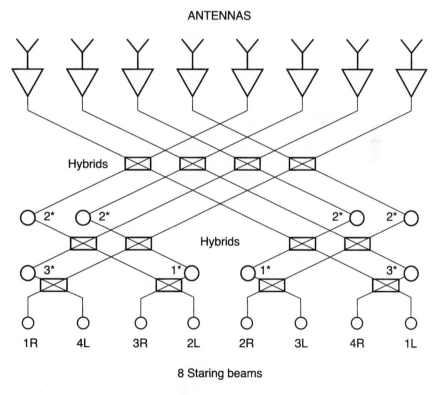

Figure 5.4 The Butler beamformer used to generate eight staring beams from an array that has as many antenna elements. Units of phase shift are $\pi/8$ radians.

5.3. THE FORMATION OF SEVERAL AGILE/STARING BEAMS

Theoretically the generation of several inertialess/agile beams affords the designer several degrees of freedom that can be used to enhance system performance or to configure more complex systems.

Satellite communications systems featuring multiple-beam LEO/MEO satellites have recently been proposed to meet the ever-growing demand for more communication channels. The use of active phased arrays on board the satellites is considered a viable means of enabling the radiation of high-power multiple beams. In this context, the BFN is expected to provide the following functions [32]:

- Precise pointing of each beam to the target area despite possible deviations of the array system components and satellite attitude.
- The formation of nulls exactly onto the neighboring area where frequency reuse could create problems.
- The provision of reconfiguration in case a failure occurs in any radiated beam.

The competing BFNs for this application are digital beamformers (DBFs), Butler matrices, and BFNs capable of deriving several agile beams simultaneously. As the beams derived by Butler matrices are staring, the matrices are unsuitable to meet the requirements and a comparison between DBFs and the beamformers we shall consider in this section will be undertaken at the end of this section.

For radar arrays the issues are more complex and application-dependent but some studies and proposals related to multiple-beam systems have been reported (e.g. [33–36]). As we have seen in Chapter 4, most current radar arrays generate one inertialess beam that scans the radar's surveillance volume with the aid of the phase-shifters located in the T/R modules.

Let us reconsider the beamformer architecture used to generate several staring beams shown in Figure 5.1 when the system operates in receive mode. If programmable vector modulators replace the resistive networks, the resulting beams can be staring or agile depending on the settings of the vector modulators [12,13]. Figure 1.1 shows the resulting BFN that operates at the system's RF. Computer commands, X, Y, induce the vector modulators to insert the required amplitude and phase weights dynamically, so that each of the resulting beams is steered independently. Alternatively, a cluster of beams can be steered in any direction within the array's surveillance volume. We have already noted that the BFN shown in Figure 1.1 is also known as a crossbar beamformer.

The steering commands for each T/R module are generated in a central computer and then distributed to each T/R module or a central computer transmits minimal information to each module, e.g. the required scan angles θ and ϕ, and the element controllers in each T/R module perform the remaining calculations and functions. The transmission of the essential information from the central computer to the T/R modules is implemented conventionally [37] or via optical lines [38]. The fiber optic link envisaged to perform the required functions [38] is small and lightweight, provides excellent isolation, has inherently wide bandwidth and is by and large immune from EMI (electromagnetic interference) and EMP (electromagnetic pulse). A compact electrooptic controller for microwave phased array antennas using nematic liquid crystal display-type technology has been demonstrated experimentally [39].

The same beamformer can operate at a convenient IF if the required downconversions are implemented. With this arrangement the beamformer can be used by a system to perform the array's designated function(s) at different RFs on a time sharing basis. It is recalled here that the accuracy attained by programmable attenuators and phase-shifters operating at IF is higher than those attained by their counterparts operating at RF. As the RF of the array is increased, the validity of this statement is strengthened.

For small arrays the increased accuracy on the amplitude/phase weights is a prerequisite for the attainment of low array sidelobes. The issues of concern here are:

- The selection of the appropriate IF, so that the instantaneous system bandwidth, B, can be easily accommodated in the BFN, i.e. $B/(\text{IF}) \approx 10\%$.
- The complexity and costs introduced by the introduction of downconverters, IF amplifiers, and the distribution of phased-locked oscillator signals to the T/R modules.

The distribution of RF/digital signals via a fiber optic link can facilitate the acceptance of IF vector modulators in phased arrays. Without a fiber optic link it is difficult to distribute a stable LO signal to the T/R modules for the required frequency downconversion. The only phased array radar employing IF beamforming is the ELRA operating by the FGAN-FFM [36]. The resulting beams can be shaped [13] and a slight variation of the same basic BFN allows the generation of hopping beams [13]. While the versatility of this BFN is phenomenal, the network has the additional attractive characteristics we shall consider after we review its many realizations.

Several realizations of beamformers operating at Ku-band [40], C-band [41] and S-band [32,.42,43] consisted of passive dividers/combiners and MMICs. The resulting beamformers are small, lightweight, and reliable and have short production times. The reported 9 × 9 beamformer, for instance had the overall dimensions of 110 mm × 13 mm × 2 mm, and weighed less than 30 g [43].

The architecture of an 81-antenna/81-beam beamformer is shown in Figure 5.5. On the basis of the work reported in reference [43], the beamformer weight is estimated to

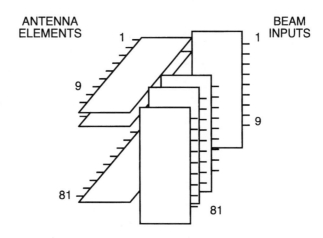

Figure 5.5 The architecture for a beamformer to derive 81 beams from an 81-antenna-element array. (From [43]; © 1996, IEEE.)

be less than 1 kg which compares favorably with 25 kg corresponding to a classical beamformer [43]. These advantages are important for beamformers operating on board any platform but are especially significant when the beamformers are designated to operate on airborne/spaceborne platforms.

The next logical step was to integrate as many functions as possible to form a mega-lithic MMIC-based beamformer. The block diagram adopted for the megalithic beamformer is shown in Figure 5.6, where M beams are formed from an array having N antenna elements [32]. The N MMICs include the input signal splitters while the M MMICs introduce the programmable phase-shifters/attenuators and the power com-biners. When $N = 32$, the resulting vector error standard deviations are within 0.38 dB rms and 2.8° rms over the bandwidth of 20 MHz at 2.5 GHz. Each of the M 'single-chip per beam' MMICs has the dimensions of 11×13 mm and all its components are of novel designs described in the references included in [32].

The other attractors of this beamformer. when compared to ADCs and digital signal processing techniques, are [32]:

* Lower DC power dissipation.
* Freedom from Nyquist sampling speed and quantization errors.

The beamformer is potentially a significant step in the ongoing development of multi-ple-carrier transmitting phased arrays in low intermodulation distortion communication channels and of GEO/LEO satellite on board active phased array transponders. The authors of the same reference coined the term microwave signal processing (MSP) architecture to describe the architecture of their beamformer. From a radar perspective, these beamformers can be used to implement adaptive beamforming/null-forming and consume modest DC powers.

As of September 1998, ten phase III GEO Defense Satellite Communication System (DSCS) satellites are used to provide high-priority secure communication channels to globally distributed military users [44]. The present DSCS satellites have an assortment of horns, reconfigurable multiple beam lens antennas, and high-gain gimbaled dish

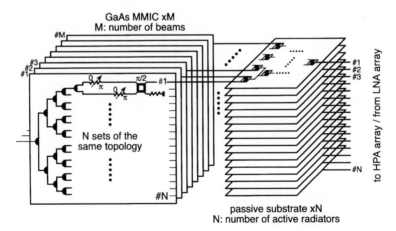

Figure 5.6 The block diagram of the megalithic beamformer that yields M beams when the array ele-ments are N. (Adapted from [32]; © 1997, IEEE.)

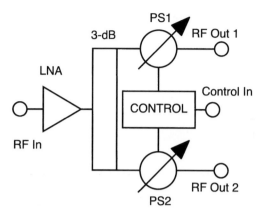

Figure 5.7 The block diagram of a receive module used to derive two independent beams over the 7–10 GHz band. (From [46]; © 1996, IEEE.)

antennas. Work toward a proof-of-concept X-band phased array antenna for potential use in future DSCS satellites is outlined in reference [45]. A crossbar BFN is used in conjunction with an 8-antenna-element array operating at X-band to derive two simultaneous independent beams and a BPF was introduced between each antenna element and its FPA.

Testing of the array included a thorough characterization of its radiation pattern in a number of beam positions. Additional tests included a series of intermodulation tests and bit error rate measurements under several scenarios. A multibeam active-aperture phased array produces intermodulation products with radiation patterns distinct from those of their originating carriers. Consequently, far-field carrier-to-intermodulation ratio characterization is more challenging for a multibeam phased array. It is therefore important to establish the non-linear far-field behavior of the antenna. The reported measurements provide experimental evidence of the non-linear far-field behavior of active phased arrays. All tests validated the concept of phased arrays for DSCS applications.

The many advantages offered by phased arrays compared to similarly reconfigurable multiple-beam antennas such as those used on the DSCS III satellites are outlined in the same reference and include high DC-to-RF efficiency and greater fault tolerance through distributed power.

Figure 5.7 is a block diagram of a receive module used to derive two independent receive beams over the frequency band from 7 to 10 GHz [46]. The wideband LNA sets up the noise figure and third-order intercept for the system and the two phase-shifters form and scan the two beams. The module has been developed to support a phased array communication antenna with dual-beam operation.

5.4. DIGITAL BEAMFORMERS

Digital beamforming can be implemented when the arrays operate in the transmit or receive mode. In the radar context, however, digital beamformers (DBFs) are mainly used on receive arrays because analog beamformers are considered adequate for transmit arrays and beamshape control and pattern nulling are not considered too critical [47]. DBF in transmit arrays may find its first application in cellular telephone networks [48,49]. The capacity of these networks can be greatly enhanced by spatial multiplexing that allows an array to transmit independent messages simultaneously in different directions at the same frequency band.

The popularity of DBFs is based on the widespread usage of digital signal processing (DSP) and the following attractive characteristics:

- Fast, adaptive null-forming; this attribute is fundamental for radar receivers operating in the presence of jammers that dynamically change position or frequency with the passage of time. In the radar context this application is the most significant driver for DBFs. Analog systems based on feedback loops used for the same application are often too slow, and their convergence is scenario-dependent. By contrast, digital open-loop systems need no feedback, are scenario-independent, and are significantly faster.
- The generation of several simultaneous beams.
- Array self-calibration, which in turn results in lower array sidelobes; this attribute is especially significant for small arrays.
- The minimization of the effects caused by mutual coupling between adjacent elements of arrays; again this attribute is significant for small arrays—see section 2.2.1.5.

While DBFs can generate several beams, this attribute is shared by analog beamformers. However, unlike the conventional beamformers based on the Butler matrix, DBFs can generate closely spaced beams. Other applications where DBFs play an important role are adaptive space-time processing and high-resolution direction finding (DF) [49].

5.4.1. Important System Parameters

A typical digital beamformer consists of elaborate analog building blocks, ADCs, and digital circuits. For the receive system, shown in Figure 5.8, the analog receive module includes the LNA, one or more frequency downconversions, and surface acoustic wave (SAW)-based filters that maintain the phase and amplitude matching between receivers. After the derivation of the in-phase, I, and quadrature, Q, components, analog-to-digital conversions take place and the complex weights and calibrations are inserted in each antenna element before the summing of powers is implemented. If I_I and Q_I and I_O, and Q_O; and I_W and Q_W represent the I and Q components of the input, output, and weight vectors respectively, the beamforming operation is characterized by

$$I_O = \sum_1^N [I_I I_W - Q_I Q_W] \tag{5.1}$$

$$Q_O = \sum_1^N [I_I Q_W + Q_I I_W] \tag{5.2}$$

where N is the number of channels used.

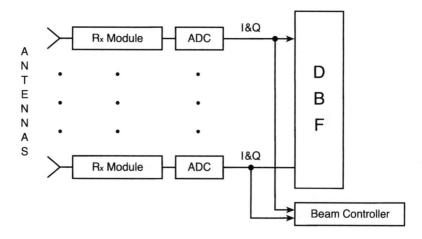

Figure 5.8 The block diagram of a digital beamformer (DBF).

For the array designer the following system parameters are important when DBFs are considered:

- Dynamic range (DR).
- Instantaneous bandwidth, B.
- Number of complex operations per second (COPS) performed by the DBF.
- Costs, weight, volume, and DC power requirements.

The characteristics of the ADCs often determine the first two important parameters; the characteristics of the beam controller are also important when adaptive pattern nulling is required. For Gaussian noise, the system's dynamic range, DR, is defined by the number of bits of the ADC, N_b, the number of parallel channels N, and the quantization noise. DR is therefore approximated by [49,50].

$$\text{DR} \approx \{6N_b + 10 \log N\} \text{ dB} \tag{5.3}$$

To digitize IF signals with components up to F_{max}, the Nyquist criterion is satisfied when the sampling frequency, $F_s \geq 2f_{max}$. If two ADCs are used after the I and Q signal components are derived, $F_s \geq f_{max}$. Usually the signals are oversampled to provide a margin for realistic finite-slope filters [49]. The 1994 state of the art supported 8-bit ADCs with sampling rates of about 2 GHz and 14-bit ADCs of approximately 20 MHz.

The number of COPS is equal to about NB [49]; thus the COPS is equal to 100×10^6 when $N = 100$ and $B = 1$ MHz , representing the parameters of a small radar array. As the system's bandwidth and/or the array's number of elements increase, the COPS become too high for current DSP technologies to meet. A 1996 estimate of the module cost is US$5000 [49], which again represents a high cost for large arrays where N ranges from 5000 to 10 000.

If a decision to use a DBF is made on the basis of system requirements, then the cost and power requirements of the ADCs and the DSP subsystems have to be considered. If costs or the power requirements cannot be met, then DBFs can be used at a subarray level. On the basis of current technology, the DBFs are eminently suitable for small arrays having a relatively narrow bandwidth where the requirements are:

- Fast nulling of jammers in a dynamically changing electronic environment.
- The generation of several simultaneous beams.
- The array sidelobes have to be minimized by the calibration of errors occurring any-where in the array. These errors can be due to array component variations and mutual coupling between antenna elements.

5.4.2. Applications

While the advantages of DBFs for radar and communications have been recognized for many years, the technology required to realize the full potential of DBFs is still evolv-ing [49,51]. In the latter reference, a 12-channel S-band digital beamforming antenna designated for radar applications is reported. The system has an agile frequency band in the 2.8–3.3 GHz range and an IF bandwidth of about 5 MHz. The signal processing is performed off-line but the feasibility of high-performance ASIC implementation of the front-end processing has also been demonstrated.

Hybrid wafer-scale integrated circuit (HWSIC) technology was used to realize an 8-channel, 4-beam, 10 MHz bandwidth beamformer in a package only 50 mm on the side [53]. Given that compensations due to amplitude and phase errors associated with each antenna element can be implemented, array sidelobes can be suppressed [52,53]. A sidelobe level of –45 dB has been reported from a 64-element phased array [52]. Often the injection of an accurate pilot tone immediately behind each antenna element is used for the implementation of the self-calibration process.

We have already considered the attractors of bistatic radar in Chapter 2. An impor-tant receiver requirement for bistatic radars is what is commonly termed a 'pulse-chasing' capability. The receiver should be able to chase the transmitted pulse through constantly changing beam angles, in the presence of jammers. This require-ment can be met by an array that yields a cluster of closely spaced low-sidelobe beams with an adaptive nulling capability. The requirements for a phased array antenna designated to operate in conjunction with a bistatic receiver at X-band have been outlined [54].

In Chapter 2 we outlined the correction procedure that is used to offset the effects of mutual coupling between adjacent antenna elements in small arrays. The Rome Laboratory's technique of implementing this correction is described in [49] and [55] and is based on the observation that, for a receive array, the individual antenna element signal has several constituents: one dominant due to the direct incident plane wave and several less dominant constituents due to scattering of the incident wave at neighbor-ing elements. These constituent contributions can be resolved and scattering can be compensated for by linear transformation, which is accomplished by a matrix multiplication performed on the element output signals. As this compensation is scan-independent, the matrix is fixed and applies for all desired patterns and scan directions [49].

Other applications of DBFs in the areas of adaptive space-time processing and super-resolution are outlined in reference [49].

A survey of early systems utilizing DBF techniques and the work in progress at the Rome Laboratory are outlined in references [56] and [49], respectively.

Testbeds for exploring the power of digital techniques applied to DBF exist in Germany and at the Rome Laboratory. In Germany, the experimental phased array radar system ELRA operating at FGAN-FFM [35,36] consists of separate circular transmit and receive arrays operating at S-band. Antenna element calibration is performed via a fixed probe in the near-field and the outputs of the double-conversion receivers are combined into 48 subarrays with digital outputs from which a cluster of beams is formed and the whole beam cluster is scanned by analog phase-shifters.

At the Rome Laboratory, work in progress includes the maintenance of acceptable array radiation patterns when antenna elements fail [57] and the development of wideband circular arrays [49].

An emerging theme is the marriage of analog and digital beamformers in systems that take advantage of the wideband characteristics of analog BFNs and the flexibility of DBFs.

5.5. PHOTONICS AND PHASED ARRAYS

Several references [e.g. 58–62] closed the knowledge gap between array designers and photonics experts and accelerated the adoption of photonic technology into phased arrays. We have already considered the most obvious application, the distribution of digital/analog signals to and from the array face.

The other area where photonics technology offers unique advantages to array designers is in the realization of volumetrically attractive, lightweight, and wideband beamformers. As optical fibers can offer true time delays, the array bandwidth can be extremely wide. Work that explores new photonics techniques by academics and communications engineers has been widely reported (e.g. [63,64]). While research in this area is growing, there is no doubt that photonics technology offers unique attractors to array designers.

5.5.1. Wideband Beamformers

The feasibility of a photonics beamformer at L- and X-bands has been demonstrated [65]. The use of one beamformer capable of operating at widely separated bands of frequency is a major achievement.

Photonics-based beamformers can have an instantaneous bandwidth extending from 3 to 6 GHz and are 75% lighter and smaller than their electronics counterparts, and the incurred loss of about 14 dB is tolerable [66]. The weight and volume advantages of photonics-based beamformers make them attractive for airborne/satellite-borne applications.

A 5-bit photonic switchable delay line for a wideband array has been reported for an airborne application [67]. The array occupies an area of $0.9 \times 2.8 \text{ m}^2$ and consists of 96 elements. It is designated to operate in the 850–1400 MHz range and follows the contour of a jumbo jet fuselage. The 96 radiating elements were divided into 24 columns with three combined into one subarray which was controlled by a 5-bit time shifter. The photonic time shifter provides the coarse delay steps ranging from 0.25 to 7.75 ns while a 6-bit electronic delay line in the T/R modules provides refined differential delays ranging from 0.01 to 0.5 ns. The array has a 50% instantaneous bandwidth and the L-band module has a 38 dB loss, but efforts are being made to reduce this.

While most wideband beamformers utilize optical fibers, other photonics-based approaches have been reported. Acousto-optic devices have been used to attain true time delays [68]. A Bragg reflection grating (BRG) has been used to construct a fiber grating prism (FGP) beamformer architecture that exhibits the following advantages over beamformers utilizing the dispersive fiber approach [69]:

- The lengths required for the grating arrays are generally much smaller.
- The entire system can be readily implemented in the form of a photonic integrated circuit.

In the same reference the first measured data of a system using Bragg gratings for the FGP are presented.

In another realization a programmable dispersion grating utilizing BFGs has been reported [70]. An integrated optic time delay unit (TDU) that exhibits time delays ranging from a few picoseconds to several nanoseconds has been reported in reference [71]. It is projected that time delays up to 10 ns can be achieved in a package size of 25 mm by 75 mm.

5.5.2. The Generation of Staring Beams

Studies for a multibeam beamformer stemmed from a proposal for a phased array-based radar that generated 1024 staring beams filling the entire surveillance volume; furthermore, the radar was capable of imaging targets of interest by using the ISAR technique [72]. The specifications called for the beamformer to be compact, lightweight, and invisible to the system.

A photonics beamformer using the beamformer architecture illustrated in Figure 5.2, has been used to realize three antenna beams from an array that had eight antenna elements [73].

The results of studies related to a beamformer capable of generating M staring beams from an array where $M = N = 1024$ have been reported [74–77]. An external modulation scheme and erbium-doped fiber amplifiers (EDFAs) have been proposed to reduce the relative-intensity noise (RIN) to a level less than -170 dB/Hz compared to the -145 dB/Hz level corresponding to direct modulation methods. With this arrangement a system noise figure of 3.5 dB resulted when the front-end noise figure was 3 dB. The system exhibited a spurious-free dynamic range [78] (SFDR) of 134 dB Hz$^{4/5}$, which corresponds to 62, 70, and 78 dB when the system bandwidth is 1, 0.1, or 0.01 GHz, respectively [75,76]. The use of EDFAs with beamformers have been proposed in several papers (e.g. [79]).

The partitioning of the array [80] and the use of a wavelength division multiplexing (WDM) scheme substantially reduced the number of calibrated fiber interconnections between the N antennas and M beams.

More explicitly, the number of interconnects K is given by

$$K = \frac{2M}{\log 2}\ \log N \approx \frac{2M \log N}{0.30103} \tag{5.4}$$

When $M = N = 1024$, the optimum number of wavelengths used is 256 and the result-ing K is equal to about 20 480, which compares favorably with 1 048 576 corresponding to the case where $(M \times N)$ accurately calibrated lengths of transmission lines/optical fibers are used.

An acousto-optic system capable of generating multiple simultaneous beams has been reported [81].

5.6. OTHER BEAMFORMERS

Most beamformers we have considered are based on linear models that in terms of sta-tistics make use of only the first- and second-order moment information of the data, e.g. the mean and variance. Natural phenomena, however, are often non-linear. In recent years, a non-linear model based on radial basis functions (RBFs) has drawn considerable attention from the signal processing community. The use of RBF was introduced by Powell [82] for multivariable interpolation. The RBF method has been used for time series prediction and the method can be implemented using a neural network structure [83].

The RBF technique can also be used in the applications of array processing and beamforming. The technique is developed on the basis of non-linear functions and it is model-independent. It therefore possesses the following desirable characteristics:

- It can estimate signals in different kinds of environments of Gaussian, non-Gaussian or colored noise.
- It is robust in terms of resolving multiple coherent signals.
- It is capable of resolving signals separated by less than an antenna beamwidth some preliminary results have been reported [84].

An analogy between the Butler matrices and neural network-based beamformers when used to perform the DF function is drawn in reference [85]. Work in progress toward DF neural beamformers at Rome Laboratory is reported in reference [49].

5.7. LOW-COST PHASED ARRAYS

Low-cost arrays having moderate sidelobe levels of the order of –25 dB or lower are required for many applications that range from communications between moving/flying and stationary platforms and the provision of global connectivity for US ships through the DSCS III satellites. In mobile satellite communications systems, for instance, the beam of a mobile platform, employing a self-focusing phased array, is always pointed to the transmitting satellite [86,87] and knowledge of the satellite's rel-ative bearings is not required. A self-focusing array can also be used on board the

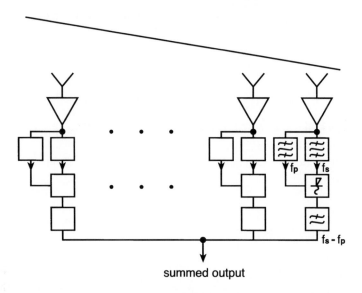

summed output

Figure 5.9 A self-cohering array when a pilot frequency, f_p, is used in conjunction with the signal frequency f_s.

satellite. Other applications envisaged include point-to-point mobile communication systems, wireless local area networks (LANs) [87] and aircraft-to-ground communications systems [86]; the possibility exists of using the same approach for aircraft-to-satellite communications.

While conventional arrays are affordable now and their cost will decrease with the passage of time, low-cost arrays that are less expensive than the affordable arrays we have considered are required.

The major cost savings result because the expensive phase-shifters and their control subsystems normally associated with conventional arrays are substituted by inexpensive subsystems that allow the array to focus and scan.

5.7.1. Self-Focusing Arrays

Consider a small linear array, shown in Figure 5.9, used to receive an incoming wavefront that contains a signal frequency, f_s, and a pilot frequency, f_p. Let $\cos(\omega_t + \phi_1)$ represent the pilot frequency with an instantaneous phase ϕ_1. The mixer at each antenna element yields the difference frequency $(f_s - f_p)$; after filtering and ignoring constants, the output voltage, E_n, at element n of the array is

$$E_n = \cos\left[(\omega_s - \omega_p) - \phi_1 + (n-1)(\psi_p - \psi_s)\right] \qquad (5.5)$$

where $\psi_s = (\omega_s/c)\, d \sin\theta$, $\psi_p = (\omega_p/c)\, d \sin\theta$, d is the distance between antenna elements, θ is the angle formed between the wavefront of the incoming signals and the horizontal, and c is the velocity of light. If the signal and pilot frequencies are close, $\psi_p - \psi_s \approx 0$ and all E_n contributions will add when the number of array elements is not too large. With this arrangement the array points toward the direction of the transmitter. In another realization [87] the two BPFs are omitted . Experimental linear arrays operating on this principle have been realized and the measurements corroborate the theoretical framework [86,87].

5.7.2. The Hughes Arrays

In reference [88] an interesting approach to low-cost, easily scannable arrays is delineated. The authors of the same reference claim that the approach was first disclosed by Welby of the Hughes Aircraft Company [89], but other references related to the same concept are given. Although a transmitting array is considered, arrays based on this approach are proposed for radars and microwave landing systems (MLSs) [90]. With reference to Figure 5.10, all array antenna elements are coupled to a transmission line and the transmitted angular frequency is ω_0. The transmission line is fed by two signals of angular frequencies ω_2 and ω_1, where ω_2 is the scanning frequency and ω_1 is selected to be equal to $(\omega_0+\omega_2)$, $(\omega_1 = \omega_0-\omega_2$ is another option). The two angular frequencies are fed into each antenna element via taps weakly coupling the two angular frequencies to a mixer that in turn selects the transmitted angular frequency for further amplification by the FPA. In the realization shown, the FPAs deliver the required power to the antennas. The array transmits power toward the direction θ_s derived from [88,90]

$$\sin \theta_s = \frac{c \, dL}{v} \frac{2\omega_1 - \omega_0}{\omega_0} \frac{1}{dx} - \frac{m\lambda_0}{dx}$$

(5.6)

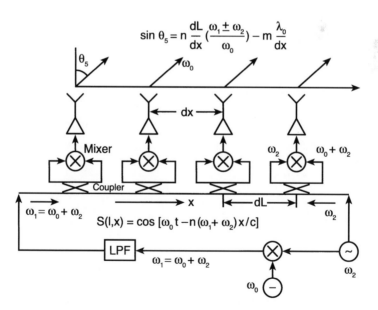

Figure 5.10 The Hughes phased array operating in the transmit mode. The transmitted angular frequency is ω_0 and the scanning is implemented by changing the angular frequency ω_2. (From [88]; ©1997, IEEE.)

where
   c = the free space wave velocity;
   v = the phase velocity of the waves;
   dL = the distance between consecutive tap points;
   dx = the distance between consecutive antennas;
   λ_0 = the free space velocity of the ω_0;
   $m = 0, \pm1, \pm2, \ldots$ is chosen so that the absolute value of the rhs of equation (5.6) is less than 1.

For the microstrip line used in the work presented in reference [90], c/v, denoted by n in Figure 5.10, is equal to 1.38. As ω_0 is constant, θ_s is a function of ω_1 only. With this arrangement the array is steered by varying ω_1, while the transmit angular frequency is invariant. A four-element array that was realized [90] corroborated the theoretical work and several extensions of the reported work are possible.

   We have already considered one solution to provide multibeam connectivity between a platform equipped with a phased array and the DSCS III satellites. To meet the requirement, an aperture of 3 m × 3 m having some 15 000 antenna elements is required. However, it is costly to derive a single beam from such a phased array and too costly to derive multiple beams with conventional BFNs [88]. The authors of the same reference propose the derivation of several beams by the use of the Hughes BFN and the mixing of more than one pair of frequencies with a photonic feed. Photonics-based beamformers are wideband, lightweight, and volumetrically attractive. Moreover, the precision offered by integrated optics makes it possible to control the phase errors and achieve a well-matched multibeam feed. The authors of the proposal demonstrated the validity of their approach by realizing two beams from a two-element transmit array [88].

5.7.3. RADANT-type Beam Scanners

A hybrid approach to low-cost arrays consists of a conventional array that scans the beam in azimuth and a scanning lens that allows the array to scan in elevation [92,93], or vice versa. Within the scanning lens, bulk phase-shifting is used instead of individual phase-shifters for each antenna element.

   A typical array having a 1° beam can be attained from an aperture $50\lambda \times 50\lambda$. Assuming that the antenna elements of an active phased array are spaced at $\lambda/2$, 10 000 antenna elements and phase-shifters are required. The relative costs of the conventional array and the hybrid array are US\$ 8–14 million and US\$3million [92], a significant saving. In more general terms, if N and M are the number of columns and rows of an array, respectively, the number of phase-shifters required by a conventional and a hybrid array are $N \times M$ and $N + M$, respectively. *Inter alia*, the applications envisaged for the hybrid arrays are for ship self-defense engagement radars [91, 5.92].

   The scanning lens can be a RADANT [93] or ferroelectric lens whose dielectric constant can be varied with an applied DC bias voltage [94]. In comparisons to the RADANT lens, the ferroelectric lens may have further advantages of smaller thickness, simpler beam-steering controls, and lower cost [91]. Figure 5.11 illustrates a ferromagnetic lens that consists of several dielectric slabs of ferromagnetic material

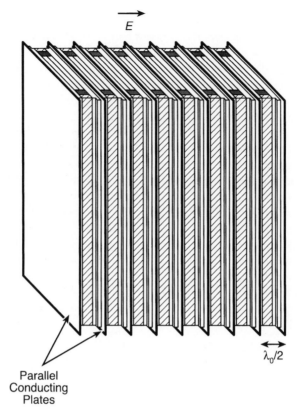

$\overrightarrow{E}$

Parallel
Conducting
Plates

$\lambda_0/2$

Figure 5.11 A ferroelectric lens used to scan the beam of an aperture along one direction. (From [91]; © 1996, IEEE.)

sandwiched between conducting plates. The dielectric constant of the ferromagnetic slabs can be changed by applying and varying the DC electric fields across each slab. If a plane wave is incident on one side of the lens with electric field E normal to the conducting plates, the beam coming out on the other side of the lens can be scanned in the E-plane if a linear phase gradient is introduced along the E-plane direction by adjusting the voltages on the ferromagnetic slabs. Plasma mirrors have also been proposed to scan the beam of an aperture [94].

5.7.4. The Beam Tagging Technique

We have already considered the research and development related to pseudo-satellites. According to the SHARP proposal, the pseudo-satellite is at an altitude of 21 km and the ground station supplies enough microwave power to maintain the satellite at this height and to power the equipment required to perform the communications and/or surveillance functions. The ground station consists of a phased array that transmits the required power, which is in turn received by the rectenna on board the pseudo-satellite. As the pseudo-satellite flies in a circle, to keep it in position against light winds, the ground phased array has to be steerable. To meet this requirement, each antenna element has a set of phase-shifters the smallest increment of which is 11.25°.

Reference [95] reviews two conventional methods used to steer the beam of a phased array and finds them inadequate for this application, while the beam tagging method used in other applications, cited in the same reference, can be adapted to meet the requirements for the SHARP system.

With reference to Figure 5.12, **R** is the vector of the field corresponding to the contributions of all antenna elements except the contribution of one antenna selected for phase alignment, designated as vector **A**. If **A** is in line with **R**, the introduction of a 90° phase lag/lead in the phase of the selected antenna rotates **A** to the positions **A′** and **A″** and the resultant field vector merely reduces the resultant vectors **R′** and **R″** by an equal amount. If, however, the phase of the vector corresponding to the field of the selected antenna forms an angle P with the resultant array vector then the introduction of a 90° phase lead or lag results in unequal vectors **R′** and **R″**. The rapid phase modulation of the field vector corresponding to the element under consideration will produce a corresponding amplitude modulation (AM) of the resulting field. Furthermore, the depth of this modulation is proportional to $\sin P$.

The phase of the signal transmitted by each antenna can be changed either by the conventional phase-shifters or by the 90° phase lag/lead modulation introduced whenever required. The other constituent parts of the system are:

- A monitor antenna located at the center of the rectenna followed by an, AM receiver, a high-pass filter (HPF), and a phase detector.

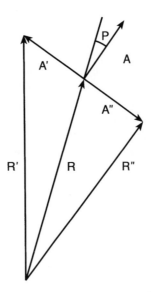

Figure 5.12 The phasor diagram of the fields at the target used to illustrate the beam tagging technique. **A** = the field from selected antenna, normal condition; **A′** = the field from selected antenna; phase advanced by 90°; **A″** = the field from selected antenna; phase retarded by 90°; **R** = combined field, all antennas except the selected one; **R′** = combined field, **A** advanced. **R″** = combined field, **A** retarded; **P** = phase error. (From [95]; © 1992, IEEE.)

- A selector switch located in the phased array that selects each antenna in turn for phase alignment.
- An altimeter or a tracking radar that derives the range between the phased array and the rectenna.
- The pseudo-satellite and the phased array, electrically connected via telemetry antennas and transceivers.
- A waveform generator that controls the sequential switching of the antennas, the application of the 90° lag/lead modulation and the introduction of the appropriate phase shifts required to phase-align the contributions from all antennas to the rectenna.

Let us assume that one antenna is selected for phase alignment and that the 90° lag/lead modulation produces a corresponding AM of the resulting field. On board the pseudo-satellite the resulting field is received by the monitor antenna that feeds the AM receiver, HPF, and phase detector. The other input of the phase detector is derived from the waveform generator located on the ground. Therefore $\sin P$ is derived from knowledge of the resulting time-averaged phase detector output and the range between the pseudo-satellite and the phased array. The $\sin P$ information is transmitted to the phased array with the aid of the telemetry system, where it is further converted to the angle P. An appropriate phase shift adjustment of the phase corresponding to the selected antenna is then applied. Once the alignment of one antenna is implemented, another antenna is selected and the same procedure is repeated until all antennas are phase-aligned. The update periods for one antenna and the array are 2 ms and 224 ms, respectively.

Computer simulations corroborated the validity of the proposed beam tagging method. The telemetry link between the array and the pseudo-satellite can operate at a different frequency from that assigned for the operation of the phased array. The proposed method results in a self-focusing and self-steering system.

5.8. CONCLUDING REMARKS

The beamformer of a typical passive array can be bulky and its insertion loss usually affects the overall array noise temperature. By contrast, the beamformer of an active array can be transparent to the system and volumetrically attractive. These two beamformer characteristics lend the designer several degrees of freedom, which we have explored in this book.

We considered the beamformers associated with conventional active arrays in Chapter 4 because vector modulators are considered as integral parts of the T/R modules. These beamformers derive one inertialess beam that can be steered anywhere within the array's surveillance volume.

Similarly, beamforming lenses were considered in Chapter 3 because some of the lenses considered, e.g. the Luneberg lenses, also perform the aperture function for a system. The beamforming lenses meet an important set of requirements, e.g. low cost, a moderate number of independent and simultaneous beams, wide-angle scanning, and wide instantaneous system bandwidth.

The crossbar beamformers explored in this chapter have unprecedented versatility, are volumetrically attractive, and are capable of deriving several simultaneous and independent agile beams. Moreover, megalithic MMIC-based realizations of these beamformers have already been reported. The additional beamformer attractors are:

- Controllable uniformity of the essential beamformer characteristics, e.g. 0.38 dB rms and 2.8° rms over a bandwidth of 20 MHz at 2.5 GHz, when the number of antennas is 32.
- Lower DC power dissipation compared to DBFs that require ADCs.
- Freedom from Nyquist sampling speed and quantization errors.

The beamformers are potentially a significant step in the ongoing development of multiple-carrier transmitting phased arrays in low intermodulation distortion communication channels and of GEO/LEO satellite onboard active phased array transponders. From a radar perspective, these beamformers can be used to implement adaptive beamforming/null-forming and consume modest DC power.

Similar beamformer architectures can be used for transmit arrays provided the non-linear characteristics of the FPAs are taken into account. Studies outlined in Chapter 3 and the measurements cited in this chapter corroborate the emerging importance of this topic.

The application of photonics to phased arrays, e.g. beamforming and the RF/digital signal distribution across the array face, represents a fertile field for continuing R&D. The attractors photonics offer to the phased array designer are compactness, light weight, and immunity against EMI and EMP. These attractors render photonics-based beamformers eminently suitable for airborne/satellite-borne platforms. Photonics beamformers offer the designer wide instantaneous bandwidths also. The same beamformer can, for instance, be used in an array operating at L- and X-bands. Furthermore, the designer can have either several staring beams or one agile beam at widely separated frequency bands.

Considerable work toward closing the knowledge gap between array engineers/scientists and photonics experts has been reported. Additionally, work that explores new photonics techniques by academics and communication engineers is found in the open literature. The evolution of integrated photonics beamformers can significantly reduce their cost.

While photonics-based and MMIC-based crossbar beamformers are eminently suitable for airborne/spaceborne applications, weight comparisons are not yet available.

Digital beamformers have the following set of attractors:

- Fast, adaptive null-forming capability; this attribute is fundamental for radar receivers operating in the presence of jamming powers that dynamically change with the passage of time. In the radar context, this application is the most significant driver for DBFs. Analog systems based on feedback loops used for the same application are often too slow and their convergence is scenario-dependent. By contrast, digital open-loop systems need no feedback, are scenario-independent, and are significantly faster.
- The generation of several simultaneous beams.
- Array self-calibration, which in turn results in lower array sidelobes; this attribute is especially significant for small arrays.
- The minimization of the effects caused by mutual coupling between adjacent elements of arrays; again this attribute is significant for small arrays.

DBFs will gain prominence as high-speed signal processing ICs evolve and become available. As can be seen, photonics-based and digital beamformers are eminently suitable for different niche applications. However, crossbar beamformers can challenge DBFs in adaptive beamforming/null-forming applications, especially in applications where only moderate DC powers are available.

Overall, there are enough conventional beamforming techniques and architectures to satisfy most requirements; indeed, some of the techniques we have explored have unprecedented versatility.

There is a niche requirement for small, low-cost arrays having moderate sidelobe levels of the order of –25 dB or lower. The applications envisaged are many and range from communications between moving/flying and stationary platforms and the provision of global connectivity for US ships through the DSCS III satellites.

While conventional arrays are affordable now and their cost will decrease with the passage of time, low-cost arrays that are less expensive than the affordable arrays that we have considered are required.

The major cost savings result because the expensive phase-shifters and their control subsystems normally associated with conventional active arrays are substituted by inexpensive subsystems that allow the array to focus and scan.

The low-cost arrays we have considered are self-coherent arrays, the Hughes arrays, and the RADANT-type beam scanners. The scanning function in a self-coherent receive array is implemented with the aid of a pilot frequency that is injected with the signal frequency and the appropriate signal processing functions are performed in the array.

In one realization, the scanning of a Hughes array is implemented by changing the frequency of a reference oscillator; this is an ingenious approach to scanning an array operating at a given RF frequency. In another realization, the basic Hughes array used in conjunction with a photonic feed can yield several simultaneous beams that are independently scannable.

The original RADANT lens scans the beam of an aperture in one direction by varying the applied DC voltage to it. A phased array can therefore scan the beam in one dimension conventionally and in the other dimension by using a RADANT lens. If N and M are the number of columns and rows of an array, respectively, the numbers of phase-shifters required by a conventional and hybrid array are $(N \times M)$ and $(N + M)$, respectively. Recently a ferromagnetic lens has been developed and in both realizations bulk phase-shifting is used instead of individual phase-shifters for each antenna element. The use of plasma mirrors to perform the same function has also been explored.

The beam tagging approach is proposed for applications where the position of a target, a pseudo-satellite, is not known and a connection has to be maintained between the satellite and phased array transmitting the required power to the satellite. This is an effective method that has been used in other applications.

Although some of the beamformers we have considered are well-established, most of those explored in this chapter are in different stages of development. While more R&D is required to fully explore their potential, there is little doubt that some promising candidates are emerging. What is more significant is that R&D in the area of beamforming has increased significantly mainly owing to the challenging requirements that emerged from recent systems and the application of MMIC/digital/photonic technologies to beamformers of active phased arrays utilizing MMIC-based T/R modules.

REFERENCES

[1]   Bracewell R. N. Radio astronomy techniques. In *Handbuch der Physik*, vol. 54 (S. Flugge, ed.). Springer-Verlag, Berlin (1962).

[2]   White W. D. Pattern illuminations in multiple beam antennas. *IRE Trans. Antennas Propag.* **T-AP62**, 430 (1962).

[3]   Allen J. L. A theoretical limitation on the formation of lossless multiple beams in linear arrays. *IRE Trans. Antennas Propag.* **T-AP62**, 350 (1961).

[4]   Stein S. On cross coupling in multiple beam antennas. *IRE Trans. Antennas Propag.* **T-AP62**, 548 (1962).

[5]   Chambers J. M., Passmore R. and Ladbrooke J. Beamforming for a multibeam radar. *Int. Radar Conf.*, **82**, p. 390 (1982).

[6]   Wallington J. R. Beamforming options for phased array radar. *Proc. Conf. Military Microwaves*, p. 379 (1986).

[7]   Hall P. S. and Vetterlein S. J. Review of radio frequency beamforming techniques for scanned and multiple beam antennas. *IEE Proc., Part H: Microwaves, Opt. Antennas* **137**(5), 293 (1990).

[8]   *IEE Colloq. Multiple Beam Antennas Beamformers* (Nov. 1989).

[9]   Mills B. Y. *et al.* The Sydney University cross-type radio telescope. *Proc. IREE Aust.* **24**, 156 (1963).

[10] Large M. I. and Frater R. H. The beam forming system for the Molonglo radio telescope. *Proc. IREE Aust.* **30**, 227 (1969).

[11] Clarke T. W., Murdoch H. S. and Large M. I. The delay line system for the Molonglo radiotelescope. *Proc. IREE Aust.* **30**, 236 (1969).

[12] Easton N. J., Bennett F. C. and Miller C. W. Analogue beamformers. *IEE Colloq. Multiple Beam Antennas Beamformers*, p. 12/1 (1989).

[13] Zagbloul A. I., Assal F. T. and Sorbello R. M. Multibeam active phased array system configurations for communication satellites. *IEEE Military Commun. MILCOM '87*, vol. 1, p. 289 (1987).

[14] Fourikis N. The Culgoora Radioheliograph–7. The branching network. *Proc. IREE Aust.* **28**(9), 315 (1967).

[15] Blum E. J. Le reseux Nord-Sud a lobes mortiples. Complement au grant interferometre de la station de Nancay. *Ann. Astrophys.* **24**(4), 359 (1961).

[16] Blass J. Multidirectional antenna-new approach top stacked beams. *IRE Int. Conv. Rec.*, Part 1, p. 48 (1960).

[17] Fassett M., Kaplan L. J. and Pozgay J. H. Optimal synthesis of ladder network array antenna feed systems. *Antennas Propag. Symp.*, Amherst, MA., p. 58 (1976).

[18] Inagaki N. Synthesis of beam forming networks for multiple beam array antennas with maximum feed efficiency. *IEE Int. Conf. Antennas Propag.*, p. 375 (1987).

[19] Wood P. J. An efficient matrix feed for an array generating overlapped beams. *IEE Int. Conf. Antennas Propag.*, p. 371 (1987).

[20] Hall P. S. and Vetterlein S. J. Advances in microstrip multiple beam arrays. *7th Int. Conf. Antennas Propag.*, vol. 1, p. 129 (1991).

[21] Hall P. S. and Vetterlein S. J. Integrated multiple beam microstrip arrays. *Microwave J.* **35**(1), 103 (1992).

[22] Butler J. and Howe R. Beamforming matrix simplifies design of electronically scanned antennas. *Electron. Des.* **9**, 170 (1961).

[23] MacNamara L. Simplified design procedure for Butler matrices incorporating 90° or 180° hybrids. *IEE Proc., Part H: Microwaves, Antennas Propag.* **134**(1), 50 (1987).

[24] Shelton J. and Hsiao J. Reflective Butler matrix. *IEEE Trans. Antennas Propag.* **AP-27**(5), 651 (1979).

[25] Guy J. R. F. Proposal to use reflected signals through a single Butler matrix to produce multiple beams from a circular array antenna. *Electron. Lett.* **28**(4), 209 (1985).

[26] Blokhia N. A. and Mishustin B. A. Design of planar beam-shaping circuits. *Radio Electron. Commun. Syst.* (Engl. transl.) **27**(2), 45 (1984).

[27] Chow P. E. K. and Davis D. E. N. Wide bandwidth Butler matrix network. *Electron. Lett.* **3**, 252 (1967).

[28] Withers M. J. Frequency insensitive phase shift networks and their application in a wide-band-width Butler matrix. *Electron. Lett.* **5**(20), 496 (1969).

[29] Levy R. A high power X-band Butler matrix. *Microwave J.*, **27**, 153 (Apr. 1984).

[30] Wallington J. R., Analysis, design and performance of a microstrip Butler matrix. *Proc. 6th Eur. Microwave Conf.*, Brussels, p. A14.3.1 (1973).

[31] Charczenco W. *et al.* Integrated optical Butler matrix for beamforming in phased array antennas. *Proc. SPIE–Optoelectron. Signal Processing Phased Array Antennas II*, p. 196 (1990).

[32] Ohira T., Suzuki Y., Ogawa H. and Kamitsuna H. Megalithic microwave signal processing for phased-array beam forming and steering. *IEEE Trans. Microwave Theory Tech.* **45**(12), 2324–2332 (Dec. 1997).

[33] Billam E. R. System aspects of multiple beams in phased array radar. *IEE Colloq. Multiple Beam Antennas Beamformers*, p. 1/1 (1989).

[34] Fourikis N. Novel shared aperture phased arrays. *Microwave Opt. Tech. Lett.* (Feb. 20), 189–192 (1998).

[35] Groger I., Sander W. and Wirth W.-D. Experimental phased array radar ELRA with extended flexibility. *IEEE AES Magazine*, (Nov.) 26 (1990).

[36] Wirth W.-D. Signal processing for target detection in experimental phased array radar ELRA. *Proc. IEE: Part F* **128**, 311 (1981).

[37] Waldron T. P., Chin S. K. and Naster R. J. Distributed beamsteering control of phased array radars. *Microwave J.*, **29**, 133 (Sept. 1986).

[38] Herczfeld P. R. *et al.* Optical control of MMIC-based T/R modules. *Microwave J.*, **38**, 309 (May 1988).

[39] Riza N. A. A compact electrooptic controller for microwave phased-array. *Proc. Photon. Processors, Neural Networks and Memories, Proc. SPIE* **2026**, 286 (1993).

[40] Gupta R., Hampsch T., Zaghloul A., Sorbello R. and Assal F. Beam-forming matrix design using MMICs for a multibeam phased-array antennas. *13th Ann GaAs IC Symposium Tech Dig.*, pp. 41–44 (1991).

[41] Estep G., Gupta R., Hampsch T., Zaharovits M., Pryor L., Chen C., Zaghloul A. and Assal F. A C-band beam-forming matrix for phased-array antenna applications. *IEEE Microwave Syst. Conf.*, pp. 189–192 (1995).

[42] Ohira T., Suzuki Y., Ogawa H. and Kamitsuna H. Megalithic microwave signal processing for phased-array beam forming and steering. *IEEE MTT-S Dig.*, pp. 587–590 (1997).

[43] Coromina F., Ventura-Traveset J., Corral J. L., Bonnaire Y. and Dravet A. New multibeam beamforming networks for phased array antennas using advanced MMCM technology. *IEEE MTT-S Dig.*, pp. 79–82 (1996).

[44] Fact Sheet, US Air Force. Defense satellite Communications System, Phase III. http://www.laafb.af.mil/SMC/PA/Fact_Sheets/dscs_fs.htm (Sept. 1998).

[45] Zagloul A. I, Sichan L., Upshur J. I., Estep G. C. and Gupta R. K. X-band active transmit phased array for satellite applications. *IEEE Int. Symp. Phased Array Systems and Technology*, pp. 272–277 (1996).

[46] Kopp B. A. and Axness T. A. Multi-chip receive module for a wide-band X-band dual-beam phased array communication antenna. *IEEE MTT-S Dig.*, pp. 1597–1600 (1996).

[47] Steyskal H. and Rose J. F. Digital beamforming for radar systems. *Microwave J.*, **32**, 121 (Jan. 1989).

[48] Zetterberg P. and Ottersten B. The spectrum efficiency of a base station antenna system for spatially selective transmission. *Proc IEEE 43rd Vehicular Technology Conf.*, Sweden (1994).

[49] Steyskal H. Digital beamforming at Rome laboratory. *Microwave J.* **39**, 100–124 (Feb. 1996).

[50] Proakis J. and Manolakis D. *Digital Signal Processing: Principles, Algorithms and Applications*, p. 419. Macmillan, New York (1992).

[51] Petterson L., Danestig M. and Sjostrom U. An experimental S-band digital beamforming antenna. *IEEE Int. Symp. Phased Array Systems and Technology*, pp. 93–98 (1996).

[52] Herd J. Experimental results from a self-calibrating digital beamforming array. *IEEE Antennas Propag. Soc. Int. Symp.*, p. 384 (1990).

[53] Langgon J. L. and Hinman K. A digital beamforming processor for multiple beam antennas. *IEEE Antennas Propag. Soc. Int. Symp.*, p. 383 (1990).

[54] Prentice A. K. A digitally beamformed phased array receiver for tactical bistatic radar. *IEE Colloq. Active Passive Components Phased Array Systems*, p. 11/1 (1992).

[55] Steyskal H. and Herd J. Mutual coupling compensation in small array antennas. *IEEE Trans. Antennas Propag.* **39**(12), p. 1971 (Dec. 1990).

[56] Steyskal H. Digital beamforming antennas: an introduction. *Microwave J.*, **30**, p. 107 (Jan. 1987).

[57] Mailloux R. J. Phased array error correction scheme. *Electron. Lett.*, **29**(7), 573 (Apr. 1993).

[58] VanBlaricum M. L. Photonic systems for antenna applications. *IEEE Antennas Propag. Mag.* **36**(5), 30 (1994).

[59] Seeds A. Microwave optoelectronics. Tutorial review. *Opt. Quantum Electron.* **25**, 219 (1993).

[60] Seeds A. Optical technologies for phased array antennas. Invited paper. *IEICE Trans. Electron.* **E76-C**(2), 198 (1993).

[61] Glista A. S. Airborne photonics, a technology whose time has come. *AIAA/IEEE 12th Digital Avion. Syst. Conf.*, DASC, p. 336 (1993).

[62] Zmuda H. and Toughlian E. N. *Photonic Aspects of Modern Radar.* Artech House, Boston, MA (1994).

[63] Benjamin R. Optical techniques for generating multiple agile antenna beams. *IEE Colloq. Multiple Beam Antennas Beamformers*, p. 11/1 (1989).

[64] Paul D. K. Optical beam forming and steering for phased array antenna. *IEEE Proc. Natl. Telesyst. Conf. Commercial Applicat. Dual-Use Technol.*, p. 7 (1993).

[65] Ng W. *et al.* Wideband fiber-optic delay network for phased array antenna steering. *Electron. Lett.* **25**(21), 1456 (Oct. 1989).

[66] Ackerman E., Wanuga S. and Kasemset D. Integrated 6-bit photonic true time delay for lightweight 3–6 GHz radar beamformer. *IEEE MTT-S Int. Microwave Symp. Dig.*, p. 681 (1992).

[67] Loo R. Y., Tangonan G. L., Yen H. W., Lee J. J., Jones V. L. and Lewis J. 5 Bit photonic time shifter for wideband arrays. *Electron. Lett.* **31**(18), 1532–1533 (31 Aug. 1995).

[68] Gessell L. H. and Turpin T. M. True time delay beam forming using acousto-optics. *Opt. Technol. Microwave Appl./Optoelectron. Signal Process. Phased Array Antennas, Proc. SPIE* **1703**, vol. 3, 592 (1992).

[69] Zmuda H., Soref R A., Payson P., Johns S. and Toughlian N. Photonic beamformer for phased array antennas using a fiber grating prism. *IEEE Photonics Tech. Lett.* **9**(2), 241–243 (Feb. 1997).

[70] Tong D. T. K. and Wu M. C. Programmable dispersion matrix using fibre grating for optically controlled phased array antennas. *Electron. Lett.* **32**(17), 1532–1533 (15 Aug. 1996).

[71] Hartman N. F. and Corey L. E. A new integrated optic technique for time delays in wideband phased arrays. *ICAP 91*, vol. 2, pp. 918–921 (1991).

[72] Fourikis N., Wehner D. R. and Eccleston K. W. A proposal for a novel surveillance and imaging phased array radar. Invited Paper, *17th Int. Conf. Infrared Millimeter Waves*, p. 146 (1993).

[73] Cardone L. Ultra-wideband microwave beamforming technique. *Microwave J.*, **28**, 121 (Aug. 1985).

[74] Alameh K. E., Minasian R. A. and Fourikis N. Photonics-based beamforming network for active phased arrays. *Proc. 18th Aust. Conf. Opt. Fibre Technol.*, p. 360 (1994).

[75] Alameh K. E., Minasian R. A. and Fourikis N. Hardware compressed photonic beamformer architecture for multi-beam active arrays. *Proc. 19th Aust. Conf. Opt. Fibre Technol.*, p. 41 (1995).

[76] Alameh K. E., Minasian R. A. and Fourikis N. High capacity optical interconnections for phased array beamforming. *IEEE J. Lightwave Technol.*, Spec. Issue **13**(6), 1116 (1995).

[77] Alameh K. E., Minasian R. A. and Fourikis N. Wavelength multiplexed photonic beamformer architecture for microwave phased array. *Microwave Opt. Technol. Lett.*, p. 84 (5 Oct. 1995).

[78] Daryoush A. S. Inter-faces for high-speed fiber optic links: Analysis and experiment. *IEEE Trans. Microwave Theory Tech.* **39**(12), 2031 (1991).

[79] Banerjee S. *et al.* A wideband microwave/photonic distribution network for an X-band active phased array antenna. *IEE Colloq. Microwave Opto-Electron. Dig.*, p. 46 (1994).

[80] Goutzoulis A. and Davies K. Compressive 2-D delay line architecture for the time-steering of phased-array antennas. *Proc. SPIE–Optoelectron., Signal Processing Phased-Array Antennas II*, **1217**, p. 270 (1990).

[81] Riza N. A. Multiple-simultaneous phased array antenna beam generation using an acousto-optic system. *Proc. SPIE–Analog Photon.* **1790**, 95 (1992).

[82] Powell M. J. D. Radial basis functions for multivariable interpolation: a review. *IMA Conf. Algorithms Approx. Funct. Data*, RMCS, Shrivenham (1985).

[83] Broomhead D. S. and Lowe D. Multivariable interpolation and adaptive networks. *Complex Syst.* **2**, 321 (1988).

[84] Lo T., Leung H. and Litva J. Nonlinear beamforming. *Electron. Lett.* **27**(4), 350 (1991).

[85] Mailloux R. J. and Southall H. L. The analogy between the Butler matrix and the neural-network direction-finding array. *IEEE Antennas Propag. Mag.* **39**(6), 27–32 (Dec. 1997).

[86] Withers M. J., Davies D. E. N., Wright A. H. and Apperly R. H. Self focusing array. *Proc IEE*, **111**(9), 1683–1688 (Sept. 1965).

[87] Gupta S. and Fusco V. F. Automatic beam steered active antenna receiver. *IEEE MTT-S Dig.*, p. 599–602 (1997).

[88] Lee J. J., Stephens R R., Tangonan G. L. and Wang H. T. A multibeam array using RF mixing feed. *IEEE Antennas Propag. Int. Symp.*, pp. 706–709 (1997).

[89] Welby W. R. US patent 3,090,928 (1963).

[90] Aamo K. Frequency controlled antenna beam steering. *IEEE Natl. Telesystems Conf.*, pp. 51–54 (1994).

[91] Rao J. B. L., Trunk G. V. and Patel D. P. Two low-cost phased arrays. *IEEE Int. Symp. Phased Array Systems and Technology*, pp. 119–124 (1996).

[92] Rao J. B. L., Hughes P. K., Trunk G. V. and Sureau J. C. Affordable phased array for ship self-defense engagement radar. *IEEE Natl. Radar Conf.*, pp. 22–37 (1996).

[93] Chekroun C., Herrick D., Michel Y. Pauchard R. and Vidal P. Radant: A new method of electronic scanning. *Microwave J.*, **24** (Feb.) pp. 45–53 (Feb. 1981). English version of the paper that appeared in *L'Onde Electronique*, **59**(12) (Dec. 1979).

[94] Mathew J., Meger R. A., Gregor J. A., Murphy D. P., Pechacek R. E., Fernsler R. F. and Manheimer W. M. Electronically steerable plasma mirror. *IEEE Int. Symp. Phased Array Systems and Technology. Revolutionary Developments in Phased Arrays*, Boston, MA, pp. 58–62 (1996).

[95] East T. W. A self-steering array for the SHARP microwave powered aircraft. *IEEE Trans. Antennas Propag.* **40**(12), 1565–1567 (Dec. 1992).

Index

Printed and bound by CPI Group (UK) Ltd, Croydon, CR0 4YY

11/05/2025

01866558-0001